Practical MMIC Design

For a complete listing of the *Artech House Microwave Library*,
turn to the back of this book.

Practical MMIC Design

Steve Marsh

ARTECH
HOUSE
BOSTON | LONDON
artechhouse.com

Library of Congress Cataloging-in-Publication Data
A catalog record for this book is available from the U.S. Library of Congress.

British Library Cataloguing in Publication Data
A catalogue record for this book is available from the British Library.

ISBN-10: 1-59693-036-5
ISBN-13: 978-1-59693-036-0

Cover design by Yekaterina Ratner

685 Canton Street
Norwood, MA 02062

10 9 8 7 6 5 4 3 2 1

For my family,
wife Sue, and sons Sam, Jon, and Toby

Contents

Foreword

This book is a valuable aid to the modern microwave engineer. Due to significant R&D funding, innovations, processing advancements and electronic circuit developments, the Microwave Monolithic Integrated Circuit (MMIC) has now become commonplace, replacing many discrete circuits with individual transistors, resistors, capacitors, inductors and element interconnections. MMIC circuits show reduced size and cost with higher reliability to meet the needs of today's markets. MMICs have replaced discrete designs, except for low volume specialty circuits, as well as very high power applications at RF and microwave frequencies. At millimeter wave frequencies, with the small wavelengths, the MMIC circuit is the primary vehicle to realize practical integrated circuits.

With technology advances, the electrical engineer must keep himself updated to the latest needs of the marketplace. Dr. Steve Marsh has worked for many years developing MMIC circuits and is now sharing his knowledge and experiences by preparing this excellent book for the engineer. This book covers all aspects of MMIC design, layout, and fabrication. The book also includes the design of resistors, capacitors, inductors, diodes, transistors, transmission lines, transmission line discontinuities, amplifiers (low noise as well as power), oscillators, mixers, switches, phase shifters, filters, couplers, and baluns.

Special thanks is extended to Dr. Steve Marsh for his countless hours in writing this book, along with his willingness to share his MMIC design experience for the engineer.

Dr. Ed Niehenke
Baltimore, Maryland, USA
September 2006

Preface

This book is also dedicated to Jim Turner, one of the founding fathers of the GaAs MMIC, who died in October 1996.

During my thirteen years as an MMIC designer at the Caswell site I learned what an important part it played in the development of today's MMICs. The electronics firm Plessey had moved its research laboratory out of London during the Second World War and located it in the cow sheds of an old hunting lodge in Northamptonshire known as Caswell. Perhaps fertilized by the previous occupants, the site grew rapidly, moving into new buildings, and by 1960 became one of the best research and development centers in Europe.

It was during this period that a young engineer at Caswell called James Aubrey Turner was pioneering the development of transistors on the new semiconductor material gallium arsenide. His group was one of three worldwide that demonstrated that GaAs FETs could amplify at microwave frequencies, and in 1974 he published a paper describing how they had made the first fully-monolithic GaAs FET MMIC. Jim Turner was also instrumental in changing these early research results into a commercial GaAs MMIC foundry, offering a 0.7-micron gate-length MESFET process in 1985, a 0.5-micron gate-length MESFET process in 1988, and a 0.2-micron gate-length pHEMT process in 1995. Jim's influence in the GaAs MMIC business was by-no-means limited to the Caswell site. He helped to set up the first symposium dedicated to GaAs IC development; the IEEE GaAs IC

Jim Turner

Symposium at Lake Tahoe, Nevada, in 1979, was awarded an MBE in 1981, the GEC Nelson gold medal in 1992, and the Heinrich Welker gold medal in 1993. Turner was also a visionary, championing the concept of European companies' collaborating on MMIC process development, rather than fighting, in order to compete in the world MMIC marketplace.

It was during just this sort of collaborative European program that I first worked closely with Jim Turner. He was managing an ESPRIT project called MANPOWER that was funded by the European Commission to develop manufacturable MESFET power amplifiers, and I was one of the MMIC designers. The project was a collaboration between many European companies and institutions, including the (now GEC-Marconi) Caswell site, Dassault, Philips, Seimens, and universities in Rome, Turin, Cantabria, and Dublin. Jim understood that collaboration only works well if you make connections with the other engineers, and there is no better way of doing this than eating and drinking together! The program was arranged with three monthly meetings rotating around the different countries and establishments, and each time we socialized together after the meetings, building friendships and trust in each other. The interaction between the companies, even though they were actually competitors, was massive, and it greatly strengthened the European MMIC industry. Friendships and contacts I made during that project have endured to this day and helped me enormously during my MMIC career. I will never forget having a beer with Jim and chatting with our European friends. To me, Jim was a scientist with drive and innovation who got things done, not by an aggressive management style but through his relationship with those working for him and his legions of friends throughout the industry. Jim was invariably cheerful at work, even though he suffered from an illness that would eventually take him from us, and he believed in finding solutions to engineering problems, rather than apportioning blame.

My admiration for Jim, the kind way he worked, and his huge achievements in developing the GaAs MMIC foundry at Caswell motivated me to write this book. The current owners of the Caswell site, Bookham Inc., closed the GaAs MMIC foundry in 2004, making me redundant and closing a chapter in the history of MMIC development. Partly because I hate to see Jim's achievements at Caswell go to waste and partly because he believed in collaboration and the continual exchange of ideas and techniques, I decided to write this book on practical MMIC design. It is based primarily on everything I learned while working at Turner's MMIC foundry, and I hope that by putting it down on paper for future MMIC engineers to access, Jim Turner's achievements at Caswell will, in part at least, live on.

I would also like to acknowledge all my other friends and colleagues with whom I worked at Caswell. In particular I would like to thank my close friends and fellow MMIC designers Andy Dearn and Chris Clifton for many hours of helpful discussions as we sat at the MMIC design computers. Thanks also to Mike Brookbanks for always being willing to give advice on physics and modeling issues, Jeff Buck for design and test issues, Dave Warner, Ian Davies, Bob Wallis, Rick Davies, and John Cockrill for processing information, and Crawford Lindsay for his annoying attention-to-detail quality assurance! On the management side, I would like to express my respect and thanks to Fred Myers, who led us through some difficult years, and to my group manager Steve Wadsworth for doing a great job supporting me. Finally thanks to all my fellow lunchtime 5-per-side soccer players at Caswell for keeping me sane!

1

Introduction

This book is a down-to-earth practical guide to the technology of monolithic microwave integrated circuits (MMICs), as shown in Figure 1.1, and to the design techniques and rules of thumb that enable the system specifications to be met correctly the first time. The aim of this book is, first, to introduce the engineer to the basic technology of MMICs and outline their advantages over other components; second, to give design engineers the foundational information and design methods for a wide range of components, which enables them to start designing immediately against their requirement specifications; third, to give insight into the layout, processing, and test constraints unique to MMICs; finally, to give technical, project, and production managers more understanding of the foundry practices and economics associated with using MMICs.

The book assumes a degree-level understanding of physics or electronic engineering and is intended for radio frequency (RF) and microwave system and component design engineers, postgraduate researchers, and technical managers with a firm knowledge of RF engineering principles who require a sound understanding of the design principles, capabilities, and production issues associated with MMICs. The book is also structured with questions and answers after most chapters to make it a suitable textbook for final-year degree students specializing in electronics, RF, microwave, and solid-state subject areas.

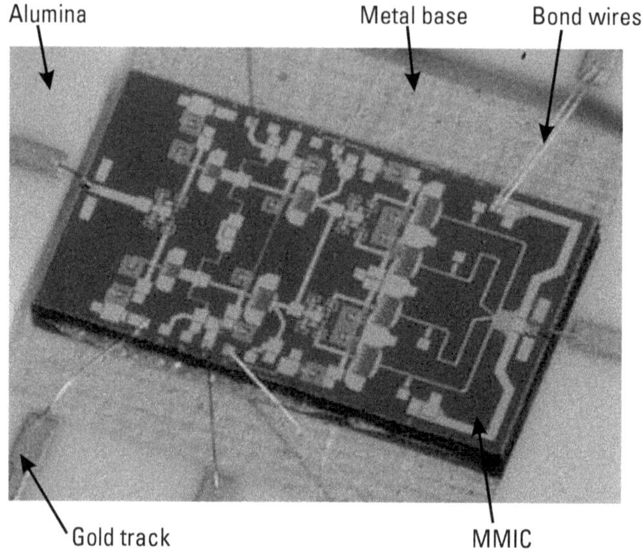

Figure 1.1 Typical MMIC mounted in an alumina substrate. (*Source:* Bookham Inc., 2006. All Rights Reserved.)

1.1 Introduction to MMICs

The acronym MMIC stands for *monolithic microwave integrated circuit.* The word *monolithic* (from the Greek) means "as a single stone" and describes the fundamental characteristic of MMICs (i.e., that they are fabricated from a single piece of semiconductor material). The word *microwave* refers to the ac signal frequency range within which they are used, which covers free-space wavelengths from 1 mm to 1m and corresponds to frequencies from 300 MHz to 300 GHz. The term *integrated circuit* (IC) indicates that the semiconductor material does not contain a single diode or transistor but consists of an electronic circuit of active devices, like transistors, and passive devices, such as capacitors and resistors, together with all their interconnections, to make up a whole system.

MMICs are utilized in most applications that involve transmitting and receiving microwave signals, ranging from cellular phones, wireless local-area networks (WLANs) and Global Positioning System (GPS) receivers at the low gigahertz end up to Earth-observation radiometers and security scanners up in the hundreds of gigahertz end. They have applications in the communications industry within optical-fiber, satellite communications, and point-to-point links; in the automotive industry within autotolling, vehicle identification, road-traffic information, and cruise-control systems; and in the military industry within electronic warfare, missile seeker heads, and phased-array radar systems.

Due to their monolithic nature, MMICs are manufactured as small parts of a whole wafer of the semiconductor material. The processing of the wafers involves forming microscopic features on their surface, so all the equipment is installed in a clean room environment to prevent dust and moisture from affecting the features. This tends to make the fabrication process very time-consuming and costly, and often the correct functionality of the circuit cannot be checked until the whole of the processing is complete. For these reasons, it is imperative that the design of the chip be right the first time or corrected before being committed to the fabrication process. Because of the importance of the design stage during the MMIC production process, this book seeks to explore MMIC design within the overall context of a foundry MMIC production run, including the technology/foundry choices, foundry characterization and modeling of the component elements, simulation and layout of the circuits, and the constraints of the processing and test methods. This book is not intended to be overly theoretical but instead to present the MMIC design techniques and practices that are used day-to-day within the MMIC design industry.

1.2 The History of MMICs

The reason that MMICs are used in systems applications instead of alternative techniques, such as waveguide or hybrid alumina circuits, is best explored by considering the history and development of today's MMICs.

The first step toward monolithic circuits was made by Jan Czochralski in 1916 when he developed a method for growing single crystals [1], but the electronics industry was not to move from vacuum-valve to solid-state technology until the transistor was invented by Bell Telephone Laboratories in 1947 [2]. Transistors, being tiny, more efficient, and much more reliable compared to valves, allowed designers to create much more complex circuits and systems, with an order of magnitude more interconnections. This multitude of interconnections brought with it its own problems because soldering so many connections was slow, costly, and, again, unreliable. This left the electronics industry searching for an economic and reliable way of making complex circuits. The solution to this problem was the IC, first patented by Jack Kilby of Texas Instruments in 1959 [3–5]. ICs can be manufactured using photolithography, which enables the multitude of components and interconnections to be printed onto the surface of a single piece of semiconductor material in just a few well-controlled process steps. Incidentally, Kilby's first integrated circuits were made of germanium (Ge), while another engineer, Robert Noyce of Fairchild Semiconductor, was awarded a patent in April 1961 for a more complex "unitary circuit" made of silicon (Si) [6, 7].

The speed of these early integrated circuits was slow by today's standards—for example, the computers aboard the Apollo space craft had only a 2-MHz clock [8]—but the extra complexity achievable in ICs enabled a single chip to perform the mathematical calculations that at the time required a large electromechanical desktop apparatus. Kilby demonstrated this with the invention of the pocket calculator [9], which became the first commercial product to use ICs and helped give impetus to the fledgling silicon IC industry.

One of the first silicon ICs operating at microwave frequencies (silicon MMIC) was an X-band transmit/receive (Tx/Rx) switch developed in 1966, but the insertion loss was too high for it to be used in the final system [10]. At microwave frequencies, the wavelength of the signals on the semiconductor become comparable with the dimensions of the circuit, so the interconnecting wires must be designed as transmission lines, and the signals interact much more with the semiconductor substrate material. If the resistivity of the substrate is low, the transmission lines are lossy, and the resulting circuits have excessive insertion loss. These early silicon MMICs suffered from what was known as inversion, where the initially high-resistivity silicon substrate exhibited low resistivity after the high-temperature processing of other components on the substrate [11]. The silicon transistors at this time were useful up to frequencies between 1 and 6 GHz [12], but the inversion problem obstructed further silicon MMIC development. One approach produced a successful 500-MHz intermediate frequency (IF) amplifier using chip transistors mounted on a monolithic circuit of passive components such as capacitors and inductors [13], but it was not a fully monolithic solution.

Meanwhile, interest in other semiconductor materials was increasing, and in 1962 the liquid-encapsulated Czochralski (LEC) method was developed for growing single crystals of materials with high vapor pressures at their melting points. This was initially applied to lead selenium (PbSe) and lead tellurium (PbTe) [14], and just as gallium arsenide (GaAs) was recognized as a suitable substrate for MMICs [15, 16], LEC was successfully applied in 1965 to produce stable high-resistivity GaAs material [17]. Also in 1965, the first GaAs field effect transistors were fabricated by Jim Turner at Plessey Research, Caswell, in the United Kingdom [18] and by C. A. Mead at the California Institute of Technology in the United States [19]. Turner's device had a gate length of ~24 μm and exhibited gain at very high frequency (VHF). In 1967, GaAs field effect transistors (FETs) showed the capability to operate at microwave frequencies [20], and Plessey produced a 4-μm-gate-length GaAs metal semiconductor field effect transistor (MESFET) with 10-dB gain at 1 MHz. This became the world's first commercial GaAs MESFET, sold as the Plessey GAT 1. The year 1968 saw the publication of the first simple monolithic microwave GaAs circuits using diodes and microstrip lines [21, 22], and in 1970, GaAs FETs were reported, for the first time, to outperform silicon transistors at microwave frequencies [23]. This ability to produce high-performance transistors on a stable high-resistivity material made GaAs the number one choice for the next few decades of MMIC development.

GaAs IC development was picking up pace, and in 1976, the first monolithic microwave integrated circuit, or MMIC, using a field effect transistor was reported by R. S. Pengelly and J. A. Turner [24], as shown in Figure 1.2. This MMIC, incorporating a single 1-μm-gate-length MESFET and a single metal layer forming single-turn loop-inductors and interdigital capacitors, exhibited a few decibels of gain from 7 to 12 GHz and gave impetus to the growth of a worldwide GaAs MMIC industry.

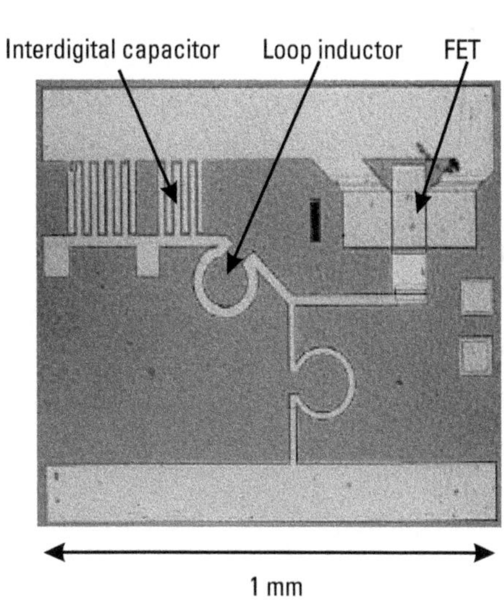

Figure 1.2 World's first GaAs FET MMIC amplifier. (*Source:* Bookham Inc., 2006. All Rights Reserved.)

This growth was accelerated when, in 1979, the Institute of Electrical and Electronics Engineers (IEEE) established the first symposium dedicated to GaAs IC developments, and in 1985, the increased optimism in the industry was demonstrated when Plessey Caswell commissioned a 0.7-μm-gate-length MESFET MMIC process on 2-in.-diameter GaAs wafers. The year 1985 also heralded the era of "band-gap engineering," which is the technique of mixing different semiconductor materials to create transistors with specific solid-state properties. This eventually led to the development of a high-electron-mobility transistor (HEMT) low-noise amplifier (LNA) MMIC in 1988 and a heterojunction bipolar transistor (HBT) power amplifier in 1989 [25]. Other milestones along the MMIC development pathway include the appearance of an indium phosphide (InP) 5–100-GHz traveling wave amplifier in 1990 [25] and the launch of Plessey's commercial 0.2-μm-gate-length pseudomorphic HEMT (pHEMT [26]) process in 1996. Further information on the history and development of MMICs can be found in [27–29].

Today, there are MMIC processes on many different semiconductor materials with transistors that exhibit gain over the whole of the microwave frequency range. Silicon MMICs are also in resurgence using transmission-line techniques that overcome the substrate-loss problems [30–32] and silicon germanium (SiGe) transistors with frequency responses comparable to GaAs [33]. The future of MMICs looks certain to continue in the direction of even more exotic semiconductor materials and ever more complicated design techniques, where the eventual capabilities of the chips will be limited only by engineers' imaginations.

1.3 The MMIC Advantage

To summarize the history of MMIC development, MMICs emerged because they combined high-performance microwave transistors with low-loss passive components and transmission lines and could be formed as complex circuits with multiple interconnections using just a few photolithographic process steps.

The microwave frequency response of the transistors also required that their dimensions be on the order of microns so that the resulting size of the chips was only a few millimeters, which is an order of magnitude smaller than the equivalent hybrid microwave integrated circuit (MIC) of packaged transistors mounted on alumina substrates. The small dimensions of MMICs also mean they have an order of magnitude less weight than their equivalent hybrid MICs. These two features of MMICs make them ideally suited to mobile electronic applications, such as mobile cell phones and portable computers, where miniaturization and lightness help to give the products a commercial advantage.

The reliability of MMICs, while always an issue with new devices on novel semiconductor materials, are well understood on Si and GaAs [34], and many MMICs and MMIC processes have such high reliability levels that they have qualified for space-borne applications [35–37]. In fact, the rugged monolithic nature of MMICs, together with their small size and weight, makes them essential components for space-borne equipment [38].

Their low cost can also be an advantage, as they cost one-third as much as the equivalent hybrid circuit [38], but this is not always clear-cut in every application. The high running cost of a semiconductor fabrication facility means that processing a wafer batch is expensive, and if used to make only a handful of chips, then each is very expensive. The low-cost advantage of MMICs is only ensured when high numbers of the chips are required, and the design produces a high-yielding circuit. This book provides the practical techniques to help the MMIC designer maintain the cost advantage by producing high-yielding MMIC designs.

1.4 Basic Design Process

The following chapters of this book are laid out in the order of the basic design process, assuming that the MMIC designer is in possession of the full requirement specification and has complete freedom of choice of MMIC processes.

The first choice that must be made is the component technology that will be used to make the chip, such as the substrate material, the type of transistor, and the available passive components. This choice will then lead to deciding which foundry process is required. At this stage, the designer will need to contact the potential MMIC foundries to discuss using their processes and get more details about their capabilities. If the requirement is for multiple functions, this is also a good point to consider the economic trade-offs between several single-function chips or one larger multifunction MMIC. The other issue that is closely connected with the foundry is the computer-aided design (CAD) simulation tools they support and for which they can supply complete model libraries of their MMIC components. This issue covers not just whether they support one particular company's CAD tools but also whether they can provide the most appropriate (e.g., nonlinear) models for the required specification that work within the CAD tools.

When the foundry is selected and the designer has acquired the CAD tools and libraries, the next step is the actual MMIC design, which is the main focus of this book. Design and layout are actually two separate chapters within this book, but the designer will find that toward the end of the design phase, they must be performed concurrently to achieve a compact and high-yielding chip design.

After completion of the design and layout, photolithographic masks are manufactured from the design data, and batches of wafers are processed with them through the foundry clean-room fabrication process. The wafers are then tested using RF-on-wafer (RFOW) equipment before being diced into individual chips, completing the MMIC production process.

References

[1] Czochralski, J., "Ein neues VerFahren zur Messung der Kristallisationsgeschwindigheit der Metalle," *Z. Phys. Chem.,* Vol. 92, No. 219, 1918 (in German).

[2] Shockley, W., and N. J. Madison, "Circuit Element Utilizing Semiconductive Material," U.S. Patent No. 2,569,347, filed June 1948, issued September 1951.

[3] Kilby, J. S., "Miniaturized Electronic Circuits," U.S. Patent No. 3,138,743, filed February 1959, issued June 1964.

[4] Stewart, R. F., "Integrated Semiconductor Circuit Device," U.S. Patent No. 3,138,747, filed February 1959, issued June 1964.

[5] Kilby, J. S., "Method of Making Miniaturized Electronic Circuits," U.S. Patent No. 3,261,081, filed February 1959, issued July 1966.

[6] Noyce, R., "Microelectronics," *Scientific American,* Vol. 23, No. 3, September 1977, pp. 63–69.

[7] Lehovec, K., "Isolation of p-n Junction," U.S. Patent No. 3,029,366, filed April 1959, issued April 1962.

[8] Hall, Eldon C., *Journey to the Moon: The History of the Apollo Guidance Computer,* Reston, VA: AIAA, 1996.

[9] Kilby, J. S., J. Merryman, and J. Van Tassel, "Miniature electronic calculator," U.S. Patent No. 3,819,921, filed in 1967, issued June 1974.

[10] Ertel, A., "Monolithic Microwave Duplexer Switch," U.S. Patent No. 3,475,700, filed December 1966, issued October 1969.

[11] Portnoy, W. M., and H. M. Leedy, "A Monolithic Single-Ended X-band Mixer Circuit," *IEEE J. Solid State Circuits,* Vol. 3, March 1966, pp. 22–23.

[12] Cooke, H. F., "Microwave Transistors: Theory and Design," *Proc. IEEE,* Vol. 59, No. 8, 1971, p. 1163.

[13] Johnson, G. D., W. H. Tulloch, and F. Opp, "Integrated 500-MHz IF Amplifier," *ISSCC Digest,* February 1966, pp. 26–27.

[14] Metz, E. P. A., R. C. Miller, and J. Mazelski, *J. Appl. Phys.,* Vol. 33, 1962, p. 2016.

[15] Uhlir, A., Jr., "Microwave Applications of Integrated Circuit Techniques," *Proc. IEEE,* Vol. 52, December 1964, pp. 1617–1623.

[16] Cronin, G. R., and R. W. Haisty, "The Preparation of Semi-Insulating Gallium Arsenide by Chromium Doping," *J. Electrochem. Soc.,* Vol. 111, July 1964, pp. 874–877.

[17] Mullin, J. B., B. W. Stranghan, and W. S. Brickell, *J. Phys. Chem. Solids,* Vol. 26, 1965, pp. 782.

[18] Arnold, J., "FET Technology for Low-Noise Front Ends," *IEEE Communication Magazine,* September 1983, pp. 37–42.

[19] Greiling, P., "The Historical Development of GaAs FET Digital IC Technology," *IEEE Trans. Microwave Theory Tech.,* Vol. 32, No. 9, September 1984, pp. 1144–1156.

[20] Hooper, W. W., and W. I. Lehrer, "An Epitaxial GaAs Field-Effect Transistor," *Proc. IEEE,* Vol. 55, July 1967, pp. 1237–1238.

[21] Mehal, E. W., and R. W. Wacker, "GaAs Integrated Microwave Circuits," *IEEE Trans. Microwave Theory Tech.,* Vol. 16, July 1968, pp. 451–454.

[22] Mao, S., S. Jones, and G. D. Vendelin, "Millimeter-Wave Integrated Circuits," *IEEE Trans. Microwave Theory Tech.,* Vol. 16, July 1968, pp. 455–461.

[23] Drangeid, K. E., R. Sommerholder, and W. Walker, "High-Speed Gallium-Arsenide Schottky-Barrier Field-Effect Transistor," *Electronics Lett.,* Vol. 6, April 1970, pp. 228–229.

[24] Pengelly, R. S., and J. A. Turner, "Monolithic Broadband GaAs FET Amplifiers," *Electronics Lett.,* Vol. 12, No. 10, May 1976, pp. 251–252.

[25] Niehenke, E. C., R. A. Pucel, and I. J. Bahl, "Microwave and Millimeter-Wave Integrated Circuits," *IEEE Trans. Microwave Theory Tech.,* Vol. 50, March 2002, pp. 846–857.

[26] Morkoc, H., et al., "Semiconductor with Strained InGaAs Layer," U.S. Patent No. 4,827,320, May 2, 1989.

[27] Pucel, R. A., "Design Consideration for Monolithic Microwave Circuits," *IEEE Trans. Microwave Theory Tech.,* Vol. 29, June 1981, pp. 513–534.

[28] Pengelly, R. S., *Microwave Field Effect Transistors—Theory Design and Applications,* New York: Wiley, 1982.

[29] Soares, R., J. Graffeuil, and J. Obregon, *Applications of GaAs MESFETs,* Dedham, MA: Artech House, 1983.

[30] Henri, H., et al., "Ultra Low Loss Transmission Lines on Low Resistivity Silicon Substrate," *IEEE MTT-Symposium Digest,* Vol. 3, June 2000, pp. 1809–1812.

[31] Leung, L. L. W., Hon Wai-Cheong, and K. J. Chen, "Low-Loss Coplanar Waveguide Interconnects on Low-Resistivity Silicon Substrate," *IEEE Trans. Packaging and Manufacturing Tech., Part A: Packaging Technologies,* Vol. 27, No. 3, September 2004, pp. 507–512.

[32] Laney, D. C., et al., "Low-Loss Microwave Transmission Lines and Inductors Implemented in a Si/SiGe HBT Process," *IEEE Bipolar/BiCMOS Circuits Tech. Meeting,* Minneapolis, MN: September 27–29, 1998, pp. 101–104.

[33] Whitaker, T., "IEDM Highlights Include SiGe HBTs Operating at 350 GHz," *Compound Semiconductor Magazine,* Institute of Physics Publishing Ltd., London, January 2003, available at http://www.compoundsemiconductor.net/articles/magazine/9/1/2/1, last access June 4, 2006.

[34] Christou, A., *Reliability of GaAs MMICs,* West Sussex, England, United Kingdom: John Wiley & Sons, 1992.

[35] Roesch, W., T. Rubalcava, and C. Hanson, "Lifetesting GaAs MMICs under RF Stimulus," *IEEE MMWMC Symposium,* Albuquerque, NM, June 1992.

[36] Potukuchi, J. et al., "Space Qualification of Monolithic Microwave Integrated Circuits," *15th AIAA Int. Comm. Sat. Systems Conf.,* San Diego, CA, March 1992.

[37] Conlon, R. F. B., and C. Lindsay, "Evaluation of a Commercial Monolithic Microwave Integrated Circuit Process for Space Applications," *Preparing for the Future,* European Space Agency, Vol. 6, No. 3, September 1996.

[38] Cazaux, J.-L., "MMICs for Space-Bourne Applications: Status and Prospectives," *GaAs IC Symposium,* 1994.

2

Component Technology and Foundry Choice

The first choice that must be made when designing an MMIC is the component technology that will be used to make the chip, and this choice will then lead to a decision as to which foundry process is required. The primary factor is the choice of substrate material as the properties of this semiconductor material have the greatest influence over the potential performance of the final MMIC. The electron mobility and peak velocity in the doped semiconductor determine how fast the electrons in the active components, such as the transistors, can react to quickly fluctuating electric fields, dictating their frequency response, and the energy band-gap of the semiconductor substrate determines how high the breakdown voltage of the transistors will be, hence their power-handling characteristics. The resistivity of the semiconductor substrate in its semi-insulating state also has an important effect on the circuit as its value determines the loss and Q-factor of the passive components created on its surface.

The next technology choice that must be made is the type of transistor that will be used as the active component to create the signal gain within the MMIC. At the simplest level, this is a choice between using a field effect transistor (FET) or a bipolar transistor, but there is a large range of device types, from traditional silicon complementary metal-oxide semiconductor

(CMOS) transistors to devices such as high-electron-mobility transistors (HEMT) with quantum well confinement of the electrons to improve their mobility.

This chapter outlines the choice of component technology available to today's designers, starting with the properties and attributes of the different semiconductor substrate materials, and going on to describe the types of transistors that are made with the active surface layers of the substrate material. This chapter also discusses the passive components that can be made with an MMIC process and how substrate technology choices can influence their characteristics. Obviously, each different technology is associated with a particular foundry, so this chapter is intended to help the MMIC designer choose the most appropriate foundry technology for a given specification.

2.1 Active Components

The active components on an MMIC are those that provide voltage or current gain to the signals on the chip—in other words, the transistors. These are known as active devices because they can use dc bias power to increase the RF signal power actively within the circuit and even to generate the signals in the first place. MMIC diodes are sometimes mistakenly classified as active components due to their nonlinear characteristics and use for frequency conversion in mixers, but they are strictly passive components and cannot convert dc power into RF power.

The elements making up the semiconductor substrate material determine the potential performance characteristics of the transistors, primarily due to the electron properties within the doped semiconductor. The faster the electron peak velocity and the higher the electron mobility (without colliding with other atoms), the more quickly the electrons can react to applied high-frequency signals and the transistor will have gain at higher frequencies. The next section details the different semiconductor substrate materials used in MMIC processes and compares their properties.

2.1.1 Substrate Material

The most commonly used substrate materials are silicon (Si) and gallium arsenide (GaAs), where Si is traditionally used for RF and lower frequencies and GaAs is used for microwave and millimeter-wave frequencies. However, with the advancement of silicon germanium (SiGe) transistors, the frequency response of circuits on Si substrates is pushing firmly into the microwave-frequency region. GaAs, on the other hand, is finding more applications further into the millimeter-wave-frequency region using transistors with gate lengths of one-tenth of a micron [1, 2]. More exotic substrate materials, such as indium phosphide (InP), are tending to take over from GaAs as frequencies extend beyond 100 GHz [3]. Other substrate materials that are increasingly being used include silicon carbide (SiC) [4] and gallium nitride (GaN) [5]. These are both wide band-gap semiconductors, which means they have much higher breakdown voltages and can operate at higher junction temperatures and higher output powers than the other semiconductor materials. The characteristics of these commonly used semiconductor materials are shown in Table 2.1.

Be aware that the SiGe semiconductor is only an epitaxial layer to provide high electron mobility in the transistors and that the substrate material

Table 2.1
Characteristics of the Most Commonly Used Semiconductor Materials

Material	Electron Mobility (cm^2/Vs)	Peak Velocity ($\times 10^7$ cm/s)	Frequency Range (GHz)	Noise Figure	Gain	Comments
Si	900–1,100	0.3–0.7	< 20	Moderate	Moderate	Mature process; 12-in. wafers
SiGe	2,000–300,000	0.1–1.0	10–40	Lower	Better	Benefits from Si; 6-in. wafers
SiC	500–1,000	0.15–0.2	15–20	Poor	Lower	4-in. wafers
GaAs	5,500–7,000	1.6–2.3	>75	Lower (F_{min} at 26 GHz = 1.1 dB)	Higher (G_{ass} at 26 GHz = 9.0)	Less mature than Si; 3-, 4- and 6-in. wafers
GaN	400–1,600	1.2–2.0	20–30	Poor	Lower	Less mature than GaAs; V_{br} > 100V; 2-in. wafers
InP	10,000–12,000	2.5–3.5	>115	Lower (F_{min} at 26 GHz = 0.9 dB)	Higher (G_{ass} at 26 GHz = 11.1)	Less mature than GaAs; 2-in. wafers

is Si, with much the same benefits and disadvantages as a purely Si process. In a similar way, InP is sometimes grown as an epitaxial layer on a gallium arsenide (GaAs) substrate to get the performance of InP transistors on a larger and better-established semiconductor wafer. This type of substrate is known as metamorphic because the lattice constant of the GaAs substrate must be gradually changed through the epitaxial layers to that of InP [6]. InP as a substrate material is more brittle and difficult to handle than GaAs. One drawback of InP transistors, on either substrate, is that they have a low Schottky barrier height, limiting the gate voltage swing and causing higher reverse diode leakage through the gate contact [7]. A graph showing the typical output power from various transistor technologies as a function of frequency is shown in Figure 2.1.

2.1.2 Transistor Type

Transistors are three terminal devices that act as switches, where a signal on one input can control the current flowing between two other connections. Incidentally, the name *transistor* comes from the fact that it is a variable re*sistor* where the voltage on one terminal is *trans*ferred to control the current

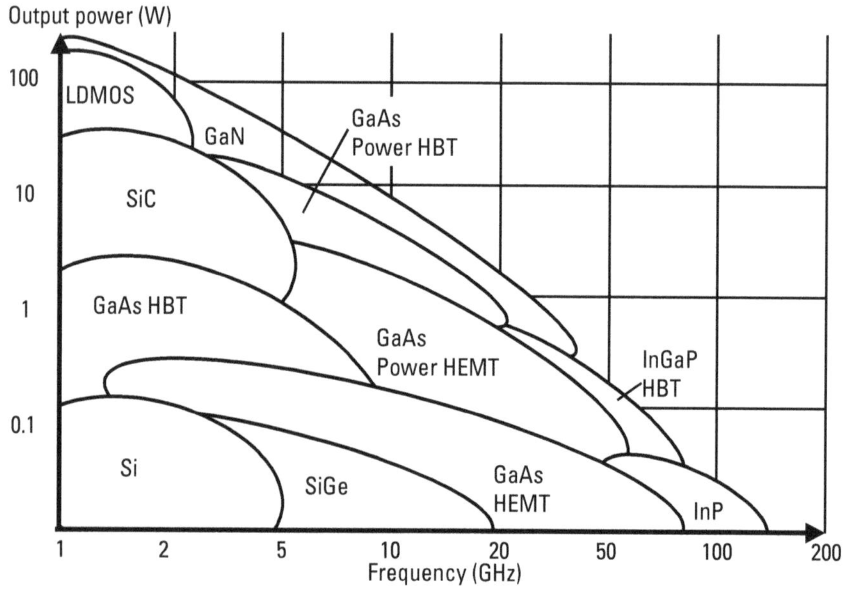

Figure 2.1 Plot of output power versus frequency for various transistor technologies.

through the other two terminals. The two basic types of active devices used in MMICs are the field effect transistor (FET) [8] and the bipolar transistor [9], and their symbols, together with the switch representation, are shown in Figure 2.2. This switching can be in a digital mode (as in computer logic), where the transistor switches between two states (off and on), or in an analog mode (most MMIC applications), where small changes in the input signal (at the gate/base terminal) are amplified between the other two connections (source and drain terminals and collector and emitter terminals).

The basic operation of the FET is that it is typically biased with a positive voltage V_{DD} (typically +5V) on the drain terminal, and the source is grounded so that current flows from the drain to the source, known as the drain current (I_D). The drain current is controlled by the voltage between the gate and the source (V_{GS}). When the gate voltage (V_{GS}) is 0, drain current can pass, and the FET is switched "on"; when the gate voltage is large and negative (typically −5V), no drain current can pass, and the FET is switched "off."

The bipolar transistor has a similar operation but requires different control voltages on the base. It is commonly biased with a positive voltage V_{CC} (typically +5V) on the collector, and the emitter is grounded, so current flows from the collector to the emitter. This current is controlled by the voltage between the base and the emitter (V_{BE}). When V_{BE} is 0, no current can pass, and the transistor is switched "off," and, typically, when V_{BE} is greater than +0.7V, the collector current can pass, and the transistor is switched "on."

A detailed description of the FET and bipolar transistor construction operation and bias voltages is given in the following sections. The types of FET include the metal semiconductor field effect transistor (MESFET) [10] and the high-electron-mobility transistors (HEMTs) [11]. The types of bipolar transistor include traditional silicon complementary metal-oxide

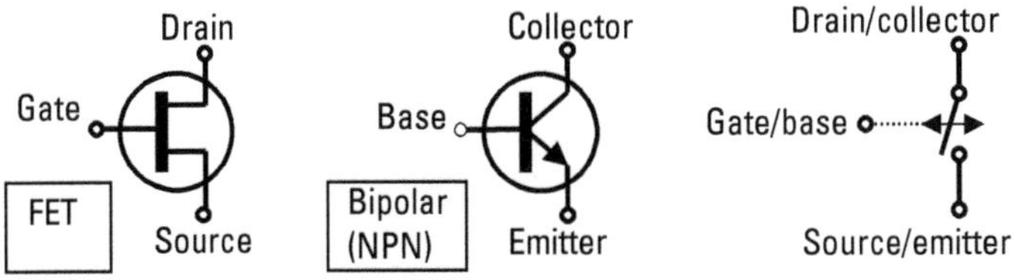

Figure 2.2 Electrical symbol and switch representation of field effect and bipolar transistors.

semiconductor (CMOS) transistors [12] and heterojunction bipolar transistors (HBTs) [13, 14]. These different transistor types are generally best suited to certain MMIC circuit types, as indicated in Table 2.2. This table, based on my experience and the literature [15–18], is not definitive but intended solely as a guide to new MMIC designers.

2.1.2.1 Field Effect Transistors

Construction and Operation of FETs

The top view of an FET, as if looking down onto the surface of an MMIC wafer, is shown in Figure 2.3. The current flows laterally across the surface of the wafer from the drain to the source and passes under the gate contact.

Table 2.2
MMIC Circuit Types Best Suited to the Different Transistor Types

	Bipolar	FET			
	CMOS	SiGe HBT	GaAs/ InP HBT	MESFET	HEMT
Oscillator	—	✓	✓	—	—
Mixer	—	✓	✓	—	—
Low-noise amplifier	—	✓	✓	—	✓
Power amplifier	—	—	✓	—	✓
Switch	—	—	—	✓	✓
Digital	✓	✓	—	—	—

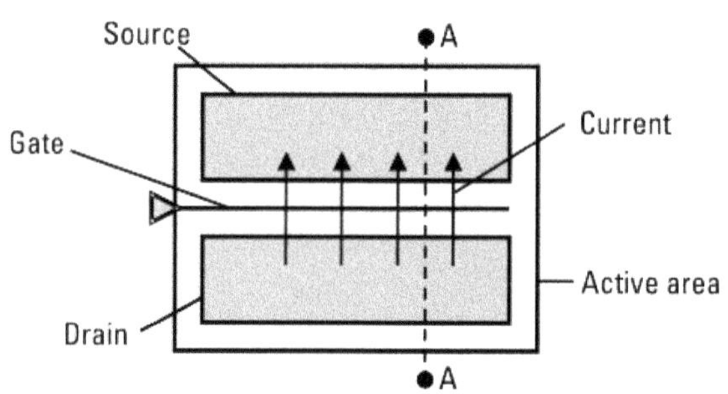

Figure 2.3 Top view of a field effect transistor.

Note that this is just a single gate FET (or unit cell), and that devices, particularly power FETs, are constructed from multiple gate fingers. (See Section 5.3.3.)

The sectional view "A-A" of the FET in Figure 2.3 is shown in Figure 2.4. The FETs are formed with a low-doped layer of the semiconductor, which forms a conducting channel below the surface of the wafer, as shown in Figure 2.4(a). The channel is normally *n*-doped so that there are free electrons to carry current in the channel. Metal source and drain terminals make contact with this conducting channel via ohmic contacts to heavily doped layers of the semiconductor. If a voltage is placed across the drain and source contacts, electric current can flow between them until all the free electrons in the channel are conducting the current; if there is zero voltage on the gate terminal, this current is known as the drain-source saturation current (I_{DSS}). This is the "on" state for a field effect transistor.

If a small negative voltage, relative to the source voltage (V_{gs}), is applied to the gate terminal, as in Figure 2.4(b), the negatively charged electrons within the channel are repelled from the gate, and an area of the channel, known as the depletion region, becomes depleted of free electrons. The effect of depleting some of the channel of its free electrons is that only the bottom part of the channel has free electrons to carry current, so the maximum current flowing through the channel is reduced. If a larger negative voltage is applied to the gate terminal (V_{gs}) as in Figure 2.4(c), the electrons are repelled even further away from the gate, and the depletion region extends all the way across the channel. When the depletion region extends all the way across the channel, there are no free electrons to carry the current; the FET is said to be pinched off, and the gate voltage at which this occurs is known as the pinch-off voltage (V_P). When the gate voltage (V_{gs}) is either set at or more negative than the pinch-off voltage, then the FET is in the "off" state.

Note that with 0V on the gate (V_{gs}), the channel is slightly pinched off by the gate Schottky-junction potential of about 0.7V, so the maximum current through the channel can actually be increased above I_{DSS} to I_{MAX} by a small amount of positive bias on the gate.

Biasing of FETs

The saturation and pinch-off gate bias points are the two extremes of the gate bias, but the FET can be biased at any gate bias voltage between the two, with the most common bias point being the gate voltage that allows a current of half the zero bias saturation current ($I_{DSS}/2$). For a GaAs MESFET, the gate bias voltage that gives $I_{DSS}/2$ is about –0.5V. A typical plot of drain

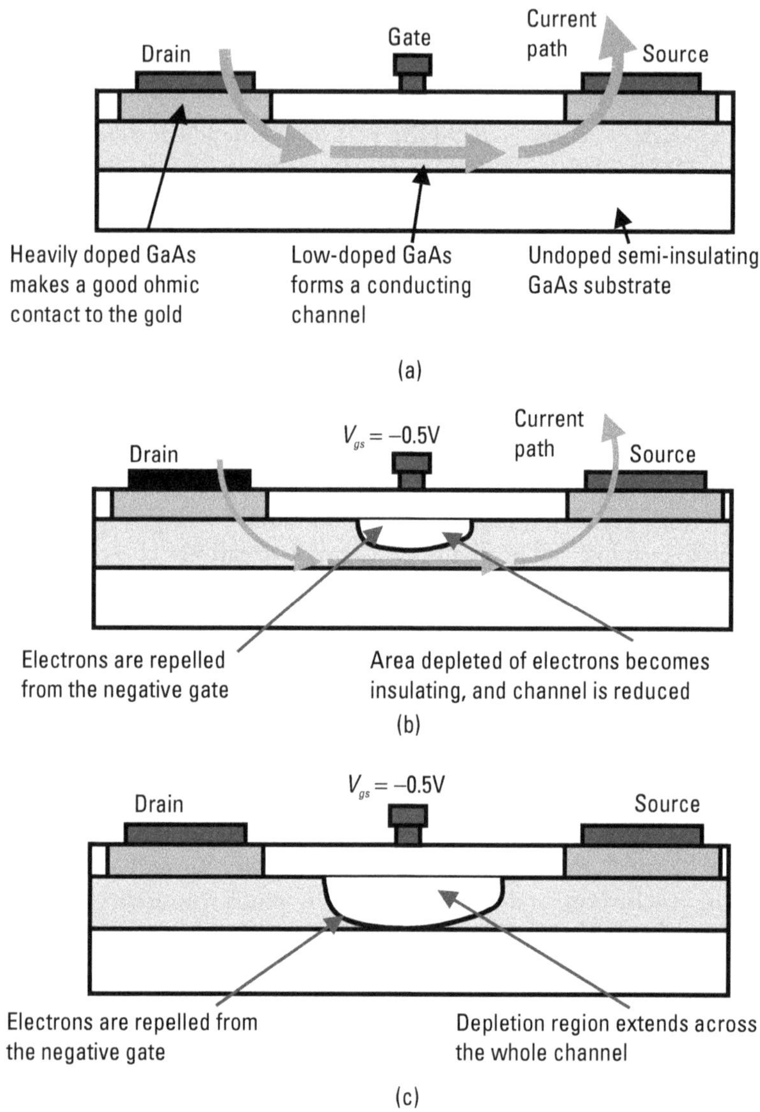

Figure 2.4 Sectional view of a field effect transistor with (a) 0V on the gate, (b) −0.5V on the gate, and (c) −1.0V on the gate, relative to the source voltage. With no voltage on the gate, current can flow from drain to source. With a small negative voltage on the gate, current flow is reduced. With a large negative voltage on the gate, current flow is stopped, and the transistor is switched off (known as pinch-off because the channel is pinched closed).

current versus drain voltage (*I*/ *V*) for several different gate bias voltages is shown in Figure 2.5.

At low drain voltages, in the linear or knee region, the drain current rises linearly with the drain voltage until all the free electrons in the channel are passing the current. At this point, the channel is saturated, so higher drain voltages do not increase the drain current. This area of the *I*/ *V* plot is known as the saturation region. At even higher drain voltages, the semiconductor material itself breaks down, and the current increases again rapidly; this is known as the breakdown region.

If the drain of the FET is loaded with a resistance R_L, varying the gate voltage will cause the drain voltage and current to follow a sloping path on the *I*/ *V* plot known as the load line, as shown in Figure 2.6. Now it can be seen that small voltage variations (RF signals) on the gate terminal create larger voltage variations at the drain terminal; in other words, they create voltage gain.

Different combinations of bias point and load resistance are used for different applications; for example, power amplifiers generally require a load line that allows the maximum swing of both drain voltage and drain current to achieve the maximum output power, while small-signal amplifiers require load impedance that gives the best small-signal gain. Similarly, the bias point is different for various applications because an FET biased near the pinch-off

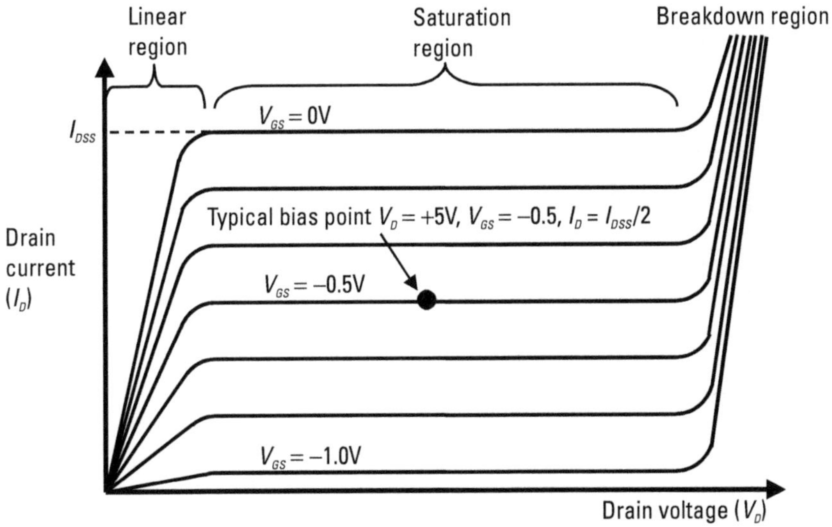

Figure 2.5 Current/voltage plot for an FET for several different gate voltages.

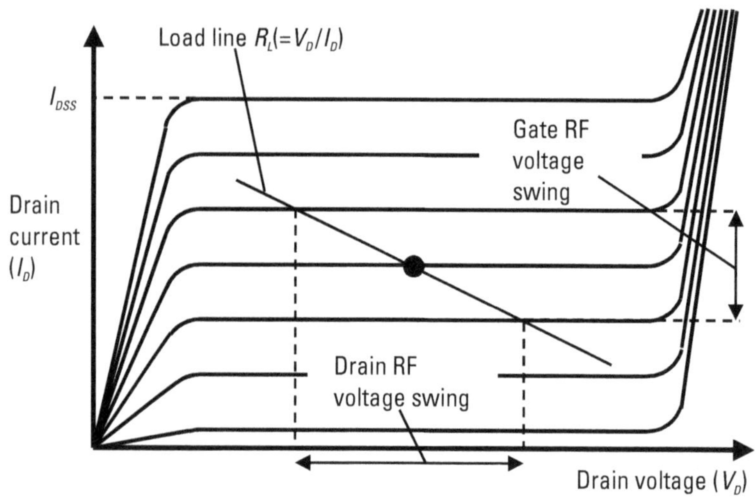

Figure 2.6 *I*/*V* characteristics of an FET loaded with resistance R_L.

voltage passes less current and has a lower noise figure applicable to low-noise amplifier design. However, near the pinch-off voltage, the voltage gain is also lower, so high-gain amplifiers are usually biased nearer $I_{DSS}/2$. This can be seen in Figure 2.7, which shows the minimum noise figure and associated

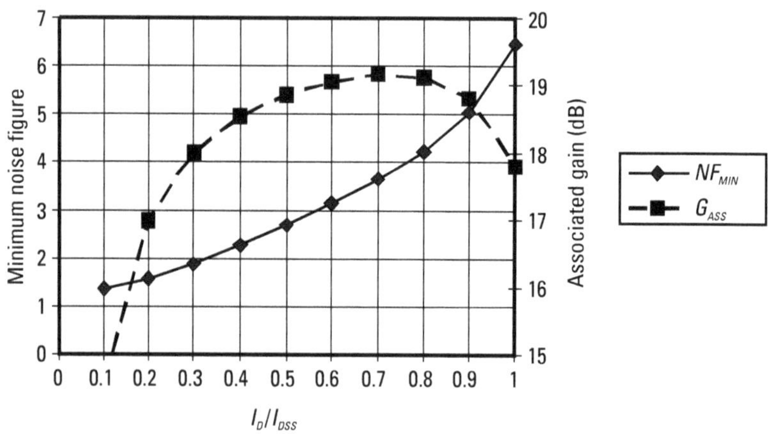

Figure 2.7 Minimum noise figure and associated gain versus drain current as a fraction of I_{DSS} for a 2 × 75 pHEMT [simulated with Agilent ADS and the PH25 Design Kit, courtesy of United Monolithic Semiconductors (UMS)].

gain for a 2 × 75–μm pHEMT as the drain current (I_{DS}) is varied from 10% to 100% of the drain saturation current (I_{DSS}).

Linearity is also an important characteristic of the active device, and it measures how linearly the drain current varies as a function of the gate voltage along the load line. Pictorially, it is a measure of how parallel and evenly spaced the *I*/*V* curves are around the bias point. This is usually optimum in the middle of the *I*/*V* plot and is primarily a function of the device technology, with GaAs MESFET and Si lateral double diffuse metal-oxide semiconductor (LDMOS) transistors usually being more linear than Si bipolar transistors for power-amplifier applications [19].

MESFET

The metal semiconductor field effect transistor (MESFET) is so called because gate contact is formed by a metal-to-semiconductor junction [10]. If the semiconductor material is only low doped, the resulting Schottky contact between the gate metal and the semiconductor makes a very-low-leakage gate contact.

HEMTs

The high-electron-mobility transistor (HEMT) operates like any other FET, except the channel is constructed from a junction of two different types of semiconductor material (known as a heterojunction) to give the free electrons in the channel higher mobility [11, 20]. For a GaAs HEMT, the other semiconductor material is typically aluminum gallium arsenide (AlGaAs). The free electrons provided by the *n*-type dopant atoms are confined very closely to the heterojunction and form what is called a two-dimensional electron gas. This two-dimensional (2D) electron gas is confined away from the lattice atoms so they will not collide with them, and this gives the electrons a higher mobility. Another consequence of the electrons' experiencing less collision is that the noise figure of the HEMT is much lower than that of ordinary FETs. An example HEMT LNA MMIC developed for a satellite receiver is shown in Figure 2.8.

HEMTs are sometimes referred to in the literature as heterostructure FETs [21] or heterojunction FETs (HFETs) [22], or even modulation doped FETs (MODFETs) [23], but they are all high-electron-mobility field effect transistors (HEMTs) because they have a heterojunction between different types of semiconductors, which increases the mobility of the carriers in the channel. Pseudomorphic HEMTs (pHEMTs) use an extremely thin layer of the different semiconductor [typically indium gallium arsenide (InGaAs)],

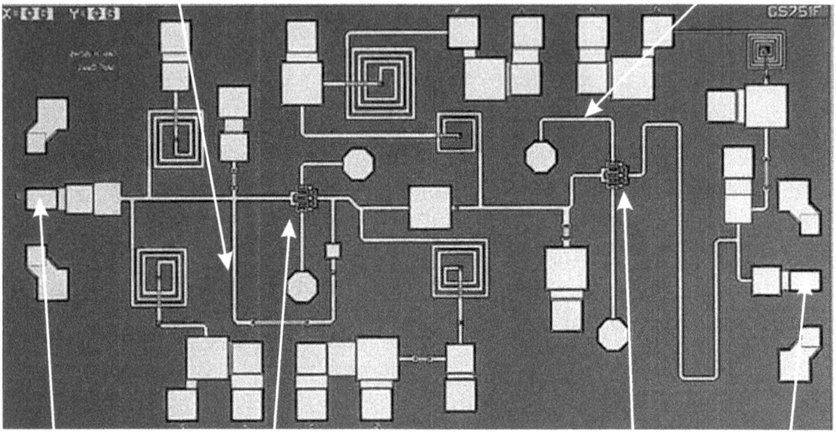

Figure 2.8 A HEMT MMIC LNA. (Courtesy of TM R&D, Malaysia).

which is strained to the lattice constant of the surrounding semiconductor (typically AlGaAs), making a pseudomorphic layer (an InGaAs layer with the lattice dimensions of AlGaAs); this gains the enhanced electron transport properties of InGaAs while the full MMIC is still being fabricated on a GaAs substrate [24]. A metamorphic HEMT (mHEMT) has a higher-mobility semiconductor layer, such as InP, grown on its surface with its own natural lattice constant. To achieve this, the lattice constant of the GaAs substrate must be gradually changed through the epitaxial layers to that of InP [6].

2.1.2.2 Bipolar Transistors

First, let me start by stating that this discussion of the operation of bipolar transistors does not suffer from the common misconception that bipolar transistors are current-controlled devices; it will explain that they are in fact voltage-controlled devices [25, 26]. One consequence of their base-emitter voltage-controlled operation is that there is a fixed ratio between the base and collector currents, and this ratio, or current gain beta (β), is often used to set the bias point, which leads to the misconception that the base current is somehow controlling the collector current.

Construction and Operation of Bipolar Transistors

The top view of a bipolar transistor, as if looking down onto the surface of an MMIC wafer, is shown in Figure 2.9. The current flows down through the

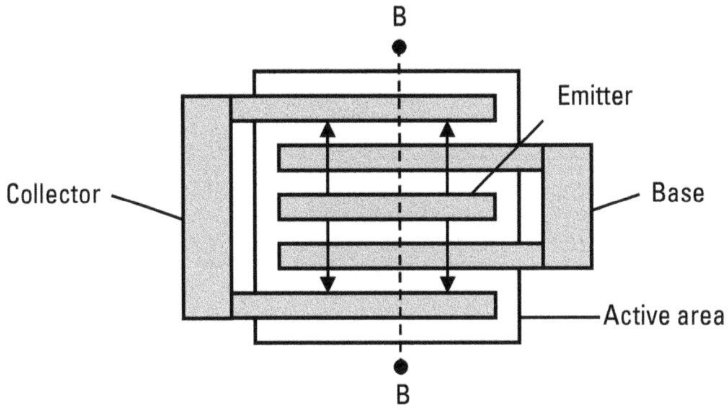

Figure 2.9 Top view of a bipolar transistor.

emitter and then laterally to the collector contacts. A scanning electron microscope (SEM) photo of an HBT is shown in Figure 2.10.

The sectional view "B-B" of the bipolar transistor in Figure 2.9 is shown in Figure 2.11. Bipolar transistors are so-called because they are formed from two polarities of doped semiconductors, *n*-type ("*n*" for negative charge), doped to have extra electrons for passing current, and *p*-type ("*p*" for positive charge), doped to have extra holes for passing current. They are usually constructed as three horizontal layers of the doped semiconductor, as shown by the NPN structure shown in Figure 2.11, and the current path is primarily vertical through the layers. This example is drawn as a

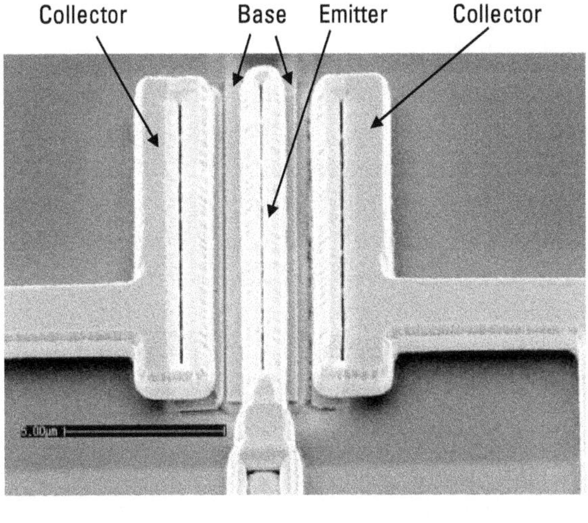

Figure 2.10 SEM photo of an HBT. (*Source:* Bookham Inc., 2006, All Rights Reserved.)

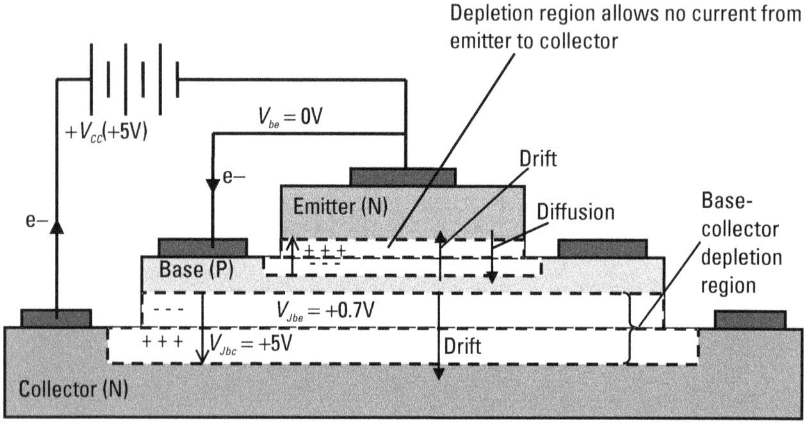

Figure 2.11 NPN bipolar transistor with zero bias on the base-emitter junction.

symmetrical device with base and collector contacts on both sides of the emitter contact.

Zero voltage bias at the base-emitter junction. In an NPN bipolar transistor, the emitter layer is usually more heavily doped than the base layer. At the base-emitter *PN* junction, there is a large number of spare electrons in the *n*-emitter, so these diffuse into the *P*-base, canceling the holes near the junction (known as diffusion current). Where the electrons have left the emitter region, and holes are cancelled in the base region, there are no longer any free charges to conduct a current (i.e., this junction area is depleted of free carriers and is known as the depletion region). This leaves exposed positively charged donor atoms on the emitter side of the junction and exposed negatively charged acceptor atoms on the base side of the junction. These exposed charge atoms create an electric field with a polarity from base to emitter (known as the junction potential, V_{JBE}), effectively forcing electrons from the base to the emitter, known as a drift current. Equilibrium is reached when the junction potential increases to a level where the drift current it creates is identical and opposite to the diffusion current, so, in total, no current passes the junction. The built-in junction potential at equilibrium is a function of the doping concentration in both types of semiconductor, but is typically +0.7V for silicon.

Base leakage current. At the base-collector *PN* junction, a depletion region forms in the same way with the junction polarity from base to collector, which forces (or drifts) electrons from the base to the collector. If a positive collector-bias voltage is applied between the collector and the emitter, this pulls more electrons away from the base-collector junction and increases the thickness of this junction depletion region until the voltage across the base-collector depletion region (V_{JBC}) is the same as the applied bias voltage. The only current in this circuit is a small leakage current of the minority carrier electrons in the base drifting over this base-collector junction into the collector. A bipolar transistor with zero or reverse bias on the base-emitter junction is in its "off" state.

Reverse-biased base-emitter junction. If a negative bias voltage is applied to the base-emitter junction, as in Figure 2.12, electrons are pulled from the emitter, and holes are pulled from the base, widening the base-emitter depletion region. The only current will be a small leakage current of the minority carrier electrons in the base drifting over the base-emitter junction.

Forward-biased base-emitter junction. If a small positive bias is applied to the base-emitter junction, as in Figure 2.13, this polarity is effectively opposing the base-emitter junction potential and reducing its value. This makes the depletion region thinner, and more electrons can diffuse into the base. Another way to view it is that the negative bias potential on the emitter pushes the electrons closer to the junction, while the positive charge on the base pushes holes toward the junction, thinning the depletion layer.

There is now a cloud of electrons that have diffused into the base layer, and these go one of two ways. Some drift into the top of the base-collector

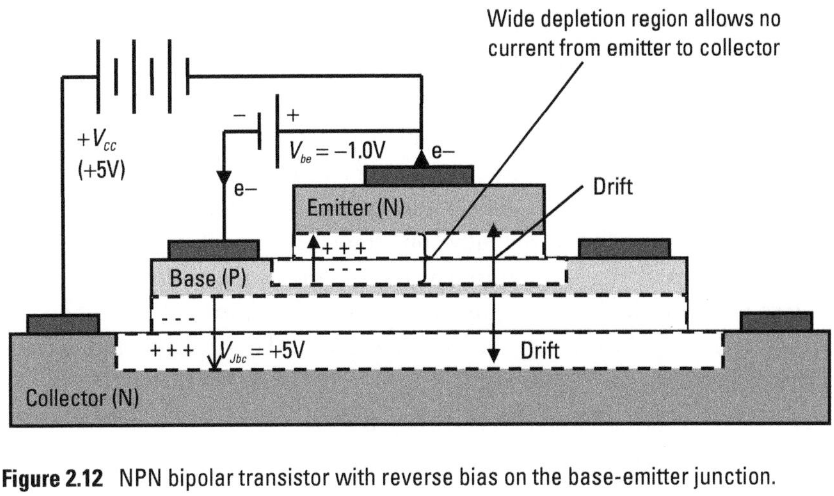

Figure 2.12 NPN bipolar transistor with reverse bias on the base-emitter junction.

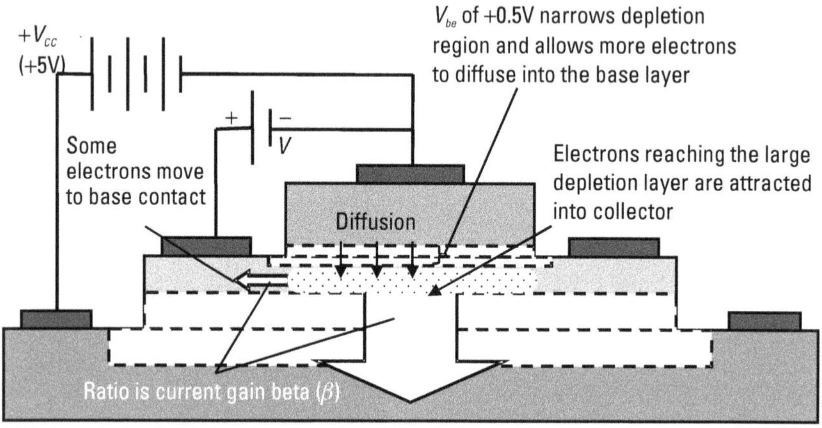

Figure 2.13 NPN bipolar transistor with base-emitter junction forward biased at +0.5V.

depletion region, experience the force of the large electric field there, and are pulled into the collector. Others recombine with the holes in the base and form a current with the low base voltage sideways to the base contact. The base layer is deliberately designed to be thin so that most of the electrons diffusing across the base-emitter junction flow straight into the collector. The ratio of the collector current to the base current is proportional to the probability of a diffused electron's meeting the base-collector junction area, compared to its meeting a hole and recombining. This ratio is fixed by the dimensions of the layers, contact areas, and doping of the semiconductor and gives the constant current gain of the bipolar transistor known as beta (β). Common values for beta for a Si bipolar are between 50 and 100.

Increasing the forward-bias voltage to the base-emitter junction continues to thin the base-emitter junction depletion region until it disappears completely (at about V_{BE} = +0.7V for silicon), as shown in Figure 2.14. With these bias conditions, a very large number of electrons can cross into the base region and be swept into the collector, creating a large forward current from emitter to collector. This is the fully "on" state for a bipolar transistor.

In summary, the base-emitter voltage controls the width of the base-emitter junction depletion region, which in turn controls the number of electrons that can diffuse into the base. The physical dimensions and doping of the base determine the ratio of electrons in the base that flow either out of the base terminal or out of the collector, giving the relatively fixed current gain beta (β).

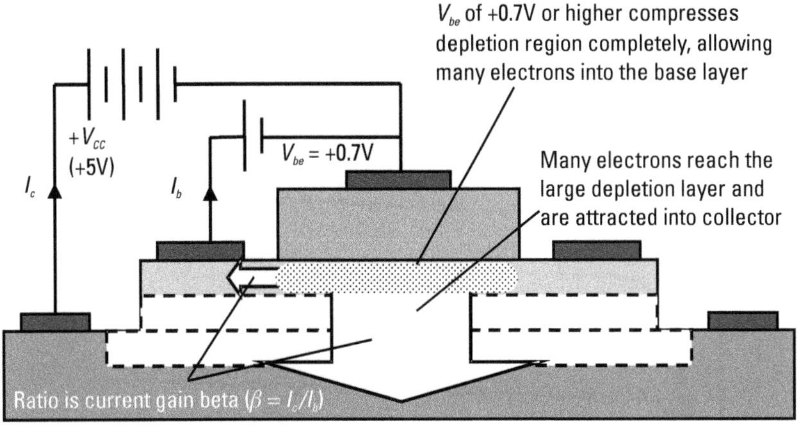

Figure 2.14 NPN bipolar transistor with base-emitter junction forward biased at greater than +0.7V.

Biasing of Bipolar Transistors

The reverse-biased and fully forward-biased base-emitter-junction cases are the two extremes of the base bias, but the bipolar transistor can be biased at any base-bias voltage between the two. Like the FET, the bipolar transistor tends to be biased around the middle of its characteristics, where it is most linear. For a Si bipolar transistor, the base-bias voltage will typically be between 0V and +1.0V, and the collector-emitter voltage (V_{CC}) will typically be between +5V and +10V. Because the base and collector currents are proportional to each other by the ratio beta (β), they are often biased to give a set base current, which in turn gives a fixed collector current. A plot of a typical collector current versus collector voltage (I/V) for several different base-bias currents is shown in Figure 2.15. The load line shows how the collector current will swing as the base current is varied.

In the forward-active region, the collector voltage is not enough to sweep all the emitter diffusion electrons arriving in the base region into the collector, so they crowd the base and prevent any more electrons from entering, giving a current roughly proportional to the collector voltage. When the collector voltage is high enough (typically, twice the base-emitter voltage), most of the electrons drifting into the base are swept into the collector, and the collector current remains at a nearly constant level for a fixed base current (actually a fixed base voltage V_{BE}). The slope of this saturation region is not quite flat, because as the collector voltage increases, the depletion region at

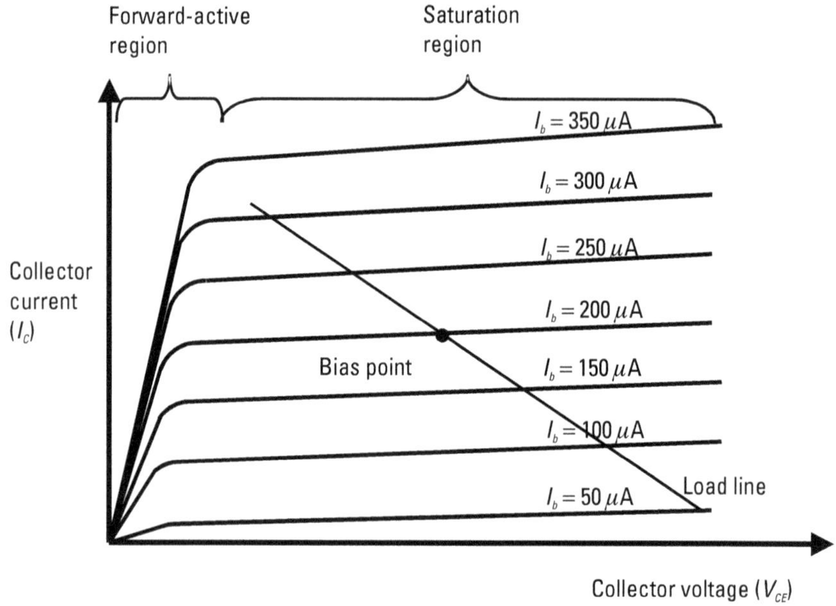

Figure 2.15 *I*/*V* characteristics of a bipolar transistor showing a typical bias point and load line.

the collector-base junction widens, and its boundary moves further into the base, effectively thinning the base layer and catching more of the diffusion electrons and attracting them into the collector (slightly increasing beta). This sloping of the saturation region is known as the Early effect, and if the slope is extrapolated down to the zero-current axis (well into negative voltage values), the voltage where it crosses is known as the Early voltage [27]. A large negative value for the Early voltage indicates a more linear device because the slope is much shallower.

HBT

The heterojunction bipolar transistor (HBT) is similar to the standard bipolar transistor except that the base-emitter junction is usually a junction of two different semiconductor materials instead of the same material with different doping [13, 14]. Examples include HBTs fabricated with an AlGaAs emitter and a GaAs base, an indium gallium phosphide (InGaP) emitter and a GaAs base, and an InP emitter and an InGaAs base [28]. Homojunction bipolar transistors maintain good injection efficiency (ratio of electrons injected from the emitter into the base over the holes injected from the base into the emitter) by having the emitter layer heavily doped and the base layer lightly doped. HBTs achieve the injection efficiency by the energy band-gap difference at the junction, which prevents holes from being injected into the emitter. This allows the base layer to be more heavily doped, which greatly reduces the base resistance, and this reduces the transit time of the device and increases its frequency response. The increased base doping also has the effect of making the base less susceptible to narrowing by increased base-collector voltages and, hence, has flatter *I* / *V* characteristics and a more negative Early voltage. Figure 2.16 shows an example of an InGaP HBT power amplifier operating at 5.8 GHz.

The SiGe HBT is different from standard silicon bipolar transistors because the base is constructed from a different semiconductor material, namely SiGe, creating heterojunctions with the silicon base and emitter layers. The SiGe base layer has a sloping concentration of germanium across its thickness, which creates an electric field to reduce the transit time of electrons moving from the base into the collector. This enables the SiGe HBT to operate at higher frequencies than silicon bipolar and allows silicon-based circuits to be considered for many microwave applications.

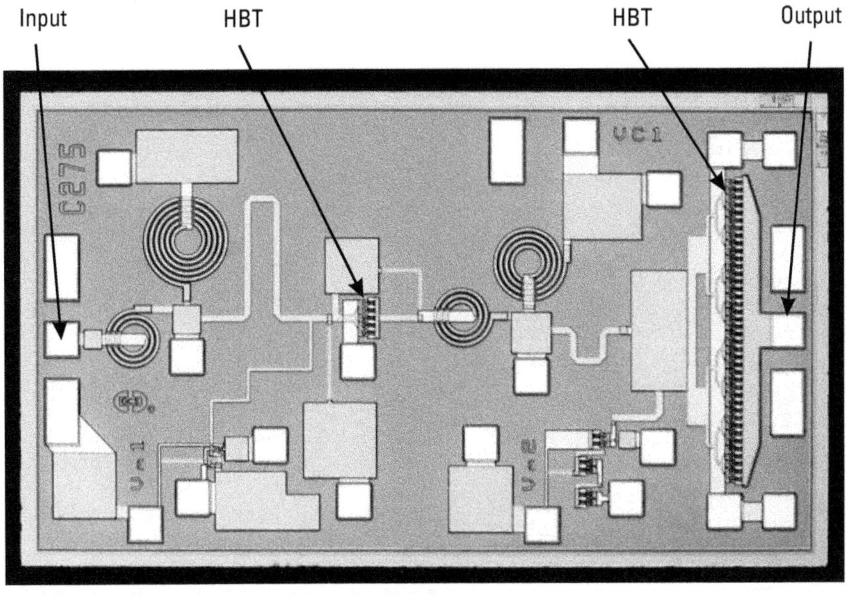

Figure 2.16 InGaP HBT power amplifier operating at 5.8 GHz. (Courtesy of Plextek Ltd, and GCS Inc. [29].)

2.1.3 Transistor Response Versus Frequency

The gain of a transistor decreases as frequency increases by about 6 dB per octave, as shown in Figure 2.17. However, a gain stage cannot necessarily be designed to operate with high, low frequency gain because it may be unstable and oscillate. J. M. Rollett determined a stability factor K [30] from the scattering parameters (*s*-parameters) of the transistor, as given in (2.1), which has a value greater than 1 when the transistor is unconditionally stable. (It will not oscillate no matter what impedances are presented to the input or output of the transistor.) In this region of the frequency range, where $K > 1$, the maximum transducer gain (G_{MAX}) is equal to the maximum available gain (MAG) of the transistor, as given by (2.2).

$$\text{Stability factor } K = \frac{1 - |S_{11}|^2 - |S_{22}|^2 + (S_{11}S_{22} - S_{12}S_{21})^2}{2|S_{12}S_{21}|} \quad (2.1)$$

$$G_{MAX} = MAG = \left(K - \sqrt{K^2 - 1}\right)\frac{|S_{21}|}{|S_{12}|}, \text{ when } K > 1 \quad (2.2)$$

At frequencies where $K < 1$ the maximum transducer gain (G_{MAX}) is equal to the maximum stable gain (MSG) of the transistor as given by (2.3).

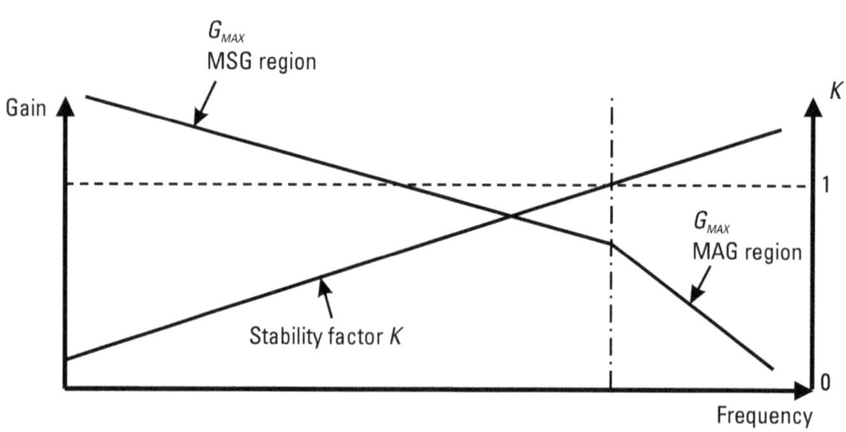

Figure 2.17 Transistor gain and stability factor versus frequency.

$$G_{MAX} = MSG = \frac{|S_{21}|}{|S_{12}|}, \text{ when } K < 1 \tag{2.3}$$

2.2 Passive Components

2.2.1 Diodes

Diodes in a MESFET MMIC process are generally created from the gate Schottky contact to the low-doped semiconductor cap layer. These diodes tend to suffer from higher series resistance compared to a dedicated Schottky-diode MMIC process [31], but this series resistance can be reduced by alternative processing methods [32]. In terms of their low-frequency $1/f$ noise, InP HBT diodes exhibit less noise than InP HEMT-based diodes [33], which is useful to note for oscillator and mixer applications. Diodes in a bipolar process can be created from either the base-emitter junction or the base-collector junction.

2.2.2 Transmission Lines

When electronic components are connected together with wires at low frequencies (<1 MHz), the distance between the components is very small compared to the wavelength of the signal, so it can be assumed that everywhere on the wire, the voltage amplitude will be the same. At RF and microwave frequencies, the distance between components, even on a tiny MMIC, can be a significant fraction of the signal wavelength; the voltage at any point of the connecting metal track will be a function of the amplitude and the phase of the signal, so the interconnecting metal must be treated as a transmission line.

The most common type of transmission line used in MMICs is called microstrip [34, 35], which consists of a metal track on a dielectric substrate with an infinite ground plane on the back surface, as shown in Figure 2.18. The track width and substrate height are finite dimensions, and the ground plane and substrate width and length are assumed to be infinite.

The electric field is confined mainly underneath the track, but it does extend a significant distance away from the edges of the track. The characteristics of a microstrip transmission line are determined primarily by the ratio (w/h) of the track width to the height of the dielectric because this is the main influence on the electric and magnetic field patterns. A track with width similar to the substrate height, as in Figure 2.18, has more parallel

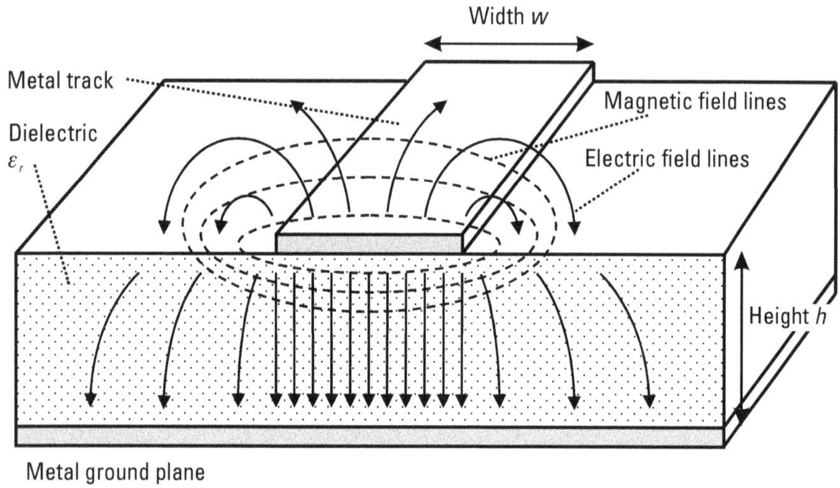

Figure 2.18 Section view through a wide microstrip transmission line showing electric and magnetic field patterns.

electric field underneath the track, similar to a parallel plate capacitor, and so looks more capacitive. A track with width much narrower than the substrate height, as in Figure 2.19, has magnetic field lines tightly packed that look more similar to a simple wire; hence, it behaves more inductively. These

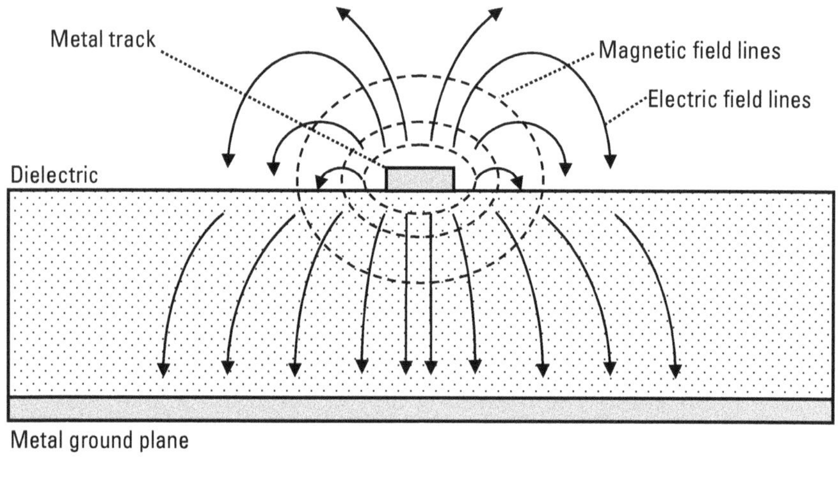

Figure 2.19 Section view through a narrow microstrip transmission line showing electric and magnetic field patterns.

field patterns also show how rules of thumb for minimum track-separation distances can be related to the track and substrate dimensions. The rule of thumb is that metal tracks should be separated, edge from edge, by either one substrate height or three track widths (whichever is the smaller) to allow reasonably close placement with a minimal level of coupling between the tracks.

The equivalent circuit of an infinitesimally short piece of uniform transmission line is shown in Figure 2.20, where R is the resistance per unit length, L is the inductance per unit length, G is the conductance per unit length, and C is the capacitance per unit length. A complete transmission line is a cascade of an infinite number of these sections. When this is applied to microstrip, R represents the conductor and dielectric losses, and G represents the finite conductance of the substrate; these are second-order effects and can be neglected in this simple transmission-line analysis.

When this is the case, the characteristic impedance of the transmission line is given by (2.4), and the velocity of propagation along the transmission line is given by (2.5) [36].

$$\text{Characteristic impedance } Z_0 = \sqrt{(L/C)} \tag{2.4}$$

$$\text{Propagation velocity } v = 1/\sqrt{(LC)} \tag{2.5}$$

Equation 2.4 also shows that narrow tracks, which are more inductive per unit length, will have higher characteristic impedance than wide tracks, which are more capacitive per unit length.

It can also be seen from Figures 2.18 and 2.19 that the electric and magnetic fields of the wave propagating along the microstrip lie within the substrate dielectric and within the air above the substrate, making the microstrip an inhomogeneous transmission line. This means that the transmission mode

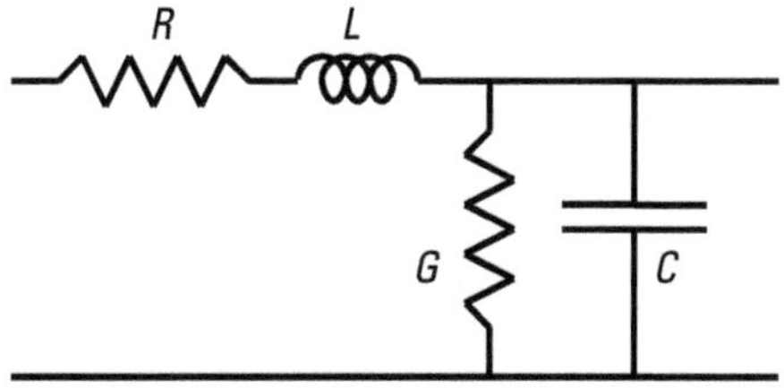

Figure 2.20 Equivalent circuit of an infinitesimally small section of transmission line.

in microstrip will not be a pure transverse-electro-magnetic (TEM) mode because the speed of propagation in the two dielectric media is different and a single TEM mode cannot be sustained. The actual mode of propagation is a hybrid of transverse electric (TE) and transverse magnetic (TM) modes, which have equal propagation velocities in each dielectric; this is known as a quasi-TEM mode [37] and can be regarded as a pure TEM mode traveling within a homogeneous dielectric with an "effective" relative dielectric constant (ξ_{eff}) that has a value between those of the substrate relative dielectric constant (ξ_r) and of air (1). The value of the effective dielectric constant for a wide track will tend toward the relative dielectric constant of the substrate material as most of the electric field will be beneath the track and in the substrate, and the value for a narrow track will tend toward the average between the relative dielectric constant of the substrate material and air as the electric fields are nearly shared equally between them both, as shown in (2.6).

$$\text{Effective relative dielectric constant range } \xi_r < \xi_{eff} < (\xi_r + 1)/2 \quad (2.6)$$

$$\text{Effective relative dielectric constant } \xi_{eff} = (\xi_r + 1)/2 + (\xi_r - 1)/2 \cdot 1\Big/\sqrt{(1 + 12\,h/w)} \quad (2.7)$$

The approximate value (within 2%) for the effective relative dielectric constant of the quasi-TEM mode traveling along the microstrip transmission line in Figure 2.16 is given by (2.7), where h is the substrate height, w is the track width, and ξ_r is the relative dielectric constant of the substrate material [34, 38]. The phase velocity of the quasi-TEM mode is slower than that of free space by the factor $\sqrt{\xi_{eff}}$, as shown by (2.8), so the wavelength of the quasi-TEM microstrip mode is shorter than that of free space and is given by (2.9).

$$\text{Phase velocity of microstrip transmission line } v_{microstrip} = c\Big/\sqrt{\xi_{eff}} \quad (2.8)$$

$$\text{Wavelength of microstrip transmission line } \lambda_{microstrip} = c\Big/\left(f\sqrt{\xi_{eff}}\right) \quad (2.9)$$

The impedance of a microstrip transmission line is given in (2.10) for when the ratio w/h is less than 1 and in (2.11) when it is greater than 1 [34, 35].

$$Z_0 = \left(60/\sqrt{\varepsilon_{eff}}\right) \cdot \ln\left(8h/w + w/4h\right) \text{ for } w/h \le 1 \quad (2.10)$$

$$Z_0 = 120\pi/\varepsilon_{eff}\left(w/h + 1.393 + 0.667\ln\left(w/h + 1.444\right)\right) \quad (2.11)$$
$$\text{for } w/h \ge 1$$

In an MMIC process, the substrate height is usually fixed at 100 μm or 200 μm, so the track impedance is set by the track width. The range of practical track widths on an MMIC process is from around 6 μm to 120 μm, which corresponds to microstrip characteristic impedances of about 120Ω down to 40Ω.

The above formulas are good for calculating approximate values for the microstrip impedance and wavelength, but in reality, the microstrip transmission lines fabricated by MMIC processes are more complicated than the simple form presented in Figures 2.18 and 2.19, and their characteristics are not easily derived from closed-form expressions. For example, Figure 2.21 shows that some interconnect metal tracks are etched slightly into the substrate material and may sit above or below multiple dielectric layers, which are far from planar around the edges of the metal track where the electric fields are most concentrated. Because of this, many foundries use two-and-a-half-dimensional (2.5D) simulation tools to find the value for the microstrip impedance and then use ring resonators manufactured on the

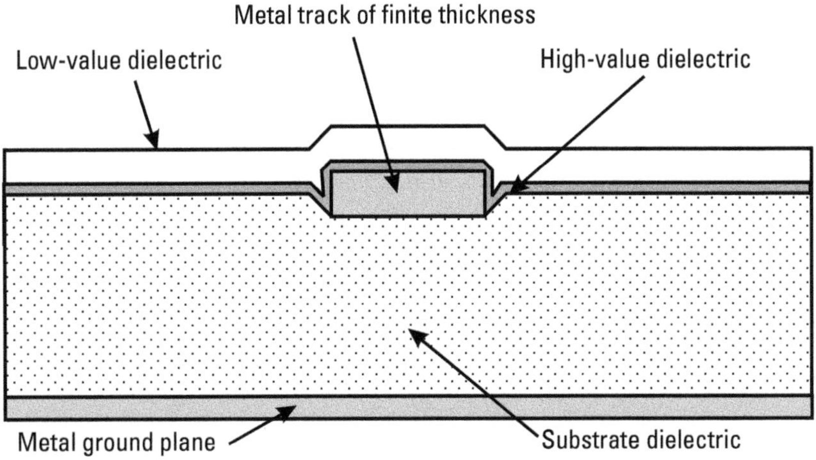

Figure 2.21 A more realistic structure for an MMIC microstrip transmission line.

MMIC process to give values for the effective relative dielectric constant and loss per unit length. The characteristic impedance for microstrip transmission lines on a 200-μm-thick substrate derived using a 2.5D simulator are shown in Figure 2.22 for upper and lower metal layer tracks as a function of the track width. A typical microstrip ring-resonator test structure is shown in Figure 2.23, and the sort of transmission response from such a resonator is shown in Figure 2.24.

The length of the resonator is designed to produce about six resonant transmission peaks across the frequency range to be characterized. The capacitive coupling to the ring resonator should be adjusted to have a maximum transmission of about −20 dB, so it is the unloaded response of the ring that is measured, while keeping the lowest transmission peak well out of the system measurement noise floor. The ring resonates in transmission when the length of transmission line making up the ring is a whole number of wavelengths of the wave propagating on the transmission line. The width of the resonant peak is determined by the Q-factor of the resonator and is used to determine the loss of the transmission line. Thus, one can fit a physical transmission-line model, with parameters of characteristic impedance (Z_0), effective relative dielectric constant (ξ_{eff}), loss per unit length (A), and physical length (L), to the measured data around each resonant peak.

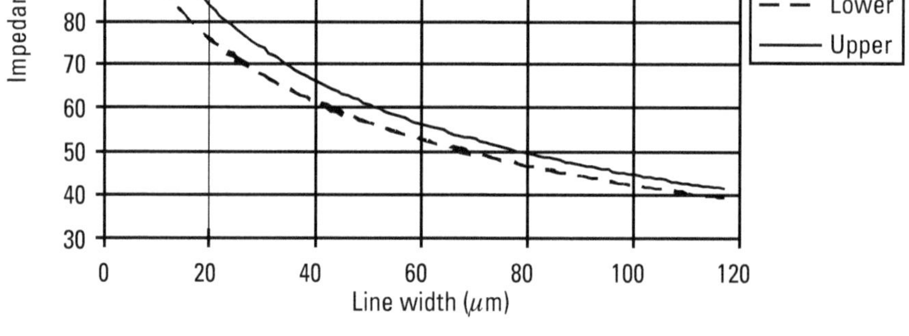

Figure 2.22 Characteristic impedance of microstrip created in upper and lower metal layers on a 200-μm-thick GaAs substrate.

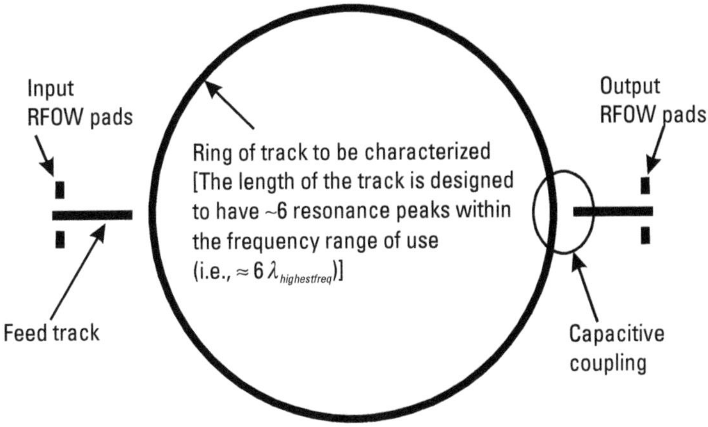

Figure 2.23 Ring resonator for microstrip transmission-line characterization.

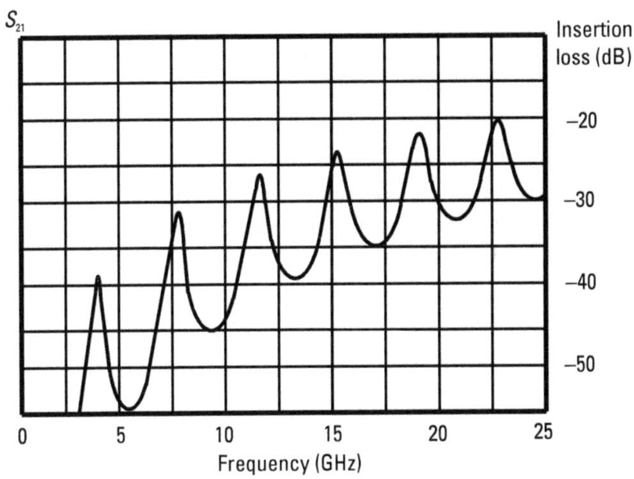

Figure 2.24 Typical ring-resonator transmission response.

In the process of fitting the model, the characteristic impedance value is derived from 2.5D simulation of the transmission-line structure, the physical length is the length of the center line of the track making up the ring resonator, and the center frequency of the resonance then determines the effective relative dielectric constant, and the width of the resonance determines the loss. Care must also be taken regarding the range of measured data around each peak that is used for fitting the model. The data should cover the −3-dB point at least and ideally also the –6-dB point on either side of the resonance to derive an accurate value for the transmission-line loss. However, data from the skirts of the resonant peak should also be discarded as they are too far from the center frequency of the resonance and tend to skew the loss value.

This characterization process gives values for the characteristic impedance (Z_0), effective relative dielectric constant (ξ_{eff}), and loss per unit length (A) at six frequency points across the frequency range and for one width of track. The process must then be repeated for several different track widths within the range allowed by the foundry process. The variation of these parameters, as a function of frequency and track width, is fairly smooth, so their values between the resonant frequencies and measured track widths can be determined by interpolation.

The typical loss (in decibels per millimeter) of 30-μm-wide transmission lines on a 200-μm GaAs substrate derived by this method are shown in Figure 2.25 for microstrip tracks constructed in lower and upper metal layers. The typical effective dielectric constant for microstrip lines on GaAs, also derived by this method, is shown in Figure 2.26.

The second most common transmission line used in MMIC is coplanar waveguide (CPW), which consists of a metal track and ground plane on the same surface of a dielectric substrate with a fixed gap between the track and the ground plane, as shown in Figure 2.27 [39]. The track width and gap are finite dimensions, and the ground plane and substrate's height, width, and length are assumed to be infinite. Note that the width of the ground plane sections of CPW also has an effect upon the characteristic impedance of the

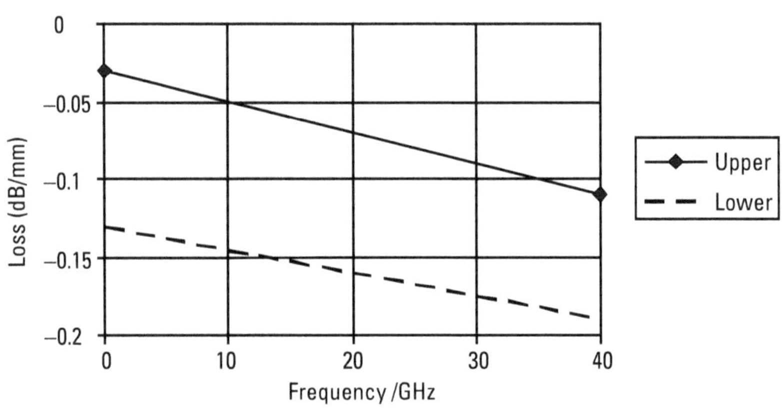

Figure 2.25 Loss of 30-μm-wide microstrip created from upper and lower metal layers on a 200-μm GaAs substrate.

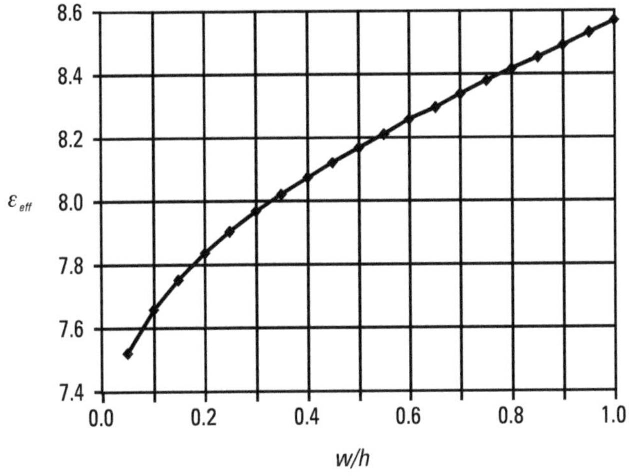

Figure 2.26 Effective dielectric constant for microstrip lines on a GaAs substrate.

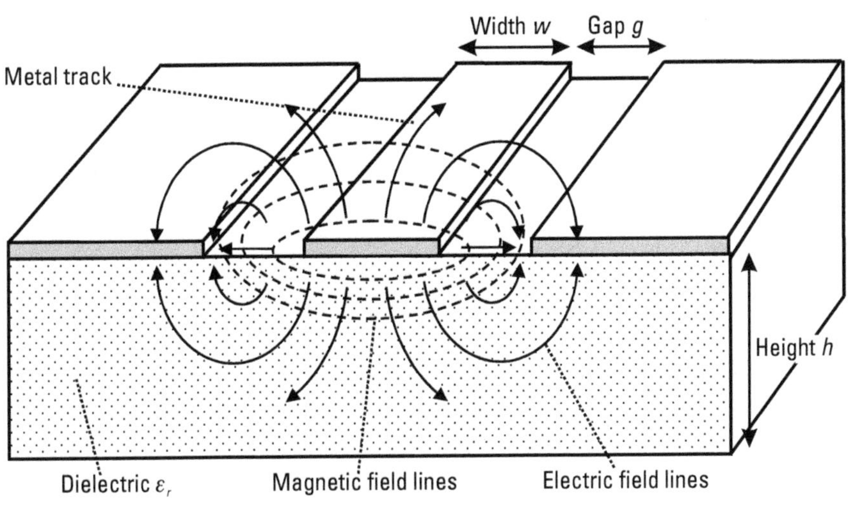

Figure 2.27 CPW transmission line.

transmission line. This is small in most cases and not important in analyzing a simple length of line. However, in densely packed MMICs where the ground plane has to be reduced to make space for other components, its effect may need to be considered. The electric and magnetic fields are in both the dielectric substrate and the air above, so the mode of propagation in CPW is also quasi-TEM [40].

The electric field is confined mainly between the track and the coplanar ground plane and does not normally extend to the bottom surface of the substrate. This means that the substrate wafers do not need to be thinned down from their bought-in thickness of over 600 μm, removing a costly process step and remaining more rugged than a thinned wafer. The back surface of the substrate can be left clear or metalized. If metal is used on the back surface, the transmission line is known as grounded coplanar waveguide (GCPW), but the characteristics are the same as ordinary CPW as long and the substrate thickness is much larger than the track width. The characteristics of a CPW transmission line are determined primarily by the ratio w/g of the track width to the gap to the ground plane because this is the main influence on the electric and magnetic field patterns. Methods of deriving the characteristic impedance of CPW have been developed [41], and CPW on a GaAs substrate requires a w/g ratio of about 2 to give 50Ω impedance.

A CPW with a higher ratio (i.e., with a track width much larger than the gap) has lower impedance because the closing of the gap increases the shunt capacitance and reduces the magnetic field density. A CPW with a lower ratio (i.e., with a track width much narrower than the gap) has less shunt capacitance, and the magnetic field lines are more tightly packed and look more similar to those of a simple wire; hence, it behaves more inductively. In theory, the actual track and gap dimensions can be scaled up or down in size, and the impedance will be the same for the same w/g ratio. However, when the dimensions become very small, the track thickness starts to play a large part in determining the impedance, and transmission losses start to become large.

CPW tends to be used for circuits operating at high microwave and millimeter-wave frequencies, such as the 44.5-GHz direct modulator MMIC shown in Figure 2.28, because microstrip can become multimode at these frequencies. For this reason, the issue of choice between microstrip and CPW can be assisted by considering the details given in Section 5.10.2.

Other transmission lines, such as coplanar strips and slotlines [41], can be used in MMICs, but their application is mainly in the area of baluns for mixers and antennas.

Transmission lines on lower-resistivity substrates suffer from several problems that can be neglected for high-resistivity substrates. These problems include slow-wave effects [42], increased transmission-line loss, and an increased difficulty in accurately characterizing the transmission line. For

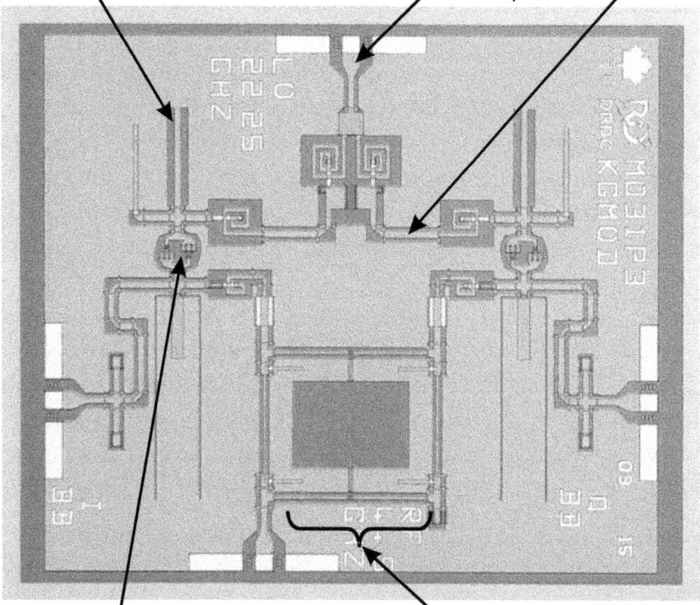

Figure 2.28 A 44.5-GHz direct modulator using CPW transmission lines. (Courtesy of Defence R&D, Canada.)

example, silicon substrates for standard CMOS processes have resistivity of the order of 1 to 20 Ω/cm; as a result, passive components and interconnects on these substrates have high loss and low Q-factors, especially at microwave and millimeter-wave frequencies. Techniques have been developed to overcome this problem, such as using high-resistivity (1,500 Ω/cm) silicon substrates [43, 44], shielding the components from the silicon substrate [45], and removing the silicon around the components with a microelectromechanical systems (MEMS) process [46]. These techniques are achieving state-of-the-art performances and are allowing low-resistivity substrates to be fully characterized [47], but they are still a long way from being compatible with commercial silicon processing steps. However, if high-resistivity substrates are available from the foundry, and the designer can use transmission lines such as CPW, then the loss can be brought down close to those levels normally associated with GaAs substrates [44]. Indeed, the increasing number of publications on high-frequency chips using silicon-based technology is forming such a body of evidence that it is now clear that silicon MMICs and radio frequency integrated circuits (RFICs) are commercial possibilities at frequencies well into the microwave region.

Transmission-line junctions, where one transmission line connects to another transmission line of a different size, to more than one other transmission line, or to a component, all tend to generate nonpropagating evanescent modes in the locality of the discontinuity in order to satisfy Maxwell's equations. These modes die away rapidly with distance from the junction but can have the effect of making the junction appear as a small lumped inductance or capacitance.

When a transmission line on one metal layer crosses over a transmission line on another metal layer, they are separated only by the dielectric layers between them. This can lead to capacitive coupling between them and can be calculated as the parallel plate capacitance of the area of overlapping metal, using the average dielectric constant of the dielectric layers between them and the distance separating the metal layers and the formula in Section 2.2.4.

2.2.3 Resistors

MMIC resistors are created either by using the active semiconductor layer under the surface of the MMIC or laying down thin films of resistive metal alloys above the surface as the resistive material. With both methods, the resistors are constructed with metal contact pads at either end of the resistive material film, as shown in Figure 2.29, and have the characteristics of typical thin-film resistors.

The resistive material is characterized in terms of its resistivity in ohms per square because a 20 × 20–μm square resistor has the same resistance as a 100 × 100–μm square resistor if it is made from the same material (resistance = resistivity × length/width = resistivity × 20/20 = resistivity × 100/100). This also makes it easy to calculate the size of a resistor quickly by counting

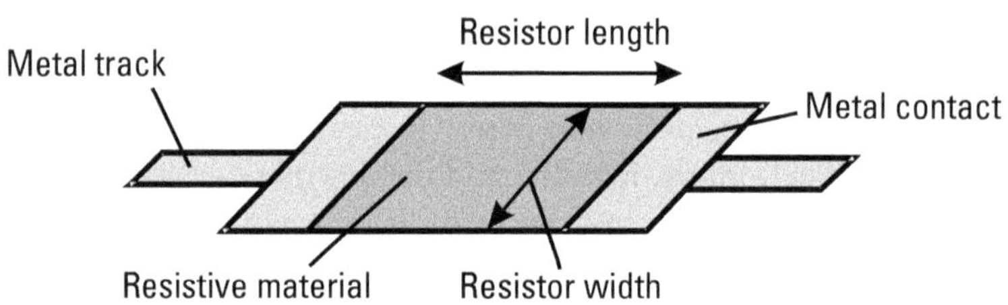

Figure 2.29 Thin-film resistor construction.

how many squares make up the resistor. For example, the resistor on the right in Figure 2.30 is made up of three squares in series, each having a resistance of 50Ω, so the total series resistance is 150Ω. The resistor on the left in Figure 2.30 can be seen as having two squares in parallel, which have half the resistance of one square in series, and thus has a total series resistance of 25Ω. Note that this principle only applies to straight resistors and does not work for resistors where the resistive film is meandered to enable a large value resistor to fit into a small area. In this situation, the straight sections have the nominal film resistivity, but the corner sections have a higher resistivity due to current crowding at the inside corner of the film.

Resistors made with the semiconductor material have the disadvantage that their resistance is a function of the electron mobility within the active semiconductor layer and so are sensitive to temperature variations. Their advantage is (1) that they usually have a resistivity of a few hundred ohms per square, which makes them useful for high-value decoupling resistors, and (2) that they are under the semiconductor surface so other interconnect metallization can be laid over them without making electrical contact. Common metal alloys used include nichrome (NiCr) and tantalum nitride (TaN), which typically have film resistivity from 20Ω to 50Ω per square, and titanium tungstosilicate (TiWSi), which can have film resistivity of 500Ω to 1,500Ω per square. Nichrome has the useful property that its resistivity is almost completely constant versus temperature variations.

If the resistor is to carry a significant current, the width of the rsistor must be designed to handle it. 50Ω/square nichrome and TaN resistors can handle typically 0.5 mA per micron of resistor width, and highly doped *p*-type polysilicon resistors can handle slightly more at 0.6 mA per micron of resistor width [48]. TaN resistors maintain their linear ohmic characteristics

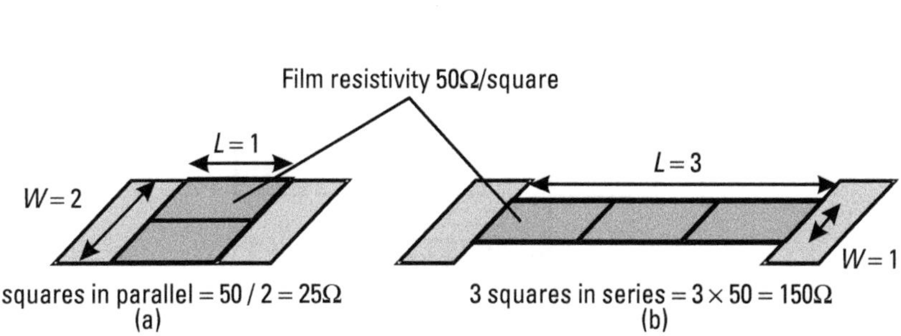

Figure 2.30 Thin-film resistor calculation of resistance using squares for (a) a 25-Ω resistor and (b) a 150-Ω resistor.

up to higher current densities than NiCr [49], but these levels of 4 mA per micron are generally higher than the reliability limits of MMIC resistors.

2.2.4 Capacitors

Capacitance is a measure of a component's ability to store electric charge, expressed as coulombs per volt, and determines its impedance to RF signals. RF signals can pass through capacitors by charging and discharging them, and the more charge storage capacity they have, the more current (charge flow) they allow per volt of signal; hence, the larger the capacitor is, the lower is its impedance (Z = volts/current).

MMIC capacitors are formed by two main methods: one uses the fringing capacitance between interdigital metal strips, and the other uses a metal-insulator-metal (MIM) sandwich. The interdigital capacitor [50], shown below in Figure 2.31, relies on the fringing capacitance between the long common-edge area of the metal fingers, which are separated by just a few microns depending on the minimum gap allowed by the foundry. These capacitors are usually formed using the same interconnect metallization as used for the transmission lines. This fringing capacitance is fairly low, so these capacitors are only able to reach capacitance values around 1 pF. However, this capacitance value is relatively insensitive to process variations and can be a useful component at millimeter-wave frequencies.

The MIM capacitor [51], shown in Figure 2.32, can achieve much higher capacitance values because they are constructed of two relatively large plates of metal separated by a much smaller distance, and the gap in between is also filled with an insulating dielectric material, which further increases the

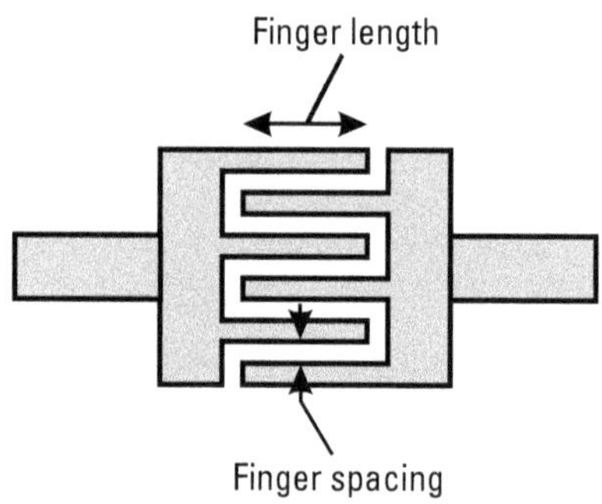

Figure 2.31 Six-finger interdigital capacitor as viewed looking down onto the wafer surface.

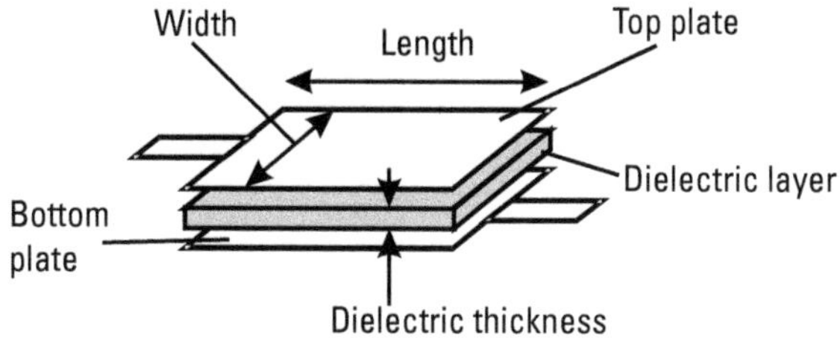

Figure 2.32 MIM capacitor.

capacitance. The size of the metal plates is typically from 20 μm × 20 μm up to 200 μm × 200 μm, and the dielectric material could be silicon nitride (SiN), silicon dioxide (SiO_2), benzocyclobutene (BCB), polyimide, or a combination of a number of these layers. The SiN dielectric thickness is typically 100 to 120 nm, the SiO_2 dielectric thickness can be as thin as 50 nm [52], and the thickness of the organic dielectrics, such as BCB or polyimide, tends to be 1,000 to 3,000 nm. Capacitance values obtained for MIM capacitors range from 50 fF up to 200 pF.

The prime capacitance (excluding fringing capacitance) of an MIM capacitor can be calculated simply from the expression $C = \xi A/D$, where ξ is the dielectric constant of the dielectric layer (equals ξ_R, the relative dielectric constant, times ξ_0, the dielectric constant of free space), A is the area of the metal plates, and D is the distance the plates are apart. The relative dielectric constant of the SiN dielectric layer is typically 6.8, the relative dielectric constant of polyimide is typically 4.5, the relative dielectric constant of BCB is typically 2.7, and the dielectric constant of free space is 8.8542×10^{-12} farad/m.

The breakdown voltage of MIM capacitors is a function of the dielectric material, the dielectric thickness, and also the quality of the surface of the bottom plate of the capacitor. Silicon nitride MIM capacitors for an MMIC process have been reported with breakdown voltages from 65V [53] to 85V [54].

The loss of an MIM capacitor is determined by the dielectric loss tangent of the insulating material used. SiN has a very low loss tangent, and the losses from this type of capacitor are often too small to be measured. Polyimide and other organic dielectric layers have higher loss tangents, which may introduce small losses to the circuit.

Note that interdigital capacitors on low-resistivity substrates will suffer from significantly increased loss compared to those manufactured on high-resistivity substrates. On the other hand, MIM capacitors on lower-resistivity substrates do not tend to suffer from increased loss because the majority of the electric field is in the insulator dielectric between the metal plates of the capacitor.

2.2.5 Inductors

Inductance is a measure of a component's ability to store current, expressed as volt seconds per amp, otherwise known as henrys. The simplest example of a component with inductance would be a narrow wire.

MMIC inductors are fabricated using lengths of interconnect metal formed as narrow transmission lines, either on their own or wound around a central point to create a spiral transmission-line inductor. The narrower the metal track in relation to the substrate height, the higher the inductance per unit length. For example, narrow lines (track width is a lot less than the substrate height) have an inductance of about 1 nH/mm of track length, while wide lines (track width is similar to the substrate height) have an inductance of about 0.6 nH/mm of track length. This is illustrated by the example tracks shown in Figure 2.33. It is a good idea to remember this property of interconnect tracks because track-width tuning can be used to tweak the performance of a circuit when it is finally laid out and track-length adjustments are unacceptable. Note also that the width of a track determines its dc current

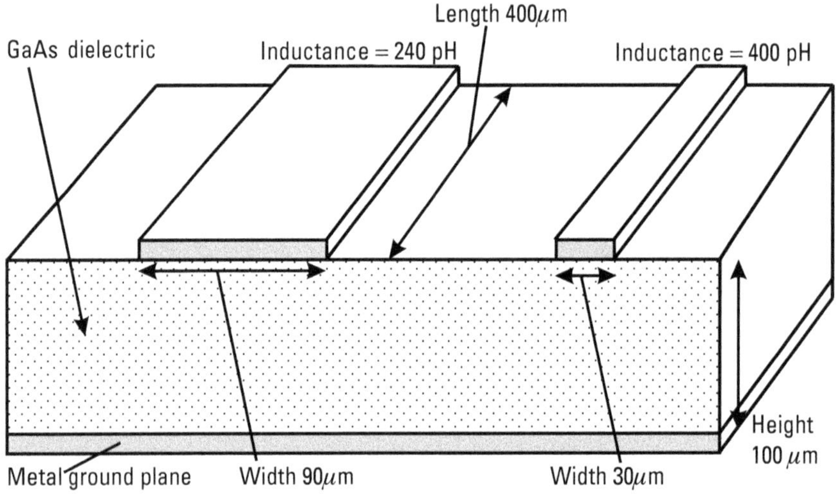

Figure 2.33 Example inductance values for narrow and wide tracks.

carrying ability. This is because high current densities in metallization can cause holes in the tracks due to electromigration. Foundries will normally quote a maximum recommended figure for current density in terms of milliamps of current per micron of track width that they have determined by reliability trials. This value can be in the region of 10 mA per micron of track width but will vary from process to process. Thus, there is a trade-off when making inductors because narrow tracks are more inductive but can carry less dc current.

Spiral track inductors have more inductance than that due to their track length alone because the magnetic fields from each turn (or complete loop) of the spiral add up, creating a larger field through the middle of the spiral and mutual inductance between all the turns. Because of this, spiral inductors are a convenient way of generating a large amount of inductance in a small area of an MMIC chip. However, the turns of metal tracks are also capacitively coupled, so at certain higher frequencies, the spiral does not look inductive and may look capacitive or even like an open circuit. Spiral inductors are constructed most efficiently when there are at least two metal interconnect levels available, as shown in Figure 2.34. One metal interconnect layer is used for the main part of the spiral, and a short length of the other metal interconnect layer is used for the underpass that returns the signal to the outer edge.

Many different types of spiral inductor are used in MMIC processes, ranging from square spirals, circular spirals, octagonal spirals, stacked spirals, to those that are in some way decoupled from the bulk substrate material. Square spirals are not ideal electrically because the square corners can lead to

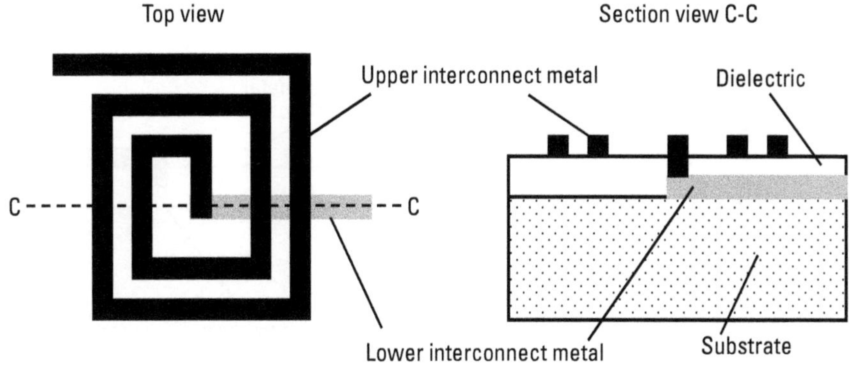

Figure 2.34 Construction of a spiral track inductor.

greater capacitive coupling between the individual turns and a slight increase in the resistive loss, but they are used by foundries because electronic design-rule checking software can quickly and easily check the separation distances between layers to ensure conformity with the foundry processing rules. Circular and octagonal spiral inductors have more ideal inductive characteristics but are much harder to check electronically for design rules. Stacked spiral inductors use full turns of metal tracks on more than one of the interconnect layers and usually interleave the turns of metal to minimize the coupling capacitance between the tracks on the different layers, as shown in Figure 2.35. Because they do exhibit increased capacitive coupling between the turns on both layers, they are used mainly at very low frequencies for bias decoupling.

Inductors on monolithic substrates are fabricated from the metal tracks or transmission lines and, as a consequence, are susceptible to losses from the substrate material, particularly if the substrate has low resistivity, such as a standard silicon CMOS substrate. This is a particular problem when the inductor is used as part of a tuned circuit because the loss has the effect of increasing the resistive part of the inductor's impedance compared to its reactive impedance and reducing the Q-factor of the inductor. The Q-factor is the ratio of the imaginary (reactive) part of the impedance over the real (resistive) part of the impedance and is a measure of how fast the component's impedance varies with frequency. A high Q-factor allows the component to be used for elements such as filters with fast cutoff responses versus frequency for applications such as communications with many closely spaced frequency channels. Q-factors of inductors produced on GaAs substrates are of a

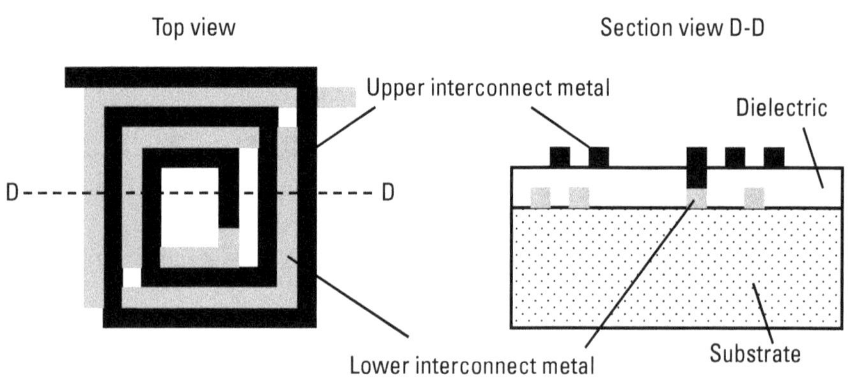

Figure 2.35 Construction of a stacked spiral inductor.

moderate level, being typically 10 to 50 at frequencies of 10 GHz, allowing them to be used for oscillator resonators at microwave frequencies but not for communications channel filters. Q-factors of inductors on low resistivity silicon substrates are commonly less than 1 and can only be used for oscillator circuits below 10 GHz. Off-chip air-core inductors are a lower loss alternative to on-chip inductors, as discussed in Section 5.3.2.1, but this is not always an acceptable solution. The on-chip solution to increase the inductor's Q-factor is to decouple it from the effects of the substrate material, and this can be done in several ways. One method is to place a solid or patterned metal ground layer between the inductor and the substrate to shield the inductor from the substrate [55]. Another alternative is to raise the metal tracks of the inductor above the surface of the chip using stand-off posts [56] or even to get the spiral inductor to self-assemble into a vertical position using MEMS [57].

2.2.6 Metal Layer Interconnects

Tracks on one metal layer can be connected to tracks on other metal layers using interconnects. In most cases, these are formed by opening a hole in the dielectric layer, which would otherwise separate the layers, and allowing the top metallization to be deposited through the hole onto the lower metal layer. If the tracks are the same width on both layers, there is seldom much RF discontinuity associated with these interconnects; however, the hole in the dielectric layer is often narrower than both tracks and could have less dc current carrying capability than the tracks themselves.

2.2.7 Bond-pads

To bring dc and RF voltages and currents onto and off of the chip, bond-pads are placed on the chip, usually around the periphery. These are used for connecting the chip to its embedding circuit using wire bonds or solder bumps and for making probe contacts during dc and RFOW testing. They are commonly constructed from the ohmic metal layer and all of the upper metal layers with no dielectric layers in between, allowing easy connection from any of the interconnect metal layers. The bond-pad does not have the final dielectric passivation layer over its surface so that external electrical connection can be made to it, and it is normally the only component that is not covered by this passivation layer. The top metal layer needs to be a reasonably thick layer of gold to allow gold wires to be compression bonded to its surface and to be wetable by solder.

2.2.8 Substrate Vias

Substrate vias are holes in the dielectric substrate that are plated with metal to provide an electrical path from the front face of the wafer to the back-face ground plane. They are usually constructed in the same way as a bond-pad on the top surface of the wafer, and the metal plating on the inside surface of the substrate hole makes the electrical connection between the back-face ground plane metal and the first metal layer of the bond-pad. These are used to provide good dc and RF ground points at different places within the circuit. At low RF frequencies of less than a few gigahertz, connections to ground at the side of the chip are usually sufficient, but at microwave frequencies and above, a substrate via is required to give a low-impedance connection to the ground plane. Substrate vias are often incorporated into the active components to ensure they have the minimum inductance in their path to ground. This is especially important for multigate-finger power devices, which sometimes have substrate vias underneath each source contact.

2.3 Questions

2.1 What characteristics of the semiconductor material determine the frequency range over which it can be used for producing MMICs?

2.2 Which semiconductor substrate materials exhibit a wide bandgap, making them suitable for high-output power applications?

2.3 What are the typical gate voltage and drain current bias settings for a GaAs MESFET?

2.4 What is the bias voltage required between the base and emitter contacts of a typical silicon bipolar transistor to overcome the built-in junction potential and switch the transistor fully on?

2.5 Why must the electrical connections between components on an MMIC be treated as transmission lines?

2.6 How does the characteristic impedance of a microstrip transmission line vary as the width of the track is increased, and is this due to the track looking more inductive or capacitive?

2.7 Figure 2.36 shows the section and plan view of a nichrome resistor. If the nichrome film has a resistivity of 50Ω/square, what is the resistance of this resistor?

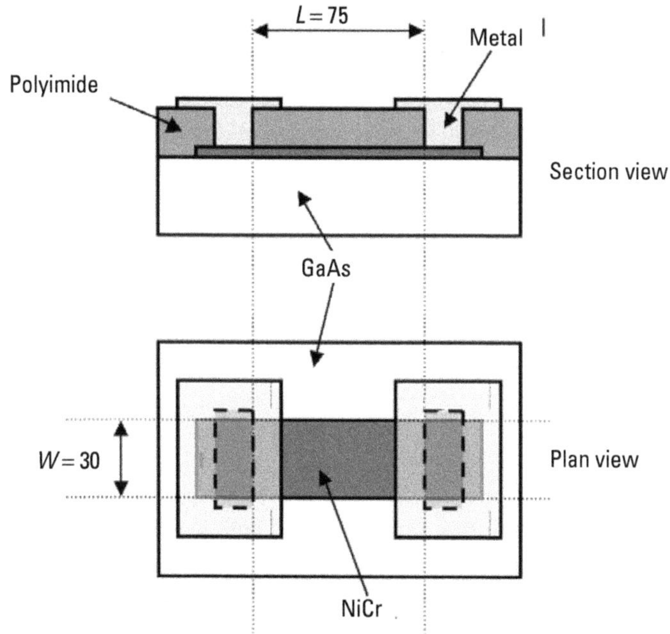

Figure 2.36 Section and plan view of a thin-film nichrome resistor.

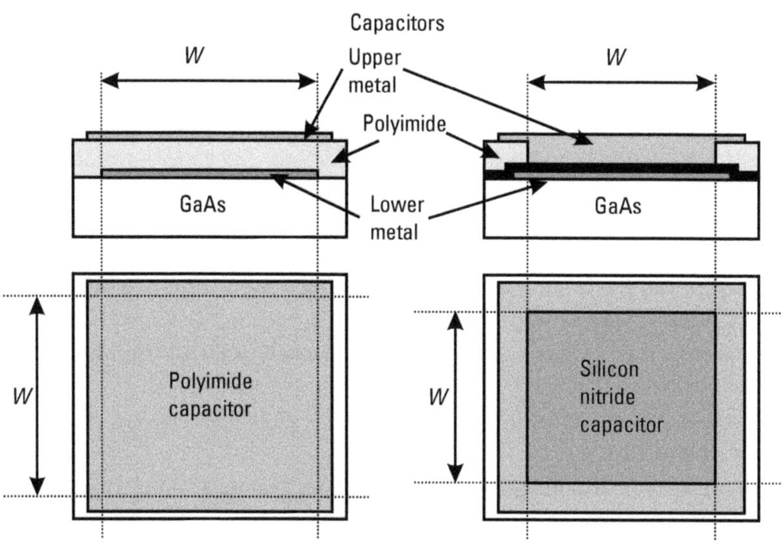

Figure 2.37 Section and plan view of polyimide and silicon nitride MIM capacitors.

2.8 Figure 2.37 shows the section and plan view of polyimide and silicon nitride MIM capacitors. Given that the silicon nitride has a thickness of 1,200Å and a dielectric constant of $\varepsilon = 7 \times 10^{-11}$ and the polyimide thickness is 1.5 μm and has a dielectric constant of $\xi = 3.6 \times 10^{-11}$, what is the capacitance of a 30-μm-square silicon nitride capacitor, and how much greater capacitance does it have than a 30-μm-square polyimide capacitor?

2.9 What limits the dc current carrying capacity of a spiral inductor?

2.10 When are stacked spiral inductors used?

References

[1] HEMT-P10S100 process, Velocium Products, Northrop Grumman, Redondo Beach, California, available at http://www.st.northropgrumman.com/velocium.

[2] Kimura, T., et al., "GaAs pHEMT Technology for Optical Communication System," *GaAs MANTECH Conference,* San Diego, CA, 2002.

[3] Uyeda, J., et al., "0.1 μm InP HEMT MMIC Fabrication on 100 mm Wafers for Low Cost, High Performance Millimeter-Wave Applications," *Proc. GaAs MANTECH Conference,* Miami Beach, FL, May 3–5, 2004.

[4] Milligan, J. W., et al., "Transition of SiC MESFET Technology from Discrete Transistors to High Performance MMIC Technology," *Proc. GaAs MANTECH Conference,* Miami Beach, FL, May 3–5, 2004.

[5] Kikkawa, T., et al., "An Over 200-W Output Power GaN HEMT Push-Pull Amplifier with High Reliability," *Proc. IEEE Microwave Theory Tech. Symposium,* Forth Worth, TX, June 6–11, 2004, pp. 1347–1350.

[6] Whelan, C. S., et al., "GaAs Metamorphic HEMT (MHEMT): an Attractive Alternative to InP HEMTs for High Performance Low Noise and Power Applications," *Proc. IEEE International Conference on Indium Phosphide and Related Materials,* Williamsburg, Virginia, May 2000, pp. 337–340.

[7] Tang, Z., et al., "Manufacturable $In_{0.8}Ga_{0.2}P/In_{0.53}Ga_{0.47}As/InP$ Doped-Channel HFETs with f_T and f_{max} over 170 GHz," *Proc. GaAs MANTECH Conference,* Vancouver, Canada, April 19–22, 1999.

[8] Lilienfeld, J. E., "Method and Apparatus for Controlling Electric Current," U.S. Patent Application No. 17,451,745, October 1926.

[9] Brattain, W. H., "Genesis of the Transistor," *The Physics Teacher,* March 1968, pp. 109–114.

[10] Mead, C. A., "Schottky Barrier Gate Field Effect Transistor," *IEEE Proceedings,* Vol. 54, 1966, pp. 307.

[11] Morkoc, H., and P. M. Solomon, "The HEMT: A Superfast Transistor," *IEEE Spectrum,* Vol. 21, No. 2, 1984, pp. 28–35.

[12] Wanlass, F., "Low Stand-By Power Complementary Field Effect Circuitry," U.S. Patent No. 3,356,858, 1963.

[13] Kroemer, H., "Heterojunction Bipolar Transistors and Integrated Circuits," *Proc. IEEE,* Vol. 70, 1982, p. 13.

[14] Asbeck, P. M., et al., "Microwave Performance of GaAs/(Ga,Al)As Heterojunction Bipolar Transistors," *Proc. IEDM 1984,* San Francisco, California, pp. 864–865.

[15] Breed, Gary, "Some Recent Trends in RFIC/MMIC Technology," *High Frequency Electronics,* Summit Technical Media, Bedford, NH, July 2005.

[16] Brehm, G. E., D. L. Green, and L. J. Mowatt, "Commercial Applications for GaAs Millimeter Wave MMICs," *Proc. GaAs MANTECH Conference,* Miami Beach, FL, May 3–5, 2004.

[17] O'Keefe, M. F., et al., "GaAs pHEMT-Based Technology for Microwave Applications in a Volume MMIC Production Environment on 150-mm Wafers," *IEEE Trans. Semiconductor Manufacturing,* Vol. 16, No. 3, August 2003, pp. 376–383.

[18] Berger, O., "GaAs MESFET, HEMT and HBT Competition with Advanced Si RF technologies," *Proc. GaAs MANTECH Conference,* Vancouver, Canada, April 19–22, 1999, pp. 7–10.

[19] Pengelly, R. S., "Improving the Linearity and Efficiency of RF Power Amplifiers," *High Frequency Electronics,* Summit Technical Media, LLC, September 2002, pp. 26–34.

[20] Smith, P. M., et al., "Advances in HEMT Technology and Applications," *Proc. IEEE Microwave Theory Tech. Symposium,* Las Vegas, NV, June 9–11, 1987, pp. 749–752.

[21] http://www.triquint.com/prodserv/divisions/foundry/docs/HFET.pdf.

[22] http://www.research.ibm.com/DAMOCLES/html_files/hfet.html.

[23] http://www.stanford.edu/class/ee328/EE328_Lecture14_5_13_04.pdf.

[24] Henderson, T., et al., "Microwave Performance of a Quarter-Micrometer Gate Low-Noise Pseudomorphic InGaAS/AlGaAs MODFET," *IEEE Electron Device Lett.,* Vol. 7, 1986, pp. 645–651.

[25] http://www.anasoft.co.uk/EE/voltagecontrolledbipolar/voltagecontrolledbipolar.html.

[26] http://www.amasci.com/amateur/transis.html.

[27] Early, J. M., "Effects of Space-Charge Layer Widening in Junction Transistors," *Proc. IRE,* Vol. 42, 1954, pp. 1761.

[28] Huber, D., et al., "InP/InGaAs Single HBT Technology for Photoreceiver OEIC's at 40 Gb/s and Beyond," *IEEE Journal of Lightwave Technology,* Vol. 18, No. 7, July 2000, pp. 992–1000.

[29] Dearn, A. W., et al., "A Universal GaAs HBT PA with Active Bias Circuitry Covering 4.9–6 GHz," *Proc. EuMW2004, GAAS Conference,* Amsterdam, Holland, October 11–12, 2004.

[30] Rollett, J. M., "Stability and Power-Gain Invariants of Linear Twoports," *I.R.E. Trans. Circuit Theory,* March 1962.

[31] Camiade, M., et al., "Full MMIC Millimeter-Wave Front-End for a 76.5 GHz Adaptive Cruise Control Car Radar," *Proc. IEEE Microwave Theory Tech. Symposium,* Anaheim, CA, June 13–19, 1999, pp. 1489–1492.

[32] Elgaid, K., et al., "Optimisation of DC and RF Performance of GaAs HEMT-Based Schottky Diodes," *Electronics Letters,* Vol. 35, No. 19, September 1999, pp. 1678–1679.

[33] Sato, K. F., et al., "Low-Frequency Noise in GaAs and InP Schottky Diodes," *Proc. IEEE Microwave Theory Tech. Symposium,* Vol. 3, Baltimore, MD, June 7–12, 1998, pp. 1695–1698.

[34] Bahl, I. J., and R. Garg, "Simple and Accurate Formulas for Microstrip with Finite Strip Thickness," *Proc. IEEE,* Vol. 65, 1977, pp. 1611–1612.

[35] Edwards, T. C., *Foundations of Microstrip Circuit Design,* New York: John Wiley, 1987 [1981].

[36] Wadell, B. C., *Transmission Line Design Handbook,* Norwood, MA: Artech House, 1991.

[37] Postoyalko, V., "Green's Function Treatment of Edge Singularities in Quasi-TEM Analysis of Microstrip," *IEEE Trans. Microwave Theory Tech.,* Vol. 34, No. 11, 1986, pp. 1092–1095.

[38] Kirschning, M., and R. H. Jansen, "Accurate Model for Effective Dielectric Constant of Microstrip with Validity up to Millimetre-Wave Frequencies," *Electronics Letters,* Vol. 18, No. 6, March 1982, pp. 272–273.

[39] Wen, C. P., "Coplanar Waveguide: A Surface Strip Transmission Line Suitable for Nonreciprocal Gyromagnetic Device Applications," *IEEE Trans. Microwave Theory Tech.,* Vol. 17, No. 12, December 1969, pp. 1087–1088.

[40] Heinrich, W., "Quasi-TEM Description of MMIC Coplanar Lines Including Conductor Loss Effects," *IEEE Trans. Microwave Theory Tech.,* Vol. 41, No. 1, 1993, pp. 45–52.

[41] Gupta, K. C., et al., *Microstrip Lines and Slotlines,* 2nd ed., Norwood, MA: Artech House, 1996.

[42] Hasegawa, H., M. Furukawa, and H. Yanai, "Properties of Microstrip Line on Si-SiO_2 Systems," *IEEE Trans. Microwave Theory Tech.,* Vol. 19, No. 11, November 1971, pp. 869–881.

[43] Hyltin, T. M., "Microstrip Transmission on Semiconductor Dielectrics," *IEEE Trans. Microwave Theory Tech.,* Vol. 13, No. 6, 1965, pp. 777–781.

[44] Suematsu, N., and M. Ono, "Silicon Based RF-ICs for Mobile Communication Terminal Applications," *Proc. International Symposium on Acoustic Wave Devices for Future Mobile Communications Systems,* Chiba University, Japan, March 2001, pp. 45–50.

[45] Guckenberger, D., and K. T. Kornegay, "Differential Distributed Amplifier and Oscillator in SiGe BiCMOS using Close-Packed Interleaved On-Chip Transmission Lines," *IEEE J. Solid State Circuits,* Bipolar/BiCMOS Circuits and Technology Meeting, October 2005, pp. 68–71.

[46] Rieh, J.-S., et al., "X and Ku-band Amplifiers Based on Si/SiGe HBTs and Micromachined Lumped Components," *IEEE Trans. Microwave Theory Tech.,* Vol. 46, 1998, pp. 685–694.

[47] Tuncer, E., and D. P. Neikirk, "Highly Accurate Quasi-Static Modeling of Microstrip Lines over Lossy Substrates," *IEEE Microwave and Guided Letters.,* Vol. 2, No. 10, October 1992, pp. 406–411.

[48] Dunn, J. S., et al., "Foundation of RF CMOS and SiGe BiCMOS Technologies," *IBM J. Research and Development,* Vol. 47, No. 2/3, 2003.

[49] Shen, H., et al., "Fabrication and Characterization of Thin Film Resistors for GaAs-Based Power Amplifiers," *Proc. GaAs MANTECH Symposium,* 2003.

[50] Esfandiari, R., D. W. Maki, and M. Siracusa, "Design of Interdigital Capacitors and Their Application to Gallium Arsenide Monolithic Filters," *IEEE Trans. Microwave Theory Tech.,* Vol. 31, No. 1, 1983, pp. 57–64.

[51] Mondal, J. P., "An Experimental Verification of a Simple Distributed Model of MIM Capacitors for MMIC Applications," *IEEE Trans. Microwave Theory Tech.,* Vol. 35, April 1987, pp. 403–408.

[52] Stein, K., et al., "High Reliability Metal Insulator Metal Capacitors for Silicon Germanium Analog Applications," *Proc. IEEE Bipolar/BICMOS Circuits and Technology Meeting,* Minneapolis, MN, September 1997, pp. 191–194.

[53] Yota, J., et al., "Development and Characterization of a 600 Å PECVD Si_3N_4 High-Density MIM Capacitor for InGaP/GaAs HBT Applications," *Proc. GaAs MANTECH Conference,* Phoenix, AZ, May 2003.

[54] De Bhola, N., and M. Shokrani, "Reliability Characterization of High Density Metal-Insulator-Metal Capacitors (MIMCAP) Fabricated by Depositing Silicon Nitride Using PECVD in Compound Semiconductor Manufacturing," *Proc. GaAs MANTECH Conference,* New Orleans, LA, April 11–14, 2005.

[55] Mohan, S. S., et al., "Bandwidth Extension in CMOS with Optimised On-Chip Inductors," *IEEE J. Solid State Circuits,* Vol. 35, No. 3, March 2000, pp. 346–355.

[56] Park, E.-C., et al., "Fully Integrated Low Phase-Noise VCOs with On-Chip MEMS Inductors," *IEEE Trans. Microwave Theory Tech.,* Vol. 51, No. 1, January 2003, pp. 289–296.

[57] Dahlmann, G. W., et al., "MEMS High Q Microwave Inductors Using Solder Surface Tension Assembly," *Proc. IEEE IMS Symposium,* Phoenix, AZ, May 20–25, 2001.

3

Foundry Use and Economics

Because of the strong link between component technology and foundry suppliers, this chapter describes the type of procedures the designers may go through when designing an MMIC with an external foundry. It also covers some of the economic issues associated with manufacturing an MMIC design through a foundry and the trade-offs when considering the design of multifunction MMICs.

3.1 Using a Foundry

Using an MMIC foundry to design and fabricate a chip to a given specification is an interactive exercise between the foundry and the customer designing the chip, as illustrated in Figure 3.1. It starts with discussions and agreements between the two parties, goes through the design process and wafer fabrication, and ends when the chips are fully tested and delivered to the customer.

3.1.1 Agreements and Discussions

Interaction with a foundry usually starts with discussions on an informal basis between the customer and foundry engineers regarding the foundry capability and available options. This will be on a very general level, discussing aspects such as the process technology available, key performance

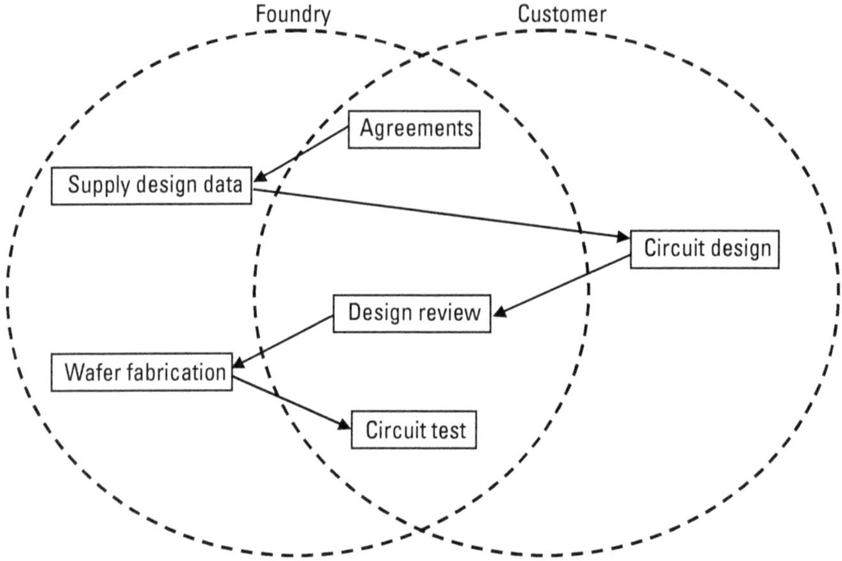

Figure 3.1 Flowchart illustrating the process of designing and fabricating an MMIC using a foundry.

benchmarks of the active devices on the process, and examples of the type of circuits that have already been produced by the foundry. This is also the time when the customer should ensure that the foundry can supply the electronic design data in a form that can be used with the customer's CAD simulation tools.

When the customer and the foundry decide there is likely to be mutual benefit in working together, they will normally enter into a nondisclosure agreement (NDA) to protect the foundry's proprietary information and the customer's design details. This will then allow the foundry to pass to the customer much more detailed process and design information so the customer can properly assess the match between the process capabilities and their design requirements. It also allows the customer to discuss their design specifications with the foundry engineers to get their experience and feedback as to the likelihood of the process's meeting the specification.

When both parties are happy, the customer and the foundry enter into a wafer agreement, which confirms the processing options required, the number of wafers or batches of wafers to be produced, and the time scales for delivery. This agreement may also include specifications for dc and RFOW testing and any chip-separation and quality-inspection requirements.

3.1.2 Delivery of Design Data

When the agreements are in place, the foundry will provide the customer with all the design data required to start designing the MMICs. This data will typically consist of a design manual and an electronic file containing the CAD simulation and layout libraries for the agreed-on MMIC process.

The design manual is often a controlled document, specific to the required process, which contains a lot of background information about the MMIC process and all the essential information needed to complement the electronic CAD libraries. It will contain a process description and sections describing all of the active, passive, and interface components available to the designer. The manual will also outline the layout rules and procedures particular to the foundry. It is also likely to contain the quality-assurance and process-control details with which the foundry operates and to describe the wafer-approval conditions that must be met before the finished MMICs are released to the customer.

The CAD simulation and layout libraries will be created specifically for one particular simulation tool and may well be particular to the hardware and operating system the CAD tool is running on. These electronic files will contain the complete library of components that can be fabricated by the process, including the schematic layouts, underlying electrical models, associated physical layouts, and possibly design-rule files.

3.1.3 Circuit Design

Some foundries offer circuit-design services and consultancy by their design engineers, but this stage is normally the responsibility solely of the customer; hence, it is the main area that this book addresses. The foundries will usually be happy to answer queries about the functioning and application ranges of their components and models but will not necessarily be willing to solve design problems. Experience has shown that even with modern schematic-to-layout synchronized CAD tools, the MMIC design process is highly complex and susceptible to simple errors at all stages. The best way to eliminate these errors is to design each chip using a team of MMIC designers who can question each others' judgments at each stage and reverse-engineer the final layout before processing starts. By reverse engineering, I mean that an independent MMIC designer takes the final layout data and recreates a schematic design in the CAD system with no input from the original design team. The RF responses generated by the independent engineer can then be compared to the original design files and to the specification. If all are in agreement, then the processing can go ahead; if there are significant discrepancies, then errors or mistakes have almost certainly been overlooked.

3.1.4 Design Review

When all the circuit designs are complete and the electronic design files have been passed back to the foundry, there is usually a process of reviewing the design. This is normally started by the foundry, which uses electronic design-rule checking on the circuit layout files to ensure that they obey all the critical rules and can be fabricated on the specified process. The foundry will then assemble all the circuit-design layouts into a reticule, as described in more detail in Section 6.4, prior to a review meeting with the customer. During the review meeting, it is common to examine extremely large plots of the reticule of circuits in order to correct visible errors and perform a critical appraisal of the designs. Foundry staff from the processing, testing, and quality-assurance departments should determine if the design conforms with the processing limitations, as well as the dc and RF testing capability, and whether small rule violations are acceptable to all involved. Minor corrections highlighted by the review process are normally corrected by the foundry, but major corrections may require redesign by the customer, followed by another review meeting. When the customer and the foundry are happy with the final reticule, an agreement is signed, and the electronic reticule data is sent to mask manufacturers, who then make the chrome-on-quartz mask plates for the photolithographic steps during the wafer-fabrication process.

3.1.5 Wafer Fabrication

Batches of wafers (numbering between 2 and 6 wafers per batch) are processed by the foundry according to the wafer agreement in place. The wafer-fabrication process can take anywhere from 6 to 16 weeks, depending on the normal process turnaround time and any priority fabrication agreements that have been made with the foundry. At any given time, there are likely to be many batches of wafers going through the clean rooms, and inevitably some batches spend time waiting for certain equipment. If quick turnaround is essential, a foundry can give priority to a particular customer's batch of wafers, but this is also likely to be at extra cost. Unplanned delays can also occur if wafer batches fail during processing, and a new batch needs to be started.

3.1.6 Test and Delivery

Before the customer's circuits are tested, the foundry will perform dc and RF measurements on foundry test structures within the drop-in arrays to ensure that the wafer has been processed correctly and that all the component parameters are within the limits guaranteed by the foundry. When this is complete, the wafers can be released to the customer, or, if arranged, the foundry can perform various RFOW measurements on the customer's circuits. Foundries may deliver the customer's circuits as complete wafers, but it is more usual for the individual chips to be separated and packed in waffle trays or gel packs. Note that the foundry releases wafers or chips to the customer based on their own measurements of the drop-in test structures and not on the basis of whether the customer's circuits work or not!

3.2 Economics

This section considers the cost of manufacturing MMICs and what can be done to minimize their cost contribution to overall system costs. Also included here is discussion of the cost/yield debate and the trade-off between using several single-function MMICs or one larger multifunction MMIC for a system requirement.

3.2.1 MMIC Production Costs

Production of any monolithic integrated circuit, including MMICs, requires clean rooms, which are maintained to a very high standard of cleanliness, temperature, and humidity control and equipped with very complex and expensive processing equipment. Skilled engineers are also required for process development and control and for defining clear processing procedures for technicians to run batches of wafers through the process. The capital expenditure to keep the technology updated is very large, and the running costs to keep the clean rooms tightly filtered and air-conditioned are very high. There is never any time when the clean rooms do not need to be maintained at these conditions, so the running costs are effectively fixed for a particular clean room setup.

The cost of the starting material (i.e., the semi-insulating wafers) is fairly low and continues to drop over time, making it less significant compared to the clean-room running costs. Therefore, the overall running costs tend to be fixed and independent of the number of wafers being produced. Thus, the cost of producing the chips can be expressed in simple terms as the overall running cost divided by the number of good chips produced. If the number of good chips or good wafers is multiplied by the area, then the cost of MMIC processing can be expressed per square millimeter of the substrate area.

Therefore, from the customer's perspective, the higher the volume of the MMIC product designed on the foundry process and the higher the overall yield, the cheaper will be the cost of the individual MMIC. In other words, a foundry process is most economically efficient when many batches of wafers are being processed, and the foundry process line is near its maximum capacity. When a process is operating near its maximum capacity, individual chip costs can only be further reduced by improving the yield of good devices produced off each wafer. For this, it is worth considering the concept of the process defect density.

3.2.2 Defect Densities

The concept of defect density was developed by the silicon industry and assumes a uniform distribution of defects across a finished wafer [1–3]. This is not generally the case for MMICs if one particular defect cause is considered, because the complexity and packing density of components in MMICs are much less than in silicon ICs. However, if the overall yield is taken to be degraded by all possible causes (e.g., processing yield, dc and RF testing yield, saw yield, visual yield), then this becomes a fair approximation.

The Murphy-Seeds rule shown in (3.1) takes the defect density (*D* defects per square centimeter) and chip area (*A* square centimeters) and predicts the probability (*p*) of getting a working device, in other words, the chip yield.

$$p = \frac{e^{-\sqrt{AD}} + \left(\frac{1 - e^{-AD}}{AD}\right)^2}{2} \tag{3.1}$$

F. A. Myers [4] shows how this rule can be useful for establishing yield and process maturity. Figure 3.2, taken from Myers, plots the chip yield versus chip area for processes with defect densities ranging from 1 to 50 defects per square centimeter. He notes that typical GaAs processes have been found experimentally to have an effective defect density in the 5 to 10 defects per square centimeter region, so a 0.1 cm^2 GaAs chip can expect to have a yield of 55%. The dashed trend line in Figure 3.2 shows how the yield of any size chip improves as a process matures and the defect density drops. In this case, the drop in defect density with process maturity is due to the operators' becoming more familiar with the process steps and the engineers' learning how to keep the process under tighter control. The dotted trend line in Figure 3.2 shows that as the MMIC designers become more confident using a process, the chip designs become more complex, and more components are packed into a smaller area, increasing the defect density and leading to chips with reduced yields. Thus, there is a trade-off between chip size and complexity to achieve the best over all yields.

This trade-off is demonstrated when one considers the economic factors associated with producing a multifunction system [4]. One option is to use a number of small, single-function chips to make up the multifunction system. Each chip must be processed, tested, possibly packaged, assembled into the system, and wire-bonded to the tracks. On the other hand, a single, but larger, chip, performing multiple functions would have fewer bond wires and package parasitics and so could perform better and would be cheaper to assemble. The larger multifunction chip would have a lower yield than the smaller single-function chips but may be an overall cheaper system solution. If the defect density of the process is known, the Murphy-Seeds equation can be used to find the cost trade-off between using two small single-function chips or one multifunction chip of twice the area. This is shown in Figure 3.3, where the cost ratio of one larger multifunction chip over two smaller single-function chips is plotted for different defect densities. Where the value of the ratio is less than one, the single, but larger, multifunction chip is the lowest cost solution.

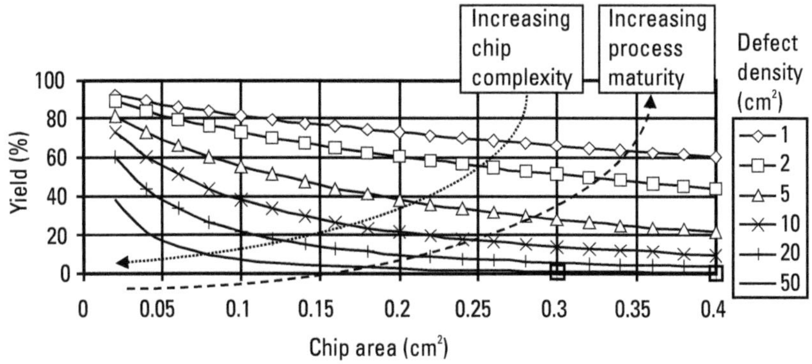

Figure 3.2 The Murphy-Seeds rule: yield as a function of chip area and process defect density [4].

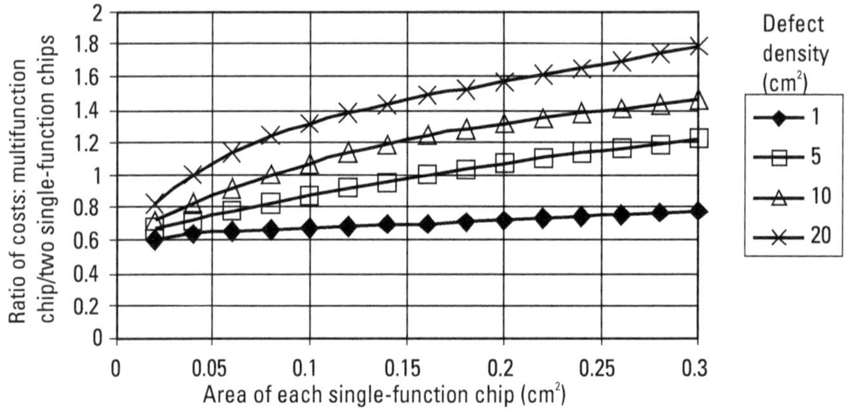

Figure 3.3 Ratio of the costs of one larger multifunction chip to two smaller single-function chips for process defect densities from 1 to 20 defects per square centimeter.

References

[1] Koren, I., and D. K. Pradhan, "Yield and Performance Enhancement through Redundancy in VLSI and WSI Multiprocessor Systems," *Proc. IEEE,* Vol. 74, May 1986, pp. 699–711.

[2] Koren, I., and C. Strapper, "Yield Models for Defect Tolerant VLSI Circuits: A Review," Department of Electronics and Computer Engineering, University of Massachusetts, Amherst, TR-89-CSE-1, February 1989.

[3] Boubekeur, A., G. Saucier, and J. Trilhe, "A Reconfigurable Wafer-Scale Memory," *IEEE J. Solid-State Circuits,* Vol. 26, No. 10, October 1991, pp. 1423–1432.

[4] Myers, F. A., "Yield Breakpoints between Single/Multi-Function MMICs," *Proc. IEEE,* 1977.

4

Simulation and Component Models

In order to design an MMIC to a given specification, the designer must be able to take ideas for the required circuit topology and accurately predict the performance of this circuit when implemented in MMIC form. This is achieved using computer-aided design (CAD) tools, such as Agilent ADS,[1] Ansoft Designer,[2] AWR Microwave Office,[3] and Cadence Virtuosos,[4] to simulate the RF and microwave characteristics. These tools allow the designer to connect together models of the individual component elements into a circuit design and simulate the MMIC circuit performance in the frequency or time domain.

High-yielding MMIC product design requires well characterized and accurately modeled MMIC components implemented within these commercial simulators. This is a time-consuming and costly activity, and it represents a significant investment by the foundry; hence, their model libraries are neither comprehensive nor perfect. Although these component models are usually hidden behind their schematic symbols, understanding how the

1. Agilent Technologies, USA, http://eesof.tm.agilent.com, last access June 5, 2006.
2. Ansoft Corporation, USA, http://www.ansoft.com/products/hf/ansoft_designer.
3. Applied Wave Research, USA, http://www.appwave.com/Products/Microwave_Office/Overview.php, last access June 5, 2006.
4. Cadence, USA, http://www.cadence.com/products/custom_ic/index.aspx, last access June 5, 2006 .

foundry may have produced the models helps the designer to understand their limitations and use them circumspectly.

The aim of this chapter is to review the concept of *s*-parameter representation of components and discus how a foundry typically goes about characterizing the individual components and develops a lumped-element equivalent circuit model for them. The chapter continues by describing the commonly used equivalent circuit topologies for both the active and passive components and how they were derived. The chapter also touches on the limitations of *s*-parameter representation of individual components and how full three-dimensional (3D) electromagnetic simulation is helping to overcome this.

4.1 Simulation

Simulation of an MMIC's microwave performance takes electrical models or *s*-parameter data files of the individual elements and calculates the overall characteristics. This can be performed on linear or nonlinear devices and derives parameters such as the input and output impedance matches, signal loss, or gain in the frequency and time domains. The interaction of closely spaced components on the chip can also be simulated using 3D simulation CAD tools.

4.1.1 The *s*-parameters

The concept of scattering parameters, or *s*-parameters, was first discussed by K. Kurokawa [1] to give a clearer and more straightforward understanding of the power relations between circuit elements connected through a multiport network. In other words, *s*-parameters give a more physical meaning to how a signal is scattered when it is incident on a circuit in terms of how much of the signal power is reflected and how much is transmitted out from the other ports of the circuit.

The *s*-parameter representation of an arbitrary circuit is called the scattering matrix, and the scattering parameters can be understood by considering the incident, reflected, and transmitted signals in terms of traveling waves. In Figure 4.1, a signal is applied to a two-port device embedded in transmission lines with characteristic impedance Z_0, and incident and reflected waves are present on the input transmission line. The generalized reflection properties of this two-port device are given in Figure 4.2, where V_{i1} is the incident voltage wave at port 1, V_{r1} is the reflected voltage wave at port

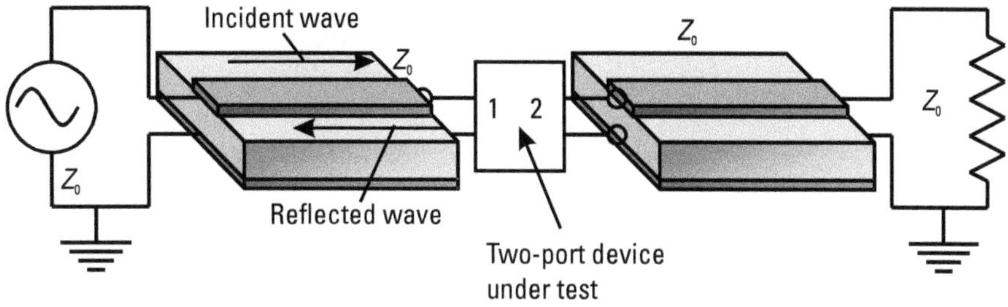

Figure 4.1 A signal applied to a two-port device embedded in transmission lines with characteristic impedance Z_0.

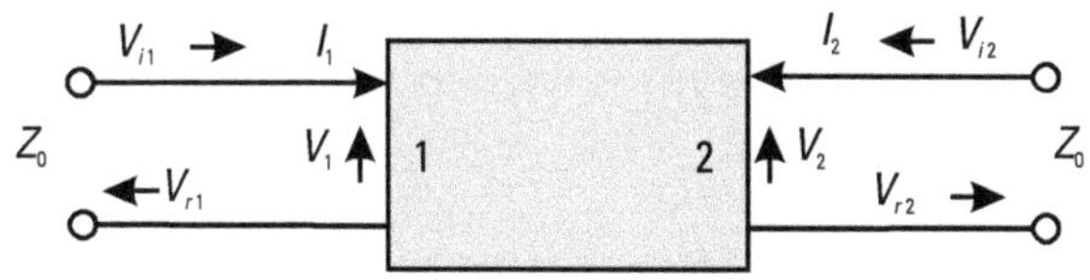

Figure 4.2 Generalization of the reflection properties of a two-port device embedded in transmission lines with characteristic impedance Z_0.

1, V_{i2} is the incident voltage wave at port 2, and V_{r2} is the reflected voltage wave at port 2.

The power incident at port 1, given in (4.1), can be used to define a new variable a_1; similarly, the power reflected from port 1, given in (4.2), can define b_1. Similarly, variables a_2 and b_2 can be defined for port 2. Rearranging these for port 1 gives the total voltage and current at port 1, as shown in (4.3) and (4.4).

$$P_{incident}(1) = \frac{|V_{i1}|^2}{Z_0} = |a_1|^2 \tag{4.1}$$

$$P_{reflected}(1) = \frac{|V_{r1}|^2}{Z_0} = |b_1|^2 \tag{4.2}$$

$$V_1 = \sqrt{Z_0}(a_1 + b_1) \tag{4.3}$$

$$I_1 = \frac{a_1 - b_1}{\sqrt{Z_0}} \tag{4.4}$$

Rearranging these equations at port 2 and solving for a_1 and a_2 as independent variables and b_1 and b_2 as dependent variables gives the scattering matrix solution for a two-port circuit, as shown in (4.5). Each scattering element S_{ij} is a complex quantity relating both the magnitude and phase of the input and output signal voltages.

$$\begin{bmatrix} b_1 \\ b_2 \end{bmatrix} = \begin{bmatrix} S_{11} & S_{12} \\ S_{21} & S_{22} \end{bmatrix} \begin{bmatrix} a_1 \\ a_2 \end{bmatrix} \tag{4.5}$$

This solution, based on the scattered waves from an arbitrary circuit, is a general solution that applies to a network with any number of ports, as given in (4.6), and the matrix $[S]$ is referred to as the scattering (S) matrix for the network.

$$[b] = [S] \times [a] \tag{4.6}$$

The individual elements of the scattering matrix have the clear physical meaning that Kurokawa discussed [1] (i.e., element S_{ij} gives the amplitude and phase of the signal voltage output from port i relative to the signal voltage input at port j). Likewise, S_{ii} is the signal voltage output from port i relative to the signal input at port i, in other words, the amount of signal reflected from port i, showing how well the port is matched to Z_0. Consequently, for a network that is perfectly matched to Z_0 at port j, the *s*-parameter S_{jj} is equal to 0. Similarly to the generalization that S_{ii} represents the match at port i, S_{ij} represents the voltage gain or loss from port j to port i, and S_{ji} represents the voltage gain or loss between port i and port j. Note that for a passive network (no elements giving voltage gain), all the S_{ij} will have a magnitude less than unity.

Going back to the example of a two-port device such as an amplifier in a 50-ohm system, S_{21} is the transmitted/incident signal known as the voltage gain, S_{11} is the reflected/incident signal at the input (i.e., the input match), S_{22} is the output/incident at output (i.e., the output match), and S_{12} is the signal transmitted out of the input/incident signal at the output, known as the reverse isolation, as shown in Figure 4.3. Note that S_{21} is the voltage gain and that the power gain is $|S_{21}|^2$.

The *s*-parameters are relatively easy to measure, as the requirement is simply that all the waves equal 0 except for the port of interest and that all the ports be terminated in the characteristic impedance Z_o. An example

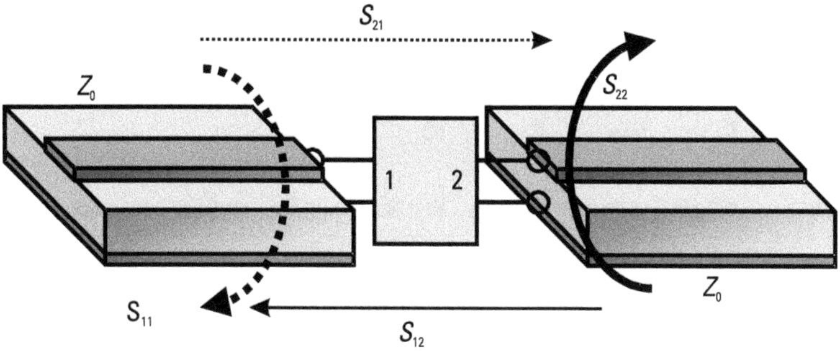

Figure 4.3 The *s*-parameters of a two-port device.

s-parameter data file in Touchstone format[5] is shown in Table 4.1, where the line starting with the "#" symbol gives the global variables for the data .

It is common to represent individual MMIC components by their *s*-parameters because the *s*-parameters of multiple components can be cascaded to predict the overall chip performance. The main limitation of *s*-parameter representation is that *s*-parameters ignore interactions with other components that are not at defined ports; for example, two closely spaced inductors will be coupled by their magnetic fields, but two-port *s*-parameter representation of the individual inductors will not take account of this. In other words, *s*-parameter representation of a component element assumes infinite spacing from other components except at the component element's defined ports. Another limitation of an *s*-parameter data file is that it is only valid over the actual frequency range of the data, and at any frequency points outside of this range, the simulator tends to extrapolate the last two data points, which are not often valid because they are commonly where the data is most noisy.

4.1.2 Component Characterization

MMIC foundries characterize the individual components' elements, such as capacitors, resistors, and spiral inductors, based on *s*-parameter measurements of the actual elements fabricated on characterization wafers. To do this, the foundry designs a mask-set that includes multiple geometries of all the different types of components from the passive devices through to the active transistor devices. Figure 4.4 shows one section from a characterization

5. http://www.eda.org/pub/ibis/connector/touchstone_spec11.pdf, last access June 5, 2006.

Table 4.1
s-Parameter Data File for a Two-Port Device

# hz	S	ma	R	50				
! freq	magS_{11}	angS_{11}	magS_{21}	angS_{21}	magS_{12}	angS_{12}	magS_{22}	angS_{22}
!								
1.00E+09	0.9934	−8.713	6.716	173.9	0.01323	82.93	0.5867	−6.543
1.10E+10	0.8508	−80.79	4.87	123.8	0.105	43.2	0.4544	−60.33
2.10E+10	0.7508	−120.4	3.252	95.21	0.1326	23.29	0.3533	−90.61
3.10E+10	0.7113	−143	2.378	75.97	0.141	12.61	0.3148	−109.2
4.10E+10	0.6951	−157.9	1.869	60.59	0.1438	5.643	0.3065	−122.3
5.10E+10	0.6888	−168.9	1.545	47.18	0.1445	0.4931	0.3134	−132.6
6.10E+10	0.6872	−177.7	1.321	34.97	0.1442	−3.607	0.329	−141.3
7.10E+10	0.6885	174.8	1.157	23.6	0.1434	−7.011	0.3502	−149.1
8.10E+10	0.6914	168.2	1.03	12.9	0.1424	−9.896	0.3751	−156.4
9.10E+10	0.6956	162.3	0.929	2.751	0.1414	−12.36	0.4026	−163.3
1.00E+11	0.7	157.4	0.8523	−5.971	0.1405	−14.28	0.4289	−169.4

hz: frequency in hertz; S: *s*-parameters as opposed to Y or Z parameters; ma: vector format is magnitude and angle rather than real and imaginary; R 50: characteristic impedance is 50Ω). Lines preceded by "!" are comments and are ignored by data-reading software.

mask-set that includes MESFETs with 1, 2, 4, 6, and 8 gate fingers with lengths of 50 μm, 100 μm, 150 μm, and 200 μm. Note that each component is placed between two RFOW probe pads and connected with a length of relatively narrow microstrip transmission line. The length of transmission line is designed to ensure that any evanescent modes created in the locality of the component are minimal by the time they reach the RFOW pads so that only the propagating microstrip mode is measured [2]. The measurement equipment is calibrated to place the data-file reference plane where this transmission line connects to the device.

Multiple wafers and batches of wafers are fabricated using the characterization mask-set over a period of time, and each component is measured RFOW, and the data is saved. This gives the foundry statistical information about the characteristics of a "typical," or average, device and the spread or range of the characteristics produced by processing variations.

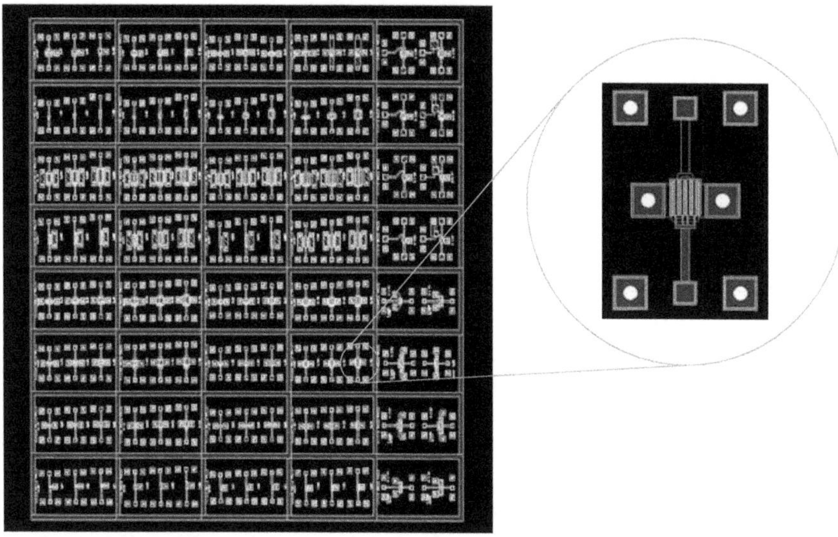

Figure 4.4 One section from a characterization mask-set, including multiple geometries of MESFET devices. (*Source:* Bookham Inc., 2006. All Rights Reserved.)

4.1.3 Model Development

Electrical models of the individual MMIC components are developed to take the vast range of measured data and make it available in a user-friendly format to the MMIC designer within microwave CAD simulation tools. The models are usually developed to work in the frequency domain but should also be capable of working in the time domain. These models, especially when provided by an MMIC foundry, commonly have component layouts linked to the electrical models to reduce the chance of errors during the layout process. Due to the complexity and specialty of current microwave CAD simulation tools, the models need to be produced specifically for each simulation tool, so foundries usually only make their model libraries for one or two of the most popular CAD tools. Within the CAD simulation tools, the models are selected from a library of components and are represented by a symbol on a schematic circuit diagram page, behind which is the compiled software code of a lumped-element equivalent circuit electrical model.

4.1.3.1 Lumped-Element Equivalent Circuit Models

Lumped-element electrical models [3] can be used for both the active and the passive components because they are usually physically small compared to the wavelength of the applied signals. The lumped-element approximation [4–6] is valid for components with dimensions less than one-tenth of the signal wavelength, so considering that the wavelength on a GaAs microstrip is about 5 mm at 20 GHz, and the components are typically smaller than 0.2 mm, this is acceptable. The gates of FETs can act as slow-wave components [7–10], greatly reducing the effective signal wavelength along them, so lumped-element models of FETs may have increased errors at high frequencies. Components, such as Lange couplers, that are designed to use the coupling effect of adjacent microstrip lines and have dimensions of one-quarter wavelength cannot be modeled with lumped-element models.

The topology of each lumped-element equivalent circuit model is generally taken from the literature [11–39] and is sometimes modified to include parasitic elements associated with the MMIC component. The parasitic capacitance from the metal making up the component through the typically 100-μm-thick substrate to the back-face ground plane is present in nearly every MMIC components and is included in the equivalent circuit model by a small capacitor to ground. Likewise, the parasitic losses within each component are taken account of in the model by a small series resistor. An example is the lumped-element equivalent circuit model for an MIM series-connected capacitor, as shown in Figure 4.5. The prime capacitance is the parallel-plate capacitance between the top and bottom plates of the capacitor; all the other elements are parasitics, such as the inductance representing the phase shift of signals as they enter one side of the capacitor and exit out the opposite side. The complete equivalent circuit model can now be fitted to the measured data of the equivalent element, which is achieved by

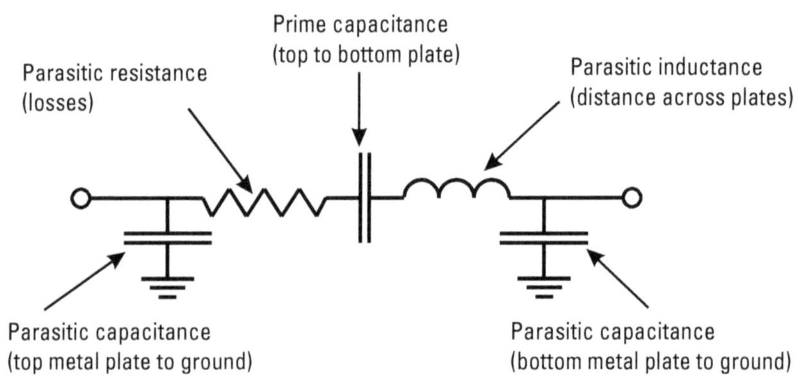

Figure 4.5 Lumped-element equivalent circuit model of a series-connected MIM capacitor.

varying the circuit elements until the measured and modeled *s*-parameters are the same. This gives the model for one particular geometry of the component, but MMIC designers prefer to use scalable models that cover the complete range of geometries.

4.1.3.2 Scalable Models

MMIC designs are very complex and often require that the design be optimized within the CAD simulation tool to get as close as possible to the required specification. The means allowing the simulation software to change the prime values of the components until the error function against performance goals are minimized. To do this, the models must be scalable against the component geometry, not just in the discrete sizes that were measured from the characterization wafers but also in a continuous manner.

A typical foundry approach is to fit the lumped-equivalent circuit model to the whole range of the geometries measured and to produce a mathematical expression [40] for how, for example, the prime capacitance of a square MIM capacitor varies with the length of the sides. The mathematical expression is normally a truncated polynomial expansion with terms appropriate to the physics of the component (i.e., in the case of the prime capacitance of a square MIM capacitor, this is dominated by the "length of side" squared term because the capacitance is proportional to the area of the plates).

The scalable models are only valid within the range of geometries that were measured and had equivalent circuit models fitted. The polynomial expressions now give valid interpolation between the measured geometries, so the scalable model can be used for geometry optimization. Extrapolation outside of the measured range of geometries is not recommended as the polynomial expressions are unlikely to be linear and could lead to increased errors.

4.1.3.3 Process Dependencies and Tolerancing

All MMIC processes produce wafers with varying physical and electrical parameters, which produce a spread in the MMIC components characteristics. For example, the thickness of the dielectric layer in MIM capacitors directly influences their prime capacitance, and the doping and etching of the active semiconductor layers influences the gain and parasitic capacitances of the transistors. Therefore, the most useful models are those that allow the designer to vary their characteristics according to the range of variation indicated by the foundry. The MMIC designer should note that the variation of the parameters across a wafer and within a batch of wafers is likely to be a lot less than the maximum and minimum specified by the foundry because their published figures must also take account of longer term drift over time.

Models for passive components such as resistors and MIM capacitors only require tolerance variables for the parameter that has greatest influence on the variation of their characteristics, such as the resistive film resistivity and the dielectric layer thickness, respectively. These parameters tend to vary independently and can be assumed to be uncorrelated when performing an overall tolerance analysis.

Models for the active devices are much more complicated and tend to exhibit characteristics that are dependant on correlated variables in their equivalent circuit models [41]. As the correlations are not entirely understood, database sampling [42, 43] is a better method to assess the effect of the active device variation during tolerance analysis.

4.1.3.4 Linked Electrical and Layout Models

Many CAD simulation tools allow a device physical layout to be linked to the electrical model, which appears on a separate layout window. This helps to minimize errors going from the design schematic to the design layout but does restrict certain aspects of the layout because the linked layouts can only be joined at predetermined points on their structure.

4.1.4 Linear Simulation

Linear simulation implicitly means small-signal analysis, and there are very few problems using the component models for linear *s*-parameter analysis.

4.1.5 Nonlinear Simulation

Nonlinear analysis involves larger signals where one or more of the components become nonlinear in their behavior and produce signal components at frequencies other than those applied. This requires techniques such as harmonic-balance analysis [44, 45], which solves the circuit at dc and at several harmonics of the applied signal, as well as at the frequency of the applied signal. There is a trade-off when using harmonic balance because the more harmonics that are simulated, the more accurate the simulation results will be, but the computation time is also much longer. Three harmonics is the minimum number that will produce meaningful results, but the designer should ideally chose five or seven if the circuit is simple enough, and the simulation time is not excessive. As a result of this, all the component models must have the correct behavior at dc when performing nonlinear analysis. Likewise, nonlinear analysis of a 30-GHz component will need to solve the circuit at 60, 90, . . . GHz. As it is unlikely that the components were measured at these high frequencies, the models must have the correct behavior when extrapolated above the frequency range of characterization.

4.1.6 2D, 2.5D, and 3D Electromagnetic Simulation

2D planar electromagnetic simulators analyze the cross-section of strip-type transmission lines and assume the propagation modes are transverse electromagnetic (TEM) or quasi-TEM. This means that all the electric and magnetic field vectors are in the plane of the cross-section, and variation along the length of the transmission line is a function of time. This type of simulator calculates the effective permeability, per-unit-length capacitance and inductances, and so forth, and can generate *s*-parameters for finite-length sections. Electric currents are allowed across the width (*x* direction) and along the length (*y* direction) of the metal strips but not vertically (*z* direction). Examples of this type of simulator include Ansoft Designer SV integrated 2D field solver [46] and LINMIC quasi-TEM and full-wave simulators [47]. 2.5D planar electromagnetic simulators such as Momentum [48] are primarily the same as 2D simulators except the extra half-dimension means that the solver can calculate currents in the vertical (*z*) direction (usually for through-substrate vias) as well as the *x* and *y* directions [49]. 3D planar electromagnetic simulators are similar, but they are also able to calculate the effects of conductor thickness and resistivity. Examples of these include Sonnet Lite [11] and EM3DS [12].

3D arbitrary geometry electromagnetic simulators define the space to be analyzed volumetrically, have no restrictions on the directions of the fields or currents, and are able to handle the effects of metal thickness and resistivity. This extra flexibility does tend to mean that the simulation time is longer than with planar simulators. Examples of these full 3D electromagnetic (EM) simulators include Sonnet's CST Microwave Studio [13] and HFSS [14].

Correct use of these various electromagnetic simulators [15] allows the MMIC designer to predict coupling effects between unconnected components, wavelength-related effects when close to the lumped-element approximation, and modes of propagation other than the assumed quasi-TEM microstrip mode, such as cavity waveguide and parallel plate modes [16–19]. This overcomes the limited *s*-parameter representation of components and is invaluable for simulating densely packed MMICs, especially at millimeter-wave frequencies. This capability must be used judiciously because computing power is not quite sufficient to simulate the whole of an MMIC within reasonable time frames, so it should be used to model small areas where coupling may be a problem and to check that performance is not degraded.

Note that if the EM-derived data is going to be used in a nonlinear harmonic-balance analysis, the frequency points at which the structure was solved must include the correct number of harmonics of the applied signals to ensure an accurate analysis. Similarly, if the EM-derived data is going to be used in a time-domain analysis, the frequency points at which the structure was solved must include time-harmonic samples (multiples) of the minimum time step to ensure convergence.

4.2 Passive Component Models

The passive components whose equivalent-circuit models are described here include capacitors, inductors, resistors, through-substrate vias, and transmission-line discontinuities.

4.2.1 Capacitors

MIM capacitors generally have the same model no matter what configuration they are used in, but interdigital capacitors require different topologies to model series and shunt configurations.

4.2.1.1 MIM Capacitors

The lumped-element equivalent circuit model for a MIM capacitor [20] is shown in Figure 4.6, along with a schematic drawing of the component showing where the reference planes of the element are normally located. The model is asymmetric because the parasitic capacitance to ground from the top plate is much less than for the bottom plate. The capacitor does not have to be square, but most foundry models assume a square geometry. This simple model assumes that the input and output connections are on opposite sides of the capacitor, with the phase change modeled by the series inductor. A more complicated model is required if the capacitor requires different or multiple connections [21, 22].

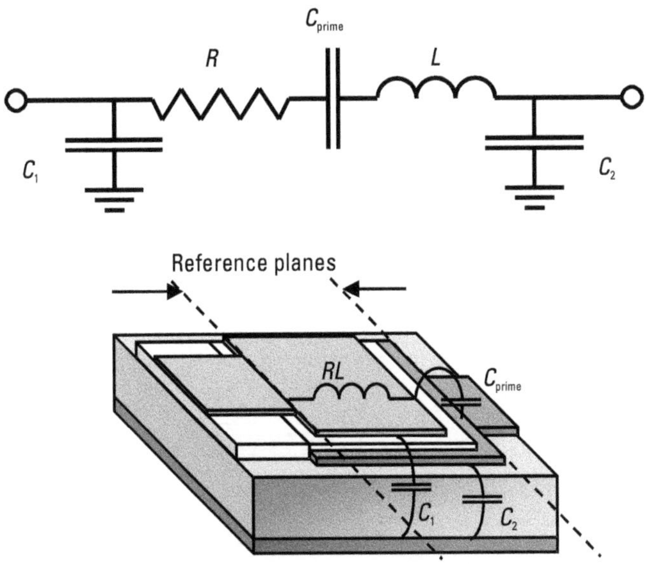

Figure 4.6 Lumped-element equivalent circuit of an MIM capacitor.

4.2.1.2 Series Interdigital Capacitors

The model for a series-connected interdigital capacitor [20], shown in Figure 4.7, is a symmetric model because the metal at each side is the same.

4.2.1.3 Shunt Interdigital Capacitors

The model for a shunt-connected interdigital capacitor is shown in Figure 4.8.

4.2.2 Inductors

The lumped-element equivalent circuit model for a spiral-track inductor [23] is shown in Figure 4.9, along with a schematic drawing showing the reference planes. The model is made up from the primary inductance of the metal track (L_{prime}) with a series resistance (R) representing the losses, capacitance from the metal tracks to the ground plane (C_1 and C_2), and a shunt capacitance (C) modeling the capacitance between the tracks. The spiral inductor is really more like a distributed component, and the use of a single feedback capacitance is only a representative model if there is only a relatively small phase change between the input and the output of the inductor. In

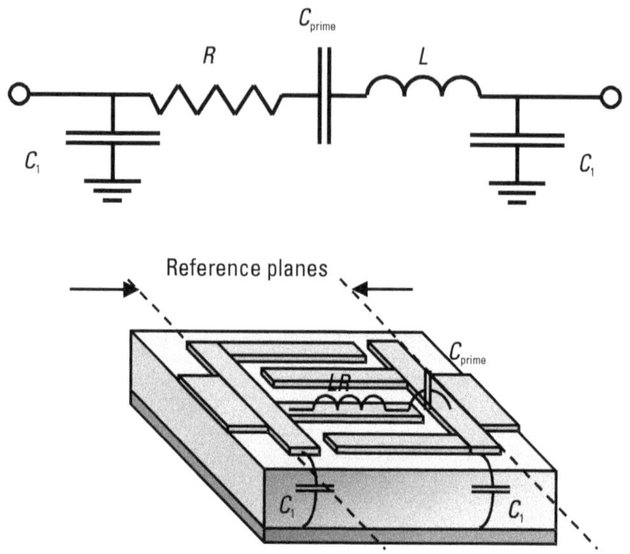

Figure 4.7 Lumped-element equivalent circuit of a series interdigital capacitor.

practice, this limits the range of applicability of the model to frequencies up to about 80% of the first resonance frequency ($f_{res} = 1/2\pi\sqrt{(L_{prime}C)}$.

When the size of a spiral inductor become large compared to the substrate thickness, the ground plane acts as a current mirror, producing an image current at twice the substrate thickness, which has the effect of reducing the primary inductance [23].

The model topology for a stacked spiral is the same as for a single-layer spiral, and the tighter coupling of the tracks significantly increases the parasitic feedback capacitance. Thus, the resonant frequency is reduced compared to the equivalent single-layer spiral, and because the feedback capacitance is much larger, the equivalent circuit model now fits well at frequencies above the resonance.

4.2.3 Resistors

Resistors cannot be modeled simply as a lumped resistance because the capacitance of the resistor contacts and thin-film resistive material to the ground plane generate significant phase change across the resistor. The best method is to model the resistor as a lossy transmission line [24] where the series resistance is no longer negligible.

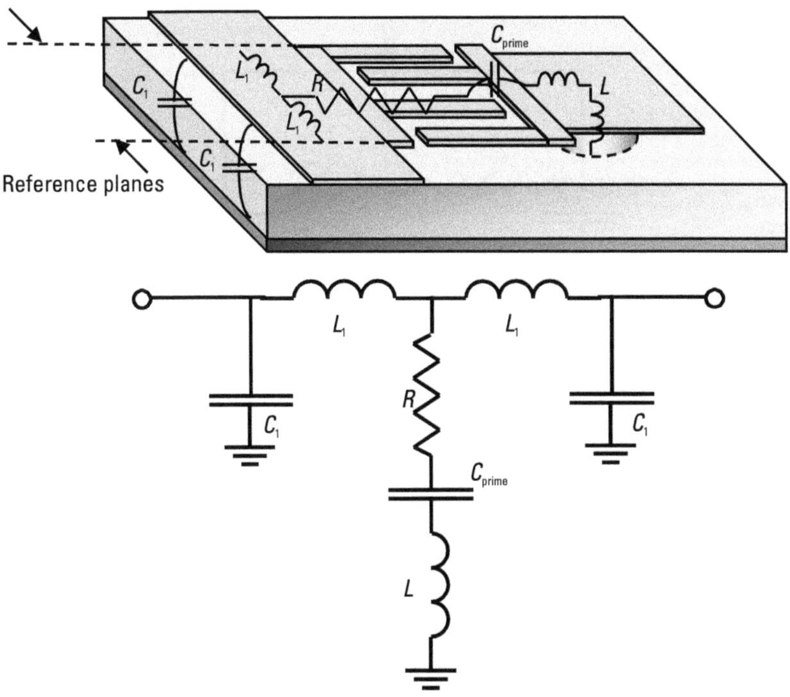

Figure 4.8 Lumped-element equivalent circuit of a shunt interdigital capacitor.

4.2.4 Vias

The equivalent circuit model for a through-substrate via is shown in Figure 4.10 [25]. The inductance is a combination of the via inductance and the distance across the via pad, and the resistance takes account of the loss due to the surface roughness.

4.2.5 Transmission-Line Discontinuities

Discontinuities in the transmission line generate higher-order and evanescent modes around the locality of the discontinuity to satisfy Maxwell's equations, and these decay rapidly with distance away from the discontinuity. The effect of the other modes can often be modeled successfully with small lumped capacitors and inductors, depending on the type of discontinuity. The effect of individual discontinuities is relatively small and difficult to assess within RFOW measurement noise, so the value of the lumped elements is normally obtained by fitting the models to 3D EM simulations of the discontinuities.

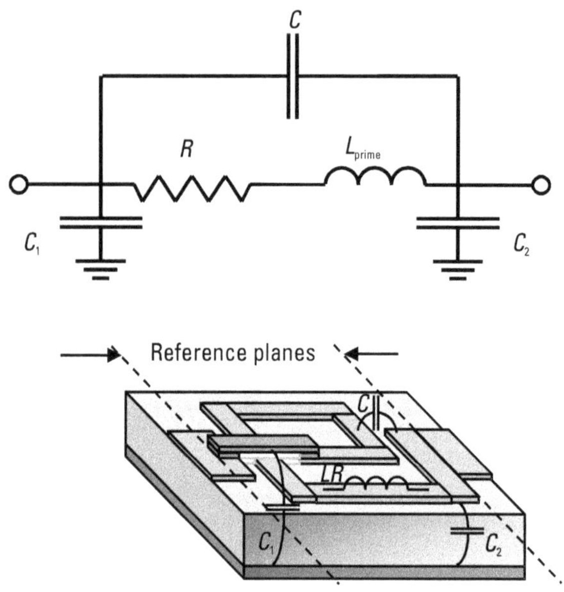

Figure 4.9 Lumped-element equivalent circuit of spiral inductor.

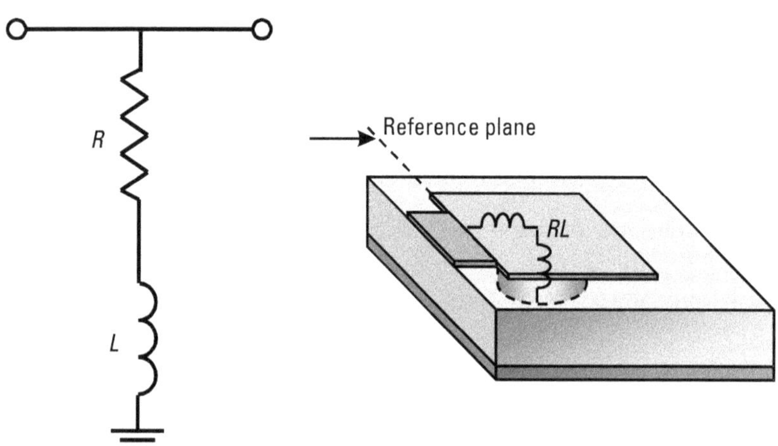

Figure 4.10 Lumped-element equivalent circuit of a through-substrate via.

4.2.5.1 Pads

dc and RF signal pads are typically modeled as parallel-plate capacitances to the ground plane, or if the track contacting them is the same width as the pad, they can be modeled as the open-end capacitance of a transmission line [26].

4.2.5.2 Bend

The lumped-element model for a microstrip bend is shown in Figure 4.11 [27], along with the reference planes for each track meeting the bend. Note that the bend is an area of metal representing the discontinuity, and a standard microstrip contacts it at the reference planes. The effective electrical length of the bend depends on the ratio of the track width to the substrate height and the frequency. Narrow tracks tend to have negative effective electrical length because the fields couple at the inside of the bend and act as if they are cutting the corner. Wide tracks tend to exhibit positive effective electrical lengths because the fields are more confined under the track and have to follow the center line of the bend. This model is applicable to chamfered or mitered bends, as well as to square bends.

4.2.5.3 Tee-junction

The equivalent circuit model for a microstrip tee-junction is shown in Figure 4.12 [28]. The topology is the same for junctions of different-width microstrip tracks, but the lumped inductor values will be different for the different track widths. A junction with more than three tracks meeting, such as a cross-junction, is modeled in the same way, except with the appropriate number of inductive arms.

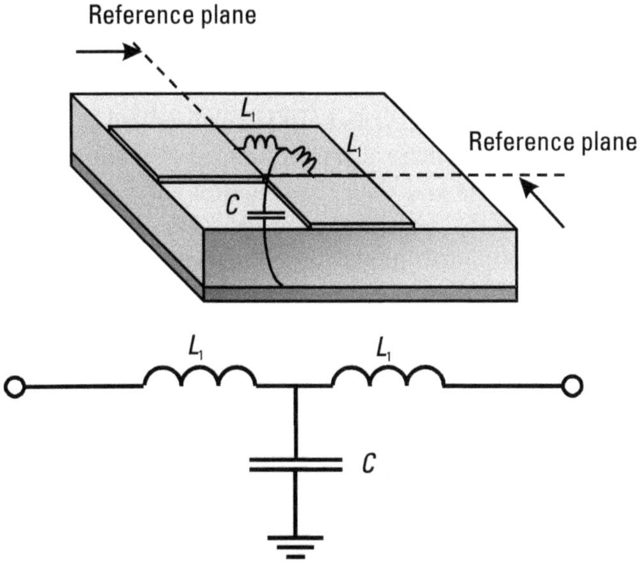

Figure 4.11 Lumped-element equivalent circuit of a microstrip bend.

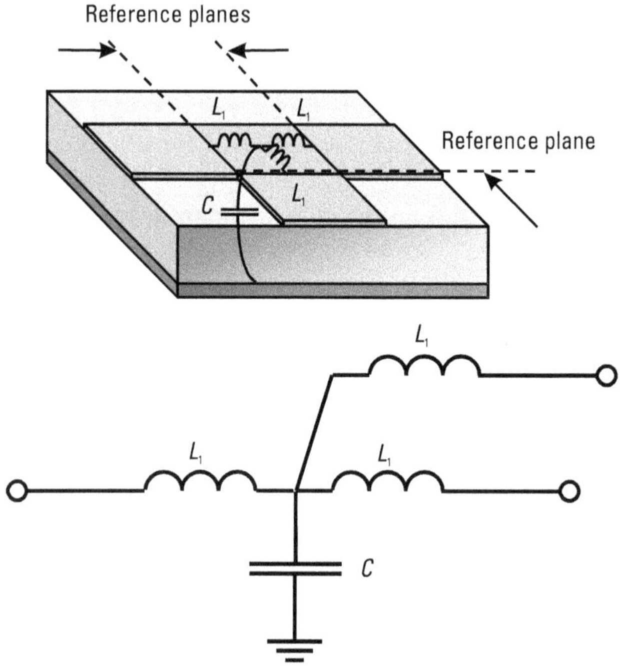

Figure 4.12 Lumped-element equivalent circuit of a microstrip tee-junction.

4.3 Active Component Models

Active components, such as FETs and HBTs, are modeled in a similar way to the passive components in that equivalent lumped-element models are fitted to measured *s*-parameters and then the values of the lumped elements are parameterized against the transistor geometry. The main difference arises because the active component models are more complicated than the passive elements, having over 20 lumped elements compared to about 5 for passive models. Also, the active models must be parameterized against their dc bias conditions, as well as their geometry, and often they must be characterized for their noise parameters, as well as their *s*-parameters. This leads to a complex and multidimensional model that requires a huge amount of data measurement and careful model fitting, which is typically why foundries only offer models for a limited number of active devices.

Another consequence of the complexity of active-device modeling is that not all models are equal. As one would expect, the more recent models for HBTs take into account more effects due to breakdown and heating than the older, but pioneering, Gummel-Poon model. Likewise, the different FET models have similar strengths or weaknesses depending upon whether they were extracted from measurements, curve-fitting, or physical-simulation models. The choice of model can be critical to the success of the MMIC design and may steer the designer to choose a foundry that happens to support the model that is most appropriate to the design requirement.

4.3.1 FET Model

The lumped-element equivalent circuit model for a MESFET or a HEMT is shown in Figure 4.13 [29]. The transconductance (G_m) is controlled by the voltage developed across the input gate-source capacitance (C_{gs}).

The associated noise model is shown in Figure 4.14 [35]. The noise sources I_g and I_d are correlated by $< I_g \cdot I_d > = \text{Cor} \cdot \sqrt{< I_g^2 >< I_d^2 >}$. The resistors R_g, R_s, and R_d are assumed to generate Johnson noise at the ambient temperature of the transistor.

Other FET models that are sometimes included in simulation CAD tools include the Curtice model [31], the Root model [32–35], the Statz model [36–38], the Angelov model [39, 50, 51], and the Triquint own model (TOM) [52].

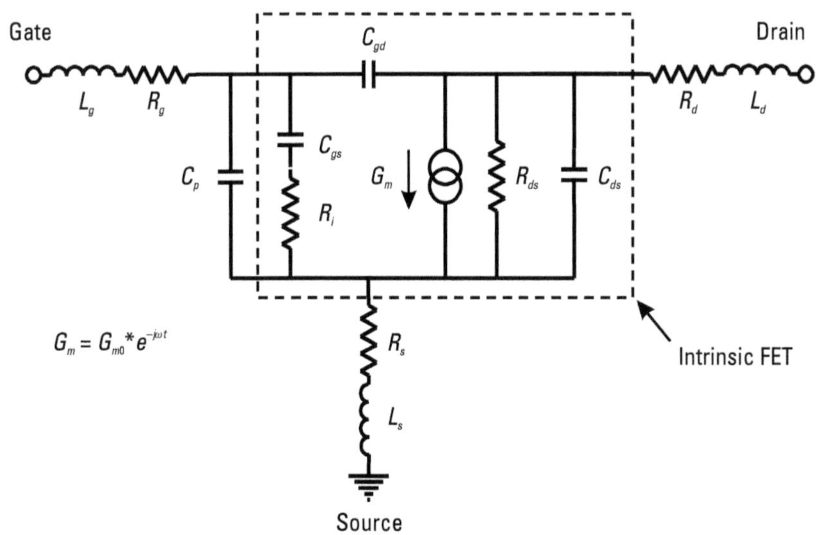

Figure 4.13 Equivalent circuit model of an FET.

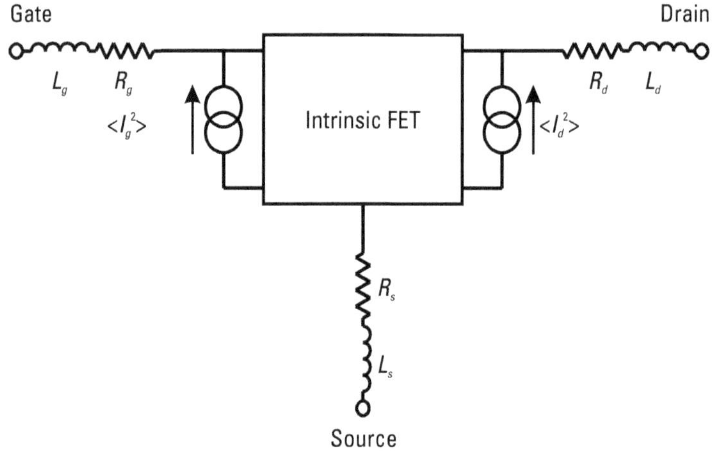

Figure 4.14 Equivalent noise model of an FET.

4.3.2 HBT Model

Bipolar transistors have historically been modeled using a Gummel-Poon model [53, 54], which works well for silicon devices. The concepts of this model have been developed further to fit GaAs and InP HBTs, such as the vertical bipolar inter-company (VBIC) [55] and high current model (HICUM) [56] models, and incorporated into CAD simulation tools, such as the Agilent HBT model [57] shown in Figure 4.15.

Further information on MMIC simulation, component models, and CAD can be found in [58].

4.4 Questions

4.1 What are the limitations of *s*-parameter representation of MMIC component elements?

4.2 How are the individual MMIC components characterized?

4.3 What is the "lumped-element" approximation?

4.4 How do the parasitic elements differ from the prime element in an equivalent circuit model?

4.5 What extra requirements does nonlinear simulation place on the models?

4.6 Why would an MMIC designer use 3D EM simulation?

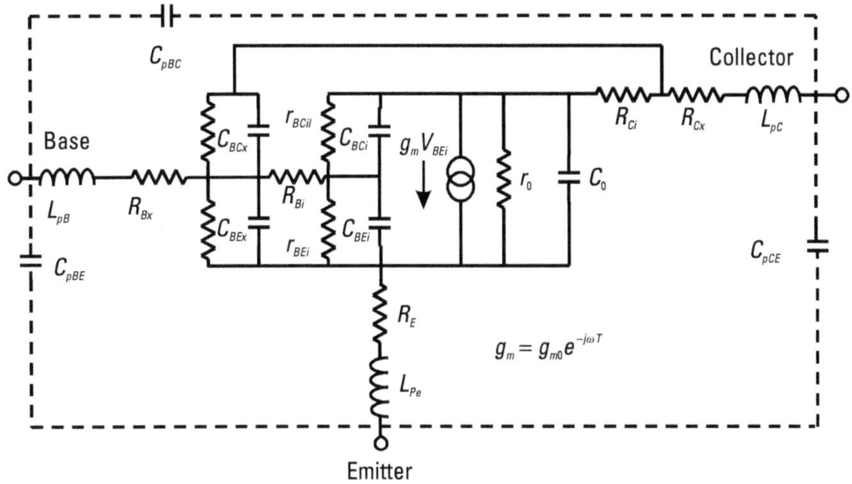

Figure 4.15 Equivalent circuit model for a heterojunction bipolar transistor.

4.7 What effect will placing a bend in a transmission line have on the electrical performance of the transmission line?

References

[1] Kurokawa, K., "Power Waves and the Scattering Matrix," *IEEE Trans. Microwave Theory Tech.,* Vol. 13, No. 2, March 1965, pp. 194–202.

[2] Zhao, A. P., A. V. Raisanen, and S. R. Cvetkovic, "A Fast and Efficient FDTD Algorithm for the Analysis of Planzar Microstrip Discontinuities by Using a Simple Source Excitation Scheme," *IEEE Microwave and Guided Wave Lett.,* Vol. 5, No. 10, October 1995, pp. 341–343.

[3] Daly, D. A., et al., "Lumped Elements in Microwave Integrated Circuits," *IEEE Trans. Microwave Theory Tech.,* Vol. 15, No. 12, December 1967, pp. 713–721.

[4] Johnson, H., "Performance Regions," in *High Speed Propagation: Advanced Black Magic,* Upper Saddle River, NJ: Prentice Hall, September 2003.

[5] Sercu, S., and L. Martens, "High-Frequency Circuit Modeling of Large Pin Count Packages," *IEEE Trans. Microwave Theory Tech.,* Vol. 45, No. 10, October 1997, pp. 1897–1904.

[6] Tsai, C.-T., "Package Inductance Characterization at High Frequencies," *IEEE Trans. Components, Packaging, Manufacturing Tech.—Part B,* Vol. 17, No. 2, May 1994, pp. 175–181.

[7] Kuvas, R. L., "Equivalent Circuit Model of FET Including Distributed Gate Effects," *IEEE Trans. Electron Devices,* Vol. 27, No. 6, 1980, pp. 1193–1195.

[8] Heinrich, W., "Distributed Equivalent-Circuit Model for Traveling-Wave FET Design," *IEEE Trans. Microwave Theory Tech.,* Vol. 35, May 1987, pp. 487–491.

[9] Fukuoka, Y., and T. Itoh, "Analysis of Slow-Wave Phenomenon in Coplanar Waveguide on a Semiconductor Substrate," *Electronics Letters,* Vol. 18, 1982, pp. 589–590.

[10] Heinrich, W., and M. Hartnagel, "Field-Theoretic Analysis of Wave Propagation on FET Electrodes Including Losses and Small-Signal Amplification," *Int. J. Electronics,* Vol. 58, No. 4, 1985, pp. 613–627.

[11] Sonnet Software Inc., http://www.sonnetusa.com/products/lite, last access June 5, 2006.

[12] EM3DS, AdvancedEM, Washington, D.C., http://advancedem.com, last access June 5, 2006.

[13] Sonnet Software Inc., http://www.sonnetsoftware.com/products/cst/index.asp, last access June 5, 2006.

[14] Ansoft Corp., HFSS, http://www.ansoft.com/products/hf/hfss.

[15] Friedrich, N., "EM Simulators Reveal Contents of Crystal Ball," *Microwaves & RF,* Penton, August 2005.

[16] Gupta, K. C., "Emerging Trends in Millimeter-Wave CAD," *IEEE Trans. Microwave Theory Tech.,* Vol. 46, June 1998, pp. 747–755.

[17] Imitiaz, S. M. S., and S. M. El-Ghazaly, "Global Modeling of Millimeter-Wave Circuits: Electromagnetic Simulation of Amplifiers," *IEEE Trans. Microwave Theory Tech.,* Vol. 45, December 1997, pp. 2208–2216.

[18] Bandler, J. W., et al., "A CAD Environment for Performance and Yield Driven Circuit Design Employing Electromagnetic Field Simulators," *Proc. IEEE Int. Symp. Circuits and Systems,* (London, England, May–June 1994), Vol. 1, pp. 145–148.

[19] Bandler, J. W., "Guest Editorial: Automated Circuit Design Using Electromagnetic Simulators," *IEEE Trans. Microwave Theory Tech.,* Vol. 45, 1997, pp. 709–711.

[20] Pettenpaul, E., et al., "CAD Models of Lumped Elements on GaAs up to 18 GHz," *IEEE Trans. Microwave Theory Tech.,* Vol. 36, No. 2, February 1988, pp. 294–304.

[21] Mondal, J. P., "An Experimental Verification of a Simple Distributed Model of MIM Capacitors for MMIC Applications," *IEEE Trans. Microwave Theory Tech.,* Vol. 35, No.4, April 1987, pp. 403–408.

[22] Bartolucci, G., et al. "MIM Capacitor Modelling: a Planar Approach," *IEEE Trans. Microwave Theory Tech.,* Vol. 43, No. 4, April 1995, pp. 901–903.

[23] Arnold, R. G., and D. J. Pedder, "Microwave Characterization of Microstrip Lines and Spiral Inductors in MCM-D Technology," *IEEE Trans. Components, Hybrids, and Manufacturing Tech.,* Vol. 15, No. 6, December 1992, pp. 1038–1045.

[24] Hammerstad, E., and O. Jensen, "Accurate Models for Microstrip Computer-Aided Design," *Proc. IEEE MTT Symposium,* Washington, D.C., May 28–30, 1980, pp. 407–409.

[25] Goldfarb, M., and R. Pucel, "Modeling via Hole Grounds in Microstrip," *IEEE Microwave and Guided Wave Letters,* Vol. 1, No. 6, June 1991, pp. 135–137.

[26] Kirschning, M., R. H. Jansen, and N. H. L. Koster, "Accurate Model for Open-End Effect of Microstrip Lines," *Electronics Lett.*, Vol. 17, No. 3, February 1981, pp. 123–125.

[27] Kirschning, M., R. H. Jansen, and N. H. L. Koster, "Measurement and Computer-Aided Modeling of Microstrip Discontinuities by an Improved Resonator Method," *Proc. IEEE MTT Symposium*, Vol. 83, No. 1, May 1983, pp. 495–497.

[28] Pucel, R. A., "Design Considerations for Monolithic Microwave Circuits," *IEEE Trans. Microwave Theory Tech.*, Vol. 29, No. 6, June 1981, pp. 513–534.

[29] Golio, J. M., et al., "Frequency Dependent Electrical Characteristics of GaAs MESFETs," *IEEE Trans. Electron Devices*, Vol. 37, May 1990, pp. 1217–1227.

[30] Cappy, A., "Noise Modeling and Measurement Techniques," *IEEE Trans. Microwave Theory Tech.*, Vol. 36, No. 1, January 1988, pp. 1–10.

[31] Curtice, W. R., "A MESFET Model for Use in the Design of GaAs Integrated Circuits," *IEEE Trans. Microwave Theory Tech.*, Vol. 28, May 1980, pp. 448–456.

[32] Root, D. E., S. Fan, and J. Meyer, "Technology-Independent Large-Signal FET Models: A Measurement-Based Approach to Active Device Modeling," *Proc. 15th ARMMS Conf.*, Bath, United Kingdom, September 1991, pp. 1–21.

[33] Root, D. E., and B. Hughes, "Principles of Nonlinear Active Device Modeling for Circuit Simulation," *Proc. 32nd ARFTG Conf.*, Tempe, Arizona, 1988, pp. 3–26.

[34] Root, D. E., S. Fan, and J. Meyer, "Technology-Independent Large-Signal Non Quasi Static FET Models by Direct Extraction from Automatically Characterized Device Data," *Proc. 21st European Microwave Conf.*, Stuttgart, Germany, September 1991, pp. 927–932.

[35] Root, D. E., and S. Fan, "Experimental Evaluation of Large-Signal Modeling Assumptions Based on Vector Analysis of Bias-Dependent *S*-Parameters Data from MESFET's and HEMT's," *Proc. IEEE MTT Symposium*, Albuquerque, NM, June 1–5, 1992, pp. 927–932.

[36] Statz, H., et al., "GaAs FET Device and Circuit Simulation in SPICE," *IEEE Trans. Electron Devices*, Vol. 34, February 1987, pp. 160–169.

[37] Divekar, D., "Comments on 'GaAs FET Device and Circuit Simulation in SPICE,'" *IEEE Trans. Electron Devices*, Vol. 34, December 1987, pp. 2564–2565.

[38] Camacho-Penalosa, C., and C. S. Aitchison, "Modeling Frequency Dependence of Output Impedance of a Microwave MESFET at Low Frequencies," *Electronics Lett.*, Vol. 21, June 1985, pp. 528–529.

[39] Angelov, I., H. Zirath, and N. Rorsmann, "A New Empirical Nonlinear Model for HEMT and MESFET Devices," *IEEE Trans. Microwave Theory Tech.*, Vol. 40, No. 12, December 1992.

[40] Sharma, R., A. Kumar, and U. C. Ray, "Model MMIC Nichrome Resistors," *Microwaves & RF*, Penton Media, February 2003, ED Online ID #5477.

[41] Filicori, F., G. Ghione, and C. U. Naldi, "Physics-Based Electron Device Modeling and Computer-Aided MMIC Design," *IEEE Trans. Microwave Theory Tech.*, Vol. 40, July 1992, pp. 1333–1352.

[42] Marsh, S. P., A. P. Long, and S. D. Wadsworth, "Statistical Techniques for MMIC Design Sensitivity and Chip Yield Analysis," *Proc. GAAS'98 Conference,* Amsterdam, Netherlands, October 1998, pp. 551–556.

[43] North, R., D. Kennedy, and M. Murphy, "Database Sampling—A Real World Solution to MMIC Yield Sensitivity Analysis," *Microwave Journal,* April 1997.

[44] Filicori, F., and V. A. Monaco, "Simulation and Design of Microwave Class-C Amplifiers Through Harmonic Balance Analysis," *Proc. IEEE MTT Symposium,* Palo Alto, CA, May 12–14, 1975, p. 362.

[45] Nakhla, M. S., and J. Vlach, "A Piecewise Harmonic Balance Technique for Determination of Periodic Response of Nonlinear Systems," *IEEE Trans. Circuits Systems,* Vol. 23, 1976, pp. 85.

[46] Ansoft Designer SV, Ansoft Corporation, http://www.ansoft.com/ansoftdesignersv.

[47] LINMIC, http://www.linmic.com/pdf/lin62.pdf, last access June 5, 2006.

[48] Momentum, http://eesof.tm.agilent.com/products/momentum_main.html, last access June 5, 2006.

[49] DeLap, J. "Planar EM Simulation of Multi-Chip Modules and BGA Packages," *Microwave Journal,* Vol. 40, No. 11, November 2001.

[50] Angelov, I., L. Bengtsson, and M. Garcia, "Extensions of the Chalmers Nonlinear HEMT and MESFET Model," *IEEE Trans. Microwave Theory Tech.,* Vol. 44, No. 10, October 1996.

[51] Canfield, P. C., "Modeling of Frequency and Temperature Effects in GaAs MESFETs," *IEEE J. Solid-State Circuits,* Vol. 25, February 1990, pp. 299–306.

[52] McCaman, A., G. McCormack, and D. Smith, "An Improved GaAs MESFET Model for SPICE," *IEEE Trans. Microwave Theory Tech.,* Vol. 38, June 1990, pp. 822–824.

[53] Getreu, I. E., *CAD of Electronic Circuits,* 1; Modeling the Bipolar Transistor, Elsevier Scientific Publishing Company, 1978.

[54] Antognetti, P., and G. Massobrio, *Semiconductor Device Modeling with SPICE,* 2nd ed., New York: McGraw Hill, Inc., 1993.

[55] McAndrew, C. C., et al., "VBIC95, the Vertical Bipolar Inter-Company Model," *J. Solid-State Circuits,* Vol. 31, No. 10, October 1996, pp. 1476–1473.

[56] Schröter, M., and T. Y. Lee, "Physics-Based Minority Charge and Transit Time Modeling for Bipolar Transistors," *IEEE Trans. Electron Devices,* Vol. 46, February 1999, pp. 288–300.

[57] ADS2004 product documentation, Agilent Technologies, USA.

[58] Gupta, K. C., R. Garg, and R. Chadha, *Computer-Aided Design of Microwave Circuits,* Dedham, MA: Artech House, 1981, pp. 213–220.

5

Design

RF and microwave design is performed primarily in the frequency domain, probably because the original sources were continuous wave (CW) devices, and different signals were separated by their frequencies. Time-domain analysis of RF and microwave components is becoming more important, especially for communication systems employing extremely complex modulation schemes, but the majority of work is still undertaken in the frequency domain. The design techniques described within this chapter will concentrate on the frequency domain.

The design of MMICs is essentially the same as that for most microwave components: one or more signals must be coupled onto the chip efficiently, then functions are performed on the signals, after which they are then coupled out of the chip and back into the system. Later sections will show that the efficient coupling of signal power requires that the impedance of the signal source match that of the chip. In fact, to pass the signal efficiently through the whole chip, the impedances of the transmission lines, the active devices, and the system it sits in must all be matched. Separation of RF signals from dc bias requires impedance mismatching, and filtering of different RF signals requires impedance changes versus frequency. Thus, MMIC design is founded on impedance matching and control across the frequency range of interest.

This chapter begins with these basis principles of impedance matching and shows how impedances may be plotted and transformed using the Smith chart. The chapter goes on to discuss the design of passive elements, such as couplers and power splitters, which may be used as subcircuits within mixer and power amplifier chips. Sections are then devoted to the different functional types of chips that can be designed, including amplifiers, oscillators, mixers, and digital circuits. The chapter concludes with a section on techniques that become important as the frequency range approaches millimeter-wave frequencies, as well as a section on design techniques to improve the yield of the chip against its specification.

5.1 Impedance Matching and the Smith Chart

Designing circuits involves the efficient transfer of signals; for example, in the case of amplifiers, it involves getting signals into active devices to amplify the signals, then transferring the higher level signals into other circuits, transmission lines, or antennas. In the early days of electric motors, it was found that to get the most efficient transfer of power from the battery (source) into the motor (load) required that the resistance of the different parts of the circuit be the same, in other words, matched; this is known as the maximum power transfer theorem [1]. The transmission lines can be made to be 50Ω, but the active devices are not normally 50Ω, so matching the active devices is one of the first design tasks.

5.1.1 Matching

The impedances of the active devices are seldom purely resistive, so the maximum power transfer must take account of the reactive part of the source and load impedances. In fact, to get maximum power transfer, the load impedance must be the complex conjugate of the source impedance. *Conjugate* means of equal value and opposite sign, so if the source has impedance $Z_S = R + jX$, then the load must have the impedance $Z_L = R - jX$ to be matched. Effectively, this means that the net phase difference is reduced to 0, and the real impedance of the source sees purely real and identical load impedance. Note that a positive reactance is inductive ($j\omega L$), a negative reactance is capacitive ($-j/\omega C$), and the conjugate condition implies that they are equal, so they resonate at a frequency of $f_{res} = 2\pi\omega_{res} = 2\pi/\sqrt{LC}$

Therefore, part of the designer's job is to design a matching circuit that will be matched to the transmission line at its input and conjugate-matched to the active device at its output, as shown in Figure 5.1.

The transfer of the signal power is dependent on impedance matching, but the performance of the active device, such as the noise output, is also dependent on the impedances presented to it [2]. For active devices such as transistors, we can define two specific impedances that maximize the performance of the circuit:

- Z_{opt} (Zee optimum) is the impedance presented to the input of the device that minimizes the noise figure. Impedances can also be expressed as reflection coefficients [described in the next section (5.1.2)], so the reflection coefficient that minimizes the noise figure is also known as Gamma opt (Γ_{opt}) [3].
- Z_{Gain} (Zee gain) is the impedance presented to the input of the device that maximizes the gain [4].

Note: These two impedances are not necessarily the same or the conjugate of S_{11} (written as S_{11}*). If the active device were unilateral [i.e., there were no feedback between the output port and the input port (S_{12}) = 0], the impedance for maximum gain would be S_{11}*. However, because S_{12} is seldom negligible, the impedance for maximum gain is slightly different from S_{11}*.

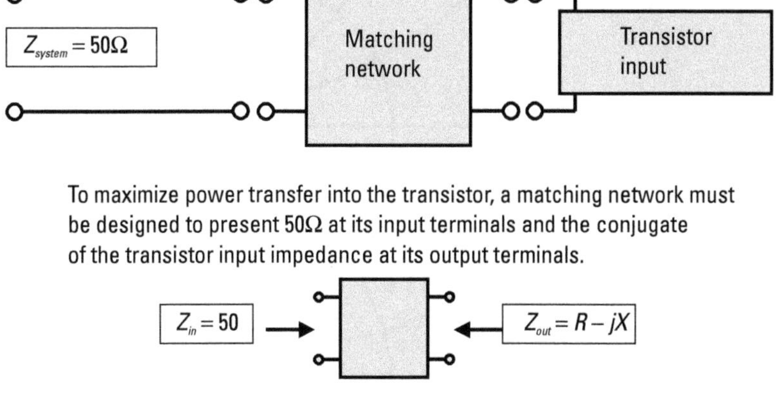

Figure 5.1 Conjugate matching the input of a transistor to 50Ω using a matching network.

5.1.2 Smith Chart

In summary, systems that the circuit will be working in generally have a characteristic impedance of 50Ω, but the active devices have arbitrary impedances that are complex. MMIC circuit design involves matching the active devices to 50Ω while presenting the ideal impedances to the active devices. The design of matching circuits and transforming impedances can be performed with complex mathematics, but this is time-consuming and error prone. In 1939, Philip H. Smith invented a pictorial method for plotting impedances and reflection coefficients known as the Smith chart [5], as shown in Figure 5.2. This is a useful tool for designing matching circuits and converting between impedance, admittance, and reflection coefficients. The advantage of the Smith chart over other graphical methods is that it allows the plotting of all possible reflection coefficients, impedances, and admittances of a passive network on a finite chart. This section describes how the Smith chart is constructed, which is a useful part of understanding how the chart is used. The following sections show how the user, by simple constructions, can determine how impedances are transformed by transmission lines and series and shunt elements, as well as understand and design complex matching networks. The Smith chart also enables the plotting of "circles of

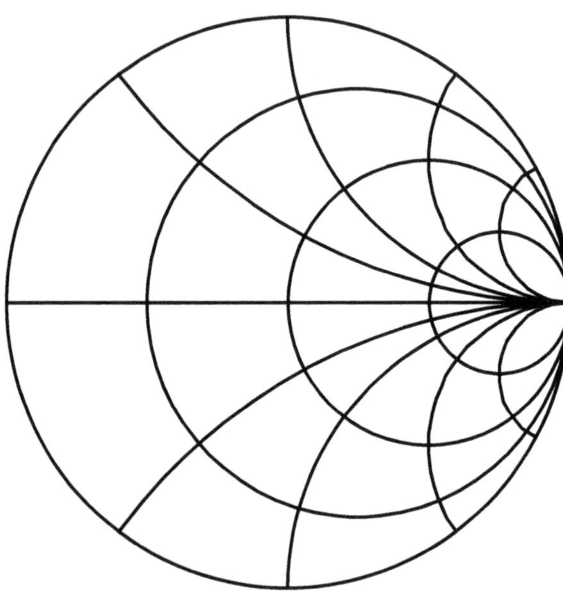

Figure 5.2 The Smith chart.

constant performance" to determine trade-offs between parameters, such as gain and noise figure or gain and output power.

Consider a transmission line with characteristic impedance Z_0 terminated by a load impedance Z_L, as shown in Figure 5.3. An open circuit $Z_L = \infty$ is difficult to plot on a finite chart; therefore, it is more convenient to plot the reflection coefficient, given in (5.1), which has a range of magnitude from 0 to 1.

$$\text{Reflection coefficient } \Gamma = (Z_L - Z_0)/(Z_L + Z_0) \tag{5.1}$$

The reflection coefficient Γ is complex (i.e., a vector with magnitude $|\Gamma|$ and angle $\angle\Gamma$) and can be plotted on a polar chart, as in Figure 5.4. From this simple polar plot, it can already be seen that the matched condition is represented by the middle of the plot, a short circuit is represented on the left-hand side, and an open circuit is represented on the right-hand side, as indicated in Figure 5.5.

Impedance values can be added to the reflection coefficient plot, and it is conventional to plot them as normalized impedances Z_N, as in (5.2) and (5.3) (i.e., relative to the characteristic impedance of the system Z_0). It follows that the reflection coefficient can also be expressed as a function of the normalized impedance in (5.4).

$$\begin{gathered}\text{Normal impedance}\\ Z_N = Z_L/Z_0 = R_L/50 + jB_L/50 \text{ (for a 50}\Omega\text{ system)}\end{gathered} \tag{5.2}$$

$$Z_N = \text{Normalized resistance} + j\,\text{normalized reactance} \tag{5.3}$$

$$\text{Reflection coefficient } \Gamma = (Z_N - 1)/(Z_N + 1) \tag{5.4}$$

If lines of constant normalized resistance are plotted on the reflection coefficient plot, they form circles shown in Figure 5.6, confirming that the

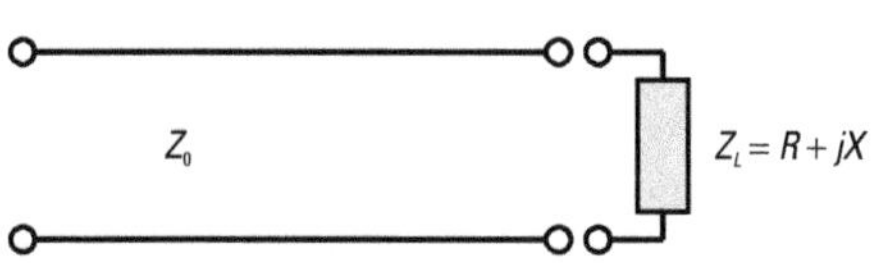

Figure 5.3 Transmission line terminated in a load impedance.

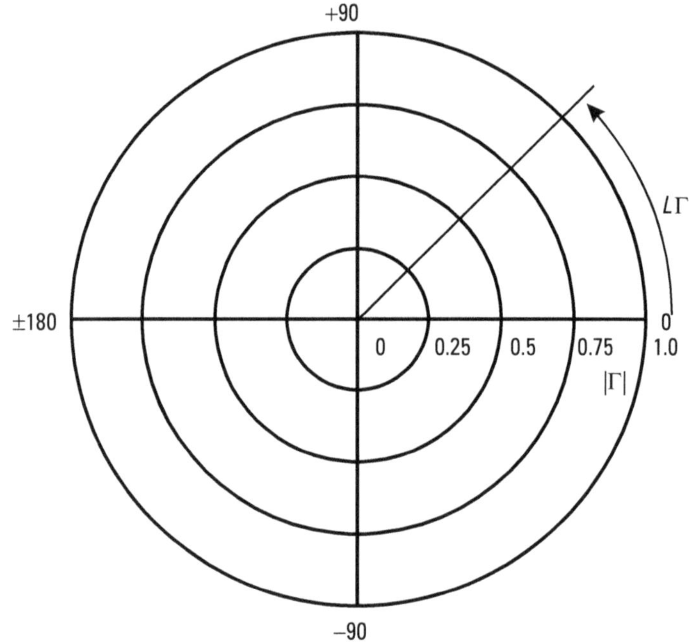

Figure 5.4 Polar plot of reflection coefficient.

short circuit (zero resistance) position is on the left side and the open circuit (infinite resistance) position is on the right side. Similarly, lines of constant normalized reactance are plotted in Figure 5.7, which shows that the center horizontal line represents purely real impedances, the top half of the plot represents inductive impedances (positive reactance), and the bottom half represents capacitive impedances (negative reactance). Putting both of the normalized impedances on the same chart gives the standard Smith chart, shown in Figure 5.8.

In addition to Smith chart representations of reflection coefficients and impedances, there are often other scales present for reading off other parameters (Figure 5.9). Additional radial scales (relating to the magnitude of the reflection coefficient) take the distance from a point on the chart to the center and read off quantities, such as the voltage standing wave ratio (VSWR) and the return loss. Other edge scales (relating to the angle of the reflection coefficient) express how impedances can be transformed around the Smith chart by transmission lines (with the characteristic impedance) as a function of their length relative to the wavelength. Some Smith charts also come with admittance lines drawn as well, as shown in Figure 5.10.

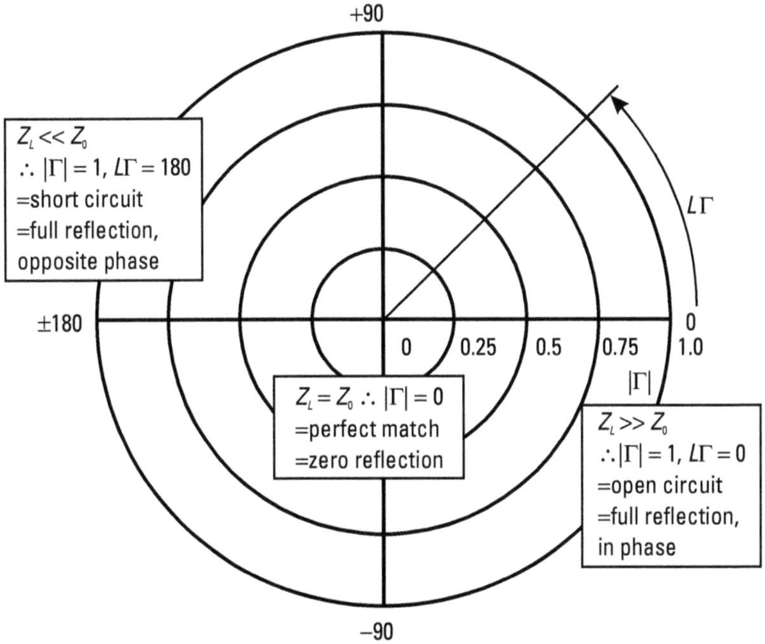

Figure 5.5 Polar plot of reflection coefficient showing short circuit, match, and open circuit positions.

With both the impedance and admittance lines plotted on the Smith chart, it becomes a very useful tool to the MMIC designer. The following paragraphs show how it can be used for performing impedance transforms, such as impedance to reflection coefficient and impedance to admittance; picturing the conjugate impedance; impedance matching using transmission lines and series and shunt elements; and plotting curves of constant Q-factor and circles of constant performance for doing trade-offs between noise, gain, and power performance.

5.1.3 Converting Impedance to Reflection Coefficient

To convert impedance to a reflection coefficient, the value of the resistance and reactance are divided by the characteristic impedance (Z_0) to normalize them, and then plotted on the Smith chart using the impedance scale (Z_N in Figure 5.11). The magnitude of the reflection coefficient ($|\Gamma|$) is calculated by measuring the radial distance from the center of the chart to the impedance point (Z_N) and reading this distance off the reflection coefficient radial

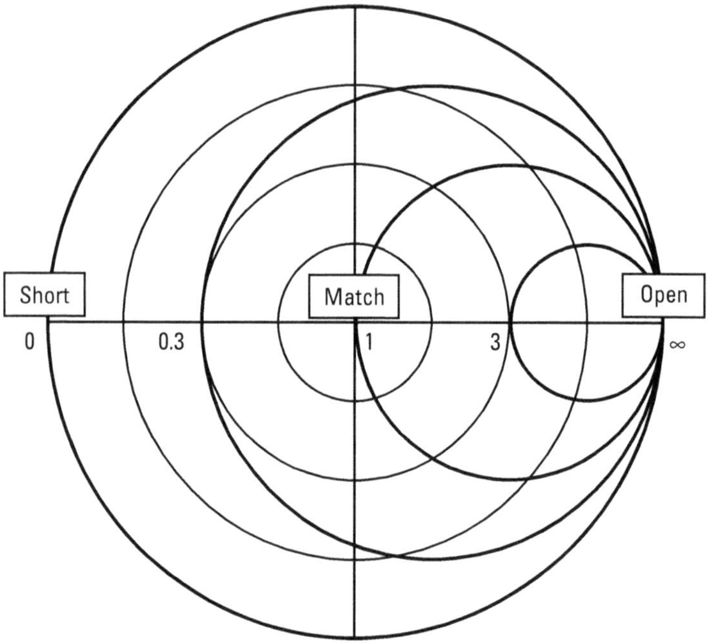

Figure 5.6 Lines of constant normalized resistance.

scale. The angle of the reflection coefficient ($\angle\Gamma$) is read by extending the radial line (from the center to the impedance point) to the edge of the chart and reading the angle from the circular scale.

5.1.4 Converting Impedance to Admittance

To convert an impedance to an admittance, it is first normalized ($/Z_0$) and plotted on the Smith chart using the impedance scale (Z_N in Figure 5.12). A second point is then constructed on the Smith chart, which on the opposite side of, and at the same distance from, the *center of the chart.* The normalized conductance (G_N) and susceptance (B_N) are read off according to where the second point sits on the admittance scales, and they are denormalized (multiplied by the characteristic admittance Y_0) to give the admittance (Y).

5.1.5 Deriving the Conjugate Impedance

As before, the impedance is normalized ($/Z_0$) and plotted on the chart using the impedance scales (Z_N in Figure 5.13). A second point is constructed on

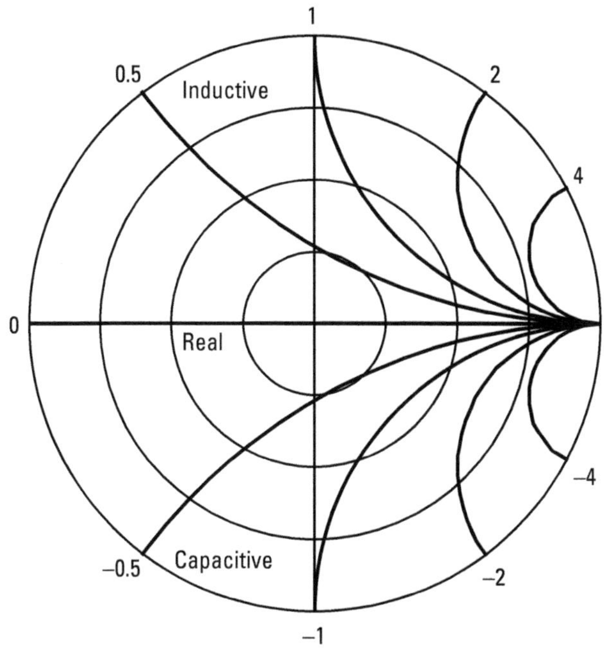

Figure 5.7 Lines of constant normalized reactance.

the chart on the opposite side of, and at the same distance from, the *center line*. The normalized conjugate impedance (Z_N*) is read off the impedance scales for this second point and denormalized (× Z_0) to give Z*.

5.1.6 Transforming a Load Impedance Along a Lossless Transmission Line

Consider the load impedance (Z_L) situated at the end of a transmission line with characteristic impedance Z_0 and length *d*, as shown in Figure 5.14. To find out what the input impedance is at the beginning of the transmission line, the load impedance must be transformed along the transmission line. To do this, the load impedance (Z_L) is first normalized (Z_L/Z_0) and plotted on the Smith chart using the impedance scales (Z_N in Figure 5.15). The transmission-line length *d* is normalized to the wavelength of the signal on the transmission line (d/λ) so that it can be represented as an angle on the "wavelengths toward the generator" edge scale of the Smith chart. The point (Z_L) is then rotated clockwise around the center of the chart by the fraction of the wavelength (d/λ) to give the normalized transmission-line input impedance ($Z_{(IN)N}$). The actual transmission-line input impedance (Z_{IN})

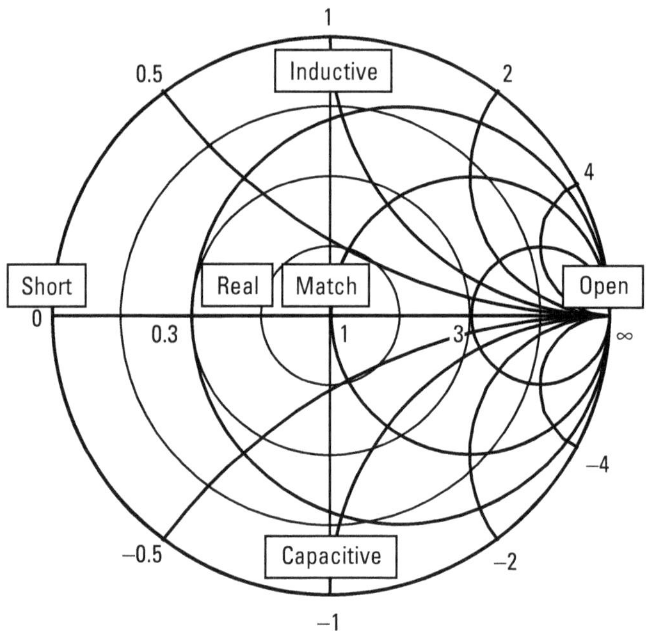

Figure 5.8 Standard Smith chart.

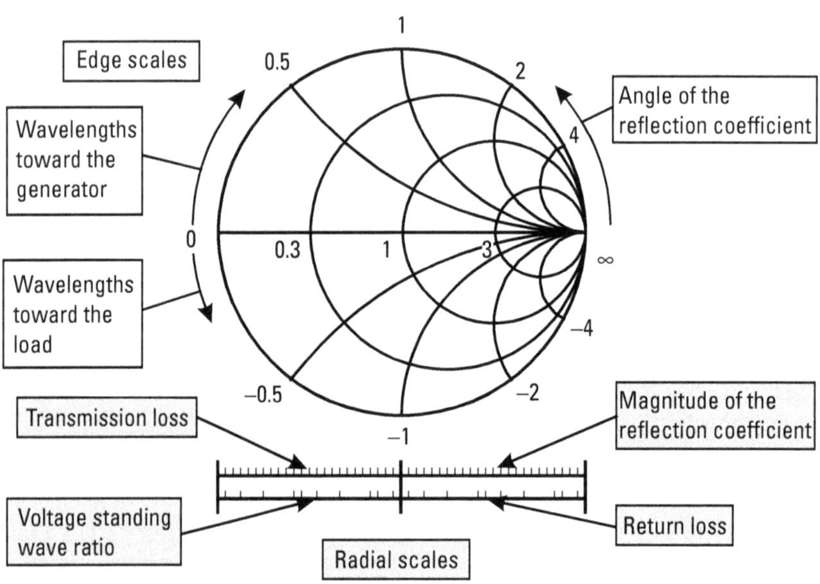

Figure 5.9 Other scales on the Smith chart.

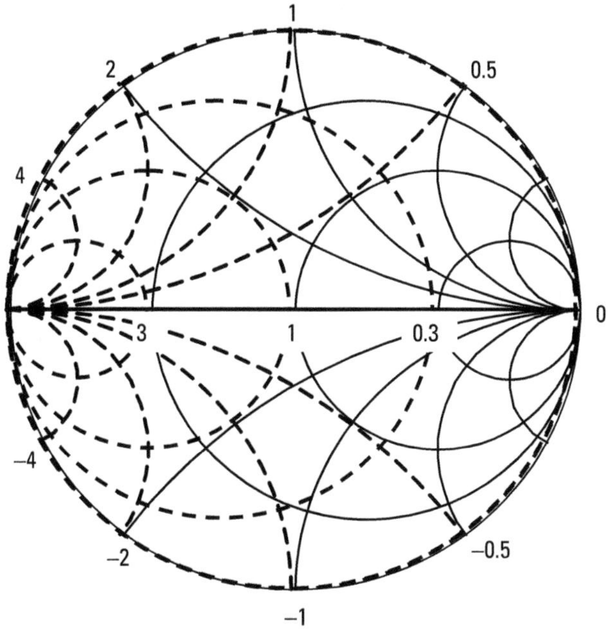

Figure 5.10 Admittance coordinates.

is then calculated by reading its position off the impedance scales and denormalizing.

The "wavelengths toward the generator" edge scale of the Smith chart is 0 at the left-hand side (short circuit position) and increases clockwise to a value of 0.25 at the right-hand side (open circuit position), then to a value of 0.5 back at the start. As a consequence, a quarter-wavelength's length of transmission line will transform a short circuit into an open circuit and vice versa, and a half-wavelength's length of transmission line will transform impedance back to the same value. This applies to transformations along transmission lines of any characteristic impedance as long as the impedance values are correctly normalized. Quarter-wavelength transmission lines are often used in circuit design to transform the value of the real part of impedance.

5.1.7 Addition of Series Reactive and Shunt Susceptance Lumped Elements

Consider the situation shown in Figure 5.16(a), where there is load impedance (Z_L), and a reactance (X) is added in series. As the components are in series, the additional reactance is simply added to the reactance of the load impedance, which has the effect of moving the impedance along the lines of

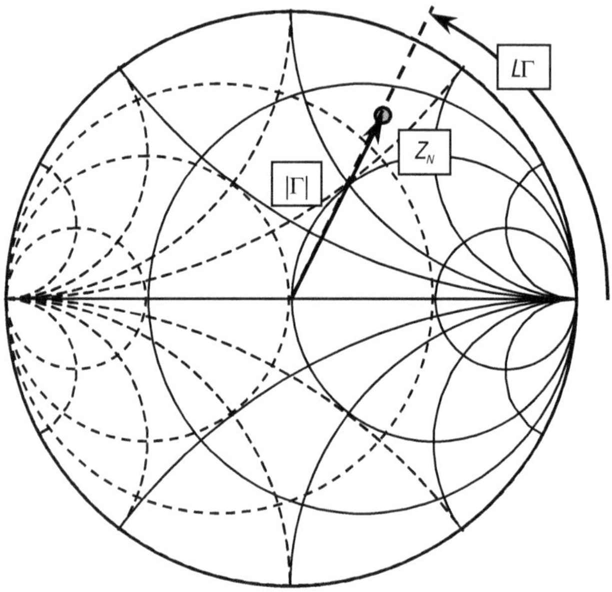

Figure 5.11 Converting impedance to reflection coefficient.

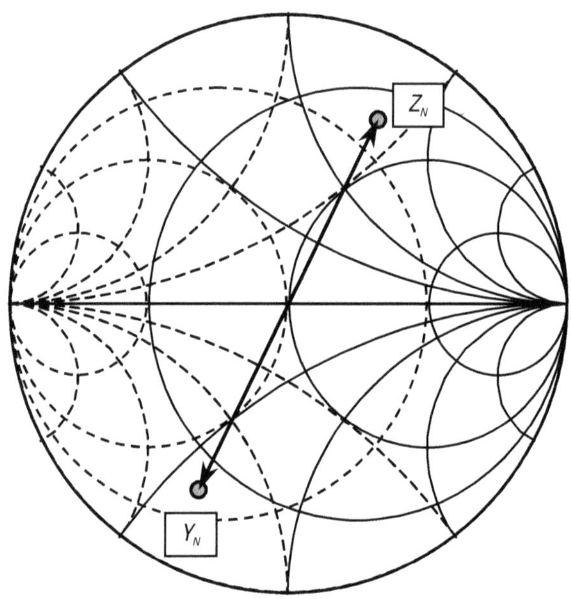

Figure 5.12 Converting impedance to admittance.

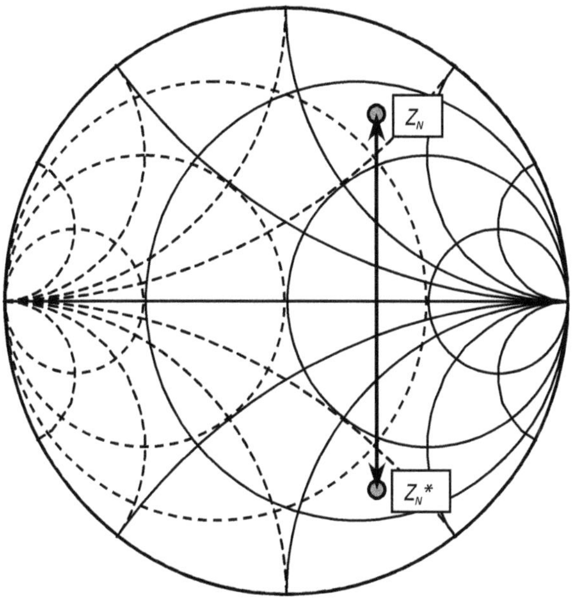

Figure 5.13 Deriving the conjugate impedance.

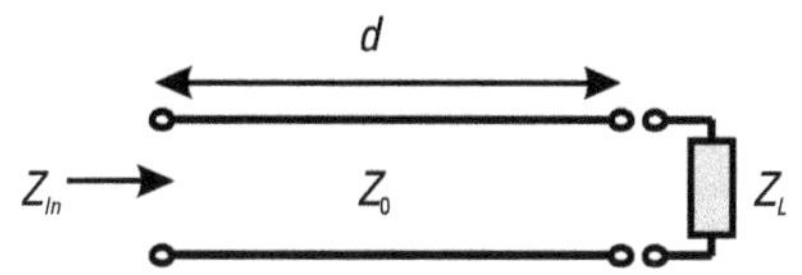

Figure 5.14 An ideal transmission line terminated in load impedance.

constant normalized resistance. If the additional reactance is inductive (positive reactance), then the point moves clockwise; if the reactance is capacitive (negative reactance), then the point moves counterclockwise. Therefore, to add a series reactance, the load impedance is normalized and plotted on the impedance scales, as shown in Figure 5.17. Then, the additional reactance value is normalized (X/Z_0), and Z_L is moved around the circle of constant resistance by this amount (X/Z_0). The impedance of this series combination [Z_{In} in Figure 5.16(a)] is calculated by reading the new position off the impedance scales and denormalizing.

Similarly, when a susceptance (B) is added in shunt to load impedance (Z_L), as in Figure 5.16(b), it changes the total susceptance but does not alter the conductance and has the effect of moving the point around the curves of

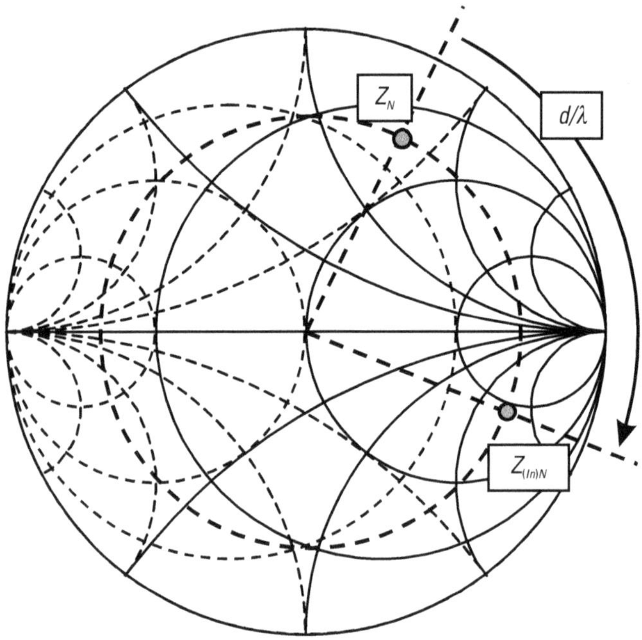

Figure 5.15 Transforming load impedance along a lossless transmission line.

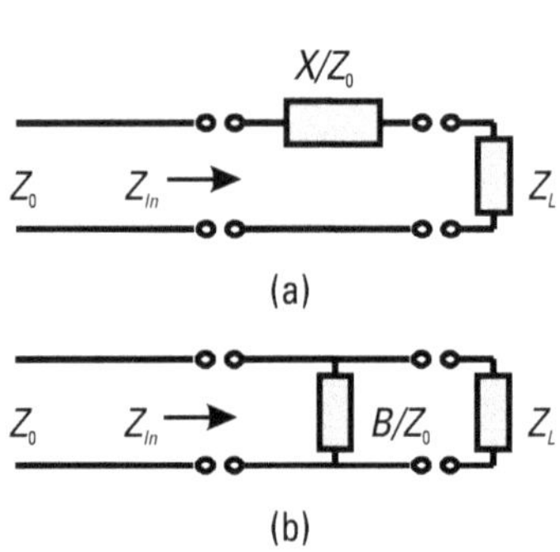

Figure 5.16 Adding (a) lumped series reactance and (b) shunt susceptance to a load impedance.

constant conductance (i.e., the circles on the admittance scale). This time, if the susceptance is inductive, the point moves counterclockwise, and if the susceptance is capacitive, the point moves clockwise. Therefore, to add shunt susceptance, the load impedance is normalized ($Z_N = Z_L/Z_0$), plotted on the impedance scales, and moved around the admittance circles by the normalized susceptance (B/Z_0), as shown in Figure 5.17. The impedance of this shunt combination [Z_{In} in Figure 5.16(b)] is calculated by reading the new position off the impedance scales and denormalizing.

5.1.8 T and Pi Matching Circuits

It is common to design lumped-element matching circuits in Pi or T configurations [6]. For example, Figure 5.18 shows how an impedance at point A on the Smith chart can be moved to the match point (50Ω) using a Pi-shaped matching network, and how an impedance point B can be moved to the match point using a T-shaped matching network.

There is a lot of information within the literature about designing matching networks [7, 8], and there are many Web-based calculators that will design networks according to the impedance transformation required.[1,2]

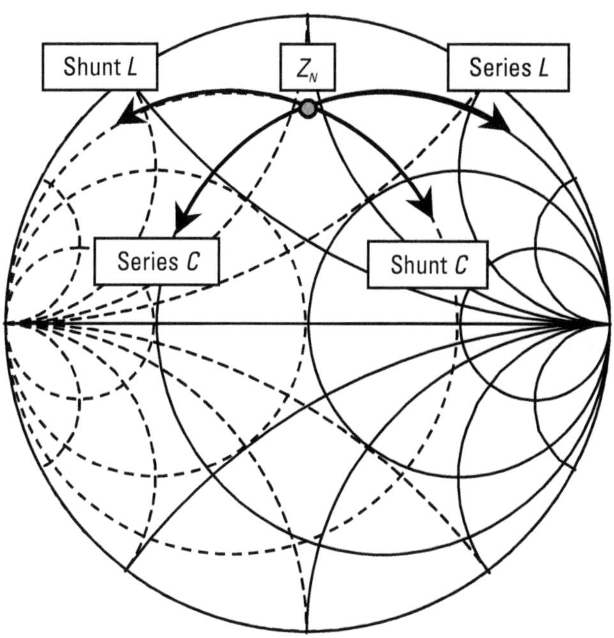

Figure 5.17 Movement of impedance on the Smith chart when adding lumped series reactance and shunt susceptance.

1. http://www.esap.com/emf.htm, last access June 5, 2006.
2. http://www.circuitsage.com/matching.html, last access June 5, 2006.

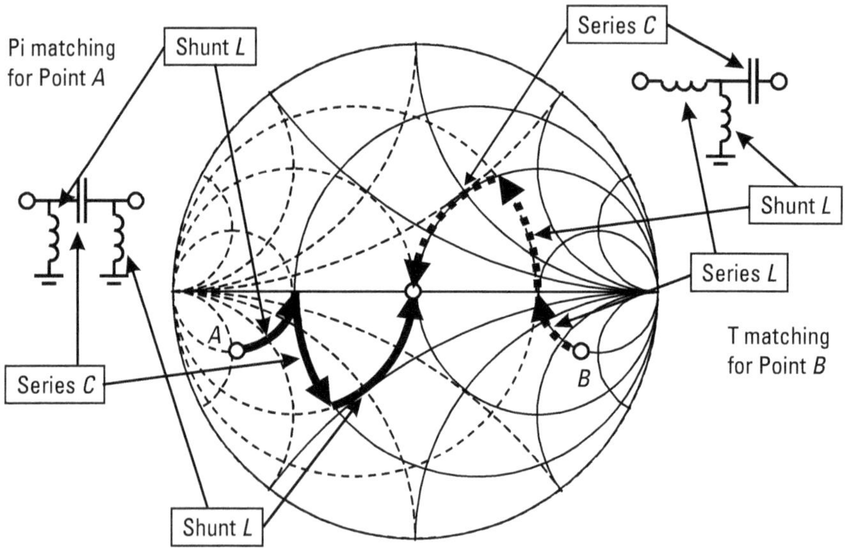

Figure 5.18 Pi and T matching circuits.

It is common practice to try several different matching-network topologies using lumped elements combined with the transistor model to investigate which give the required bandwidth and which allow easy biasing of the transistor.

5.1.9 Curves of Constant Q-factor

Each impedance point ($Z = R + jX$) has an associated Q-factor ($Q = |X/R|$), and these can be plotted on the Smith chart. For example, in Figure 5.19, $Z_1 = 0.33 + j2$ with a Q-factor of 6 ($Q = 2/0.33 = 6$), and $Z_2 = 1 + j1$ with a Q-factor of 1 ($Q = 1/1 = 1$). Curves can then be drawn on the chart (Figure 5.19) linking impedances that have the same Q-factor.

Consider now the load impedance (Z_L), which has a matching network of shunt inductors and series capacitors in front of it that transform the impedance to Z_0 along the path indicated by the arrows in Figure 5.19. At each impedance along the path, the Q-factor can be calculated, but it is clear that they all lay within the $Q = 1$ curve, so the highest Q-factor of this matching network is 1. As the Q-factor is a measure of how quickly the impedance value changes with frequency, this low value of 1 means that the matching network will be tend to be broadband. Therefore, the lines of constant

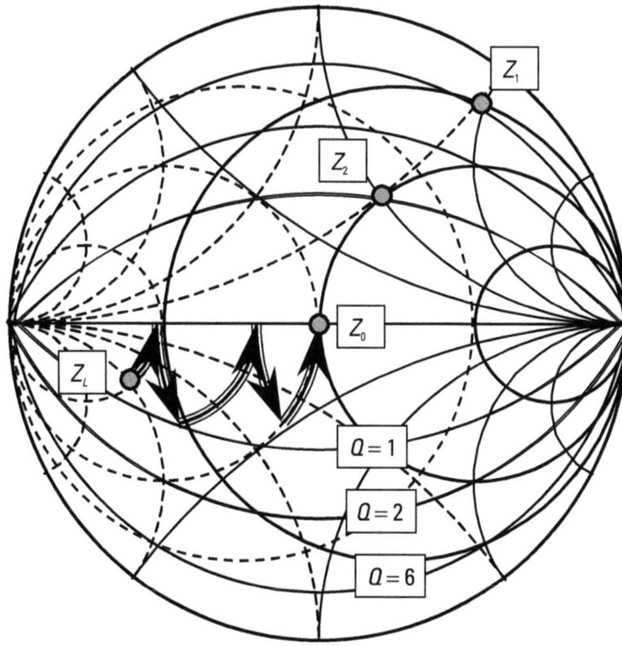

Figure 5.19 Curves of constant Q-factor.

Q-factor can be used as a guide to maintain a broadband matching network, or, in the opposite case, the designer can deliberately take the impedance through a higher Q-factor region to improve the frequency selectivity of the network in applications such as filters.

5.1.10 Circles of Constant Performance

Similarly to the curves of constant Q-factor, other contour lines can be plotted on the Smith chart to map impedances that, when presented to the input of a certain active device, for example, will result in the same performance characteristic of that device. This assumes the designer has the appropriate model of the active device or transistor under investigation and a simulator that can calculate the circuit performance over the full range of input impedances. The type of performance characteristics that are commonly plotted include the signal gain, the noise figure, the output power, and the stability factors.

For example, Figure 5.20 shows typical noise figure circles for a transistor, or, in other words, circles of impedance (or reflection coefficient) that,

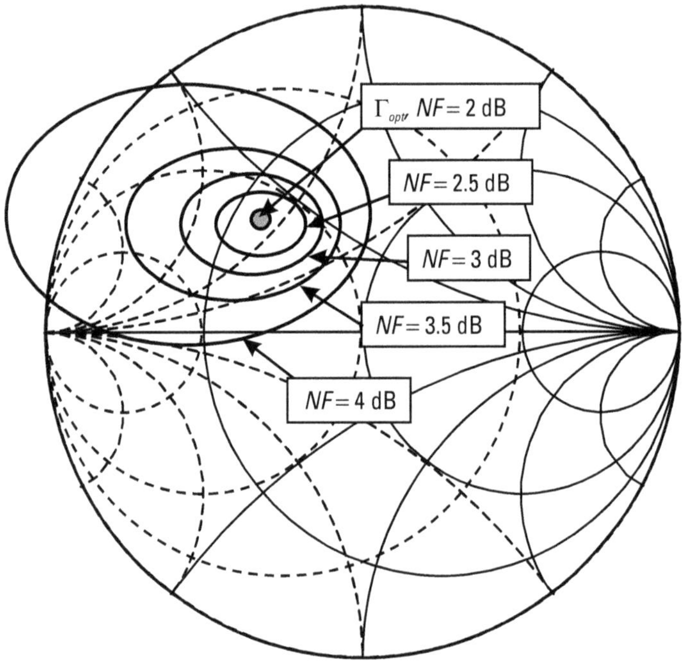

Figure 5.20 Circles of constant noise figure.

when presented to the transistor input, will give the same circuit noise performance. The optimum reflection coefficient (Γ_{opt}) is shown in the middle of the circles, and this optimum impedance results in a noise figure of 2 dB. Impedances away from this optimum value give a poorer noise figure indicated by the circles labeled 2.5 dB, 3 dB, and so forth. If the designer has a 3-dB noise figure specification, the circles show that the matching network must be designed to keep the input impedance presented to the transistor within the 3-dB circle.

Similarly, Figure 5.21 shows typical gain circles for a transistor, or, in other words, circles of impedance (or reflection coefficient) that, when presented to the transistor input, will give the same circuit gain performance. The optimum impedance for maximum gain is shown in the middle of the circles, and this optimum impedance results in a gain of 10 dB. Impedances away from this optimum value give a lower gain indicated by the circles labeled 9 dB, 8 dB, and so forth. If the designer has a 9-dB gain specification, the circles show that the matching network must be designed to keep the input impedance presented to the transistor within the 9-dB circle.

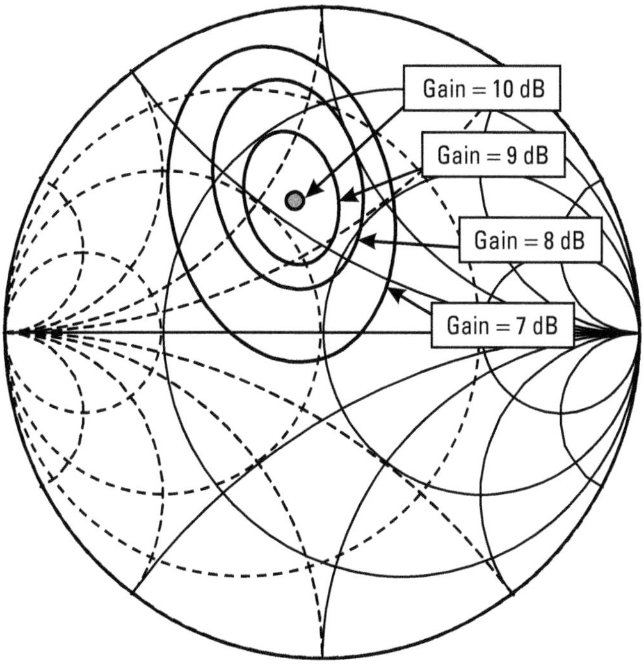

Figure 5.21 Circles of constant gain.

If both the noise figure and gain circles for the transistor are plotted on the same chart, the designer can see the trade-offs between the noise and gain performance according to which impedance is chosen to be presented to the input of the transistor. For example, Figure 5.22 shows that there is one input impedance that will cause the transistor to have a 2-dB noise figure and a gain of 8 dB, and another that will cause the transistor to have 3.5-dB noise figure and a gain of 10 dB. The designer must then decide which trade-off best meets the required specification.

In summary, the Smith chart, when properly understood, is the MMIC designer's friend and an invaluable tool for impedance manipulation and matching-network design. It allows the plotting of all possible reflection coefficients, impedances, and admittances of a passive network on a finite chart and allows the user, by simple constructions, to determine how impedances are transformed by transmission lines and series and shunt elements. It also enables the designer to plot "circles of constant performance" and to determine trade-offs between parameters, such as gain and noise figure or gain and output power.

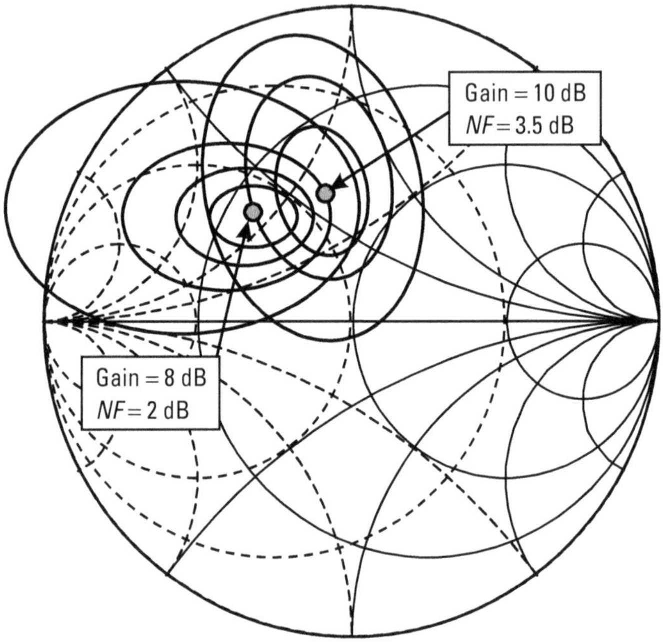

Figure 5.22 Noise figure and gain circles plotted to show performance trade-off with input impedance.

5.2 Passive Elements

The type of passive elements, whose design principles are described in this section, are those that generally form subcircuits within larger MMICs but could in themselves form stand-alone circuits. For example, baluns are often used in mixers; splitters and combiners are often used in power amplifiers; and all of them may be used in transmit and receive (Tx/Rx) chips. They form an important part of the practical MMIC designers' toolbox because, not being based on the active devices, they tend to be less sensitive to processing variations and, hence, a more reproducible component.

5.2.1 Open Circuit and Short Circuit Stubs

Designers can utilize the distributed nature of the transmission lines to transform known impedances, such as short and open circuits, into a wide range of shunt impedances, as discussed in Section 5.1.6. When used as matching elements, they are known as open and short circuit stubs. The open circuits are formed from open-ended transmission lines, and microstrip short circuits are formed from through-substrate vias at low frequencies and from radial stubs at millimeter-wave frequencies. One-quarter-wavelength open circuit and short circuit stubs are often used in multiplier circuits because they have opposite effects on the signals at f_0 and $2f_0$, as shown in Figure 5.23.

The open circuit stub transforms the open circuit to a short circuit where it meets the through transmission line. This is a useful filter circuit for multipliers because it shorts frequencies at f_0 and is open circuit to frequencies at $2f_0$. The short circuit stub transforms the short circuit to an open circuit where it meets the through transmission line. This is also useful for multiplier circuits (primarily doublers) where there are signals at harmonics of the applied signal because it shorts frequencies at $2f_0$, is open circuit to frequencies at f_0, and provides a possible dc bias return path to ground.

5.2.2 Radial Stubs

At millimeter-wave frequencies, the inductance of a through-substrate via is significant enough to make the connection to ground reactive and not a very good low-impedance short circuit to ground. The alternative is to use a

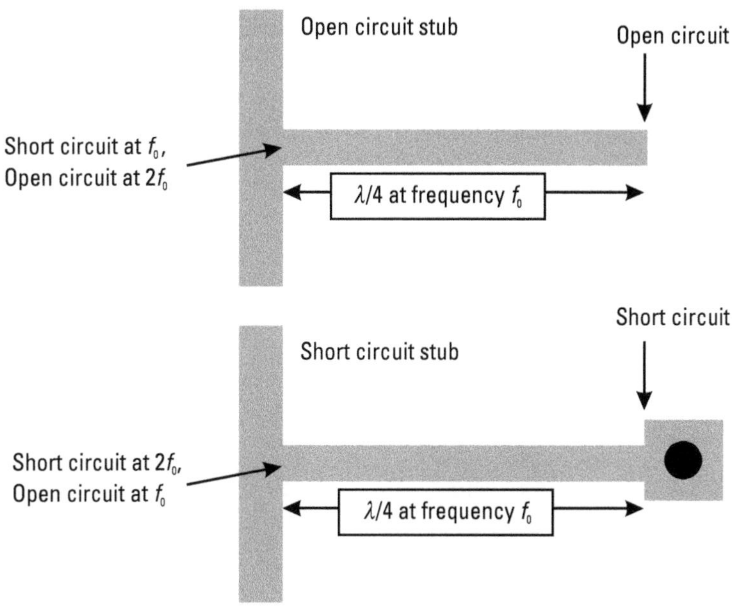

Figure 5.23 One-quarter-wavelength open and short circuit stubs.

microstrip radial stub [9, 10], shown in Figure 5.24, which transforms the open circuit at the radial edge to a good high-frequency short circuit at the apex. Because they are designed with one-quarter-wavelength dimensions, they are usually too large to use on MMICs below 30 GHz, but substrate vias are adequate for lower-frequency use. Figure 5.25 shows a radial stub used on a 38-GHz–76-GHz multiplier MMIC to decouple the drain bias sufficiently at 76 GHz.

5.2.3 Couplers, Splitters, and Combiners

Power splitters separate a signal at their input port into two or more parts with equal or unequal magnitude in each output port. Couplers perform the same function, usually with only two output ports, and tend to couple only small amounts of power into the second output port. Couplers are typically used to sample small amounts of power from a signal path where their coupled output is either connected to power meters to monitor the main signal level or fed into mixers to phase-lock the signal generator. Power splitters are used to split the power between several transistors in power amplifiers or to send the signal down more than one path in a more complex system. As only passive elements are being considered here, the splitters and couplers are reciprocal devices, so when they are used in the opposite direction, they act as power combiners and combine one or more signals together into one output port.

The simplest way to make an MMIC coupler is to use two transmission lines that are close enough together for their field patterns to interact with

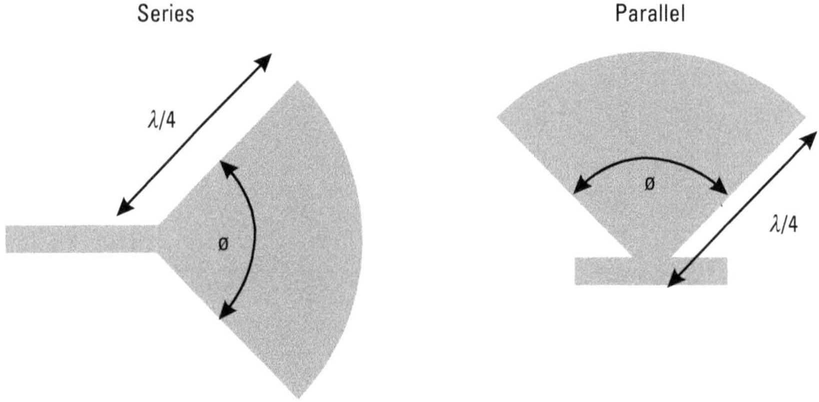

Figure 5.24 Radial stub.

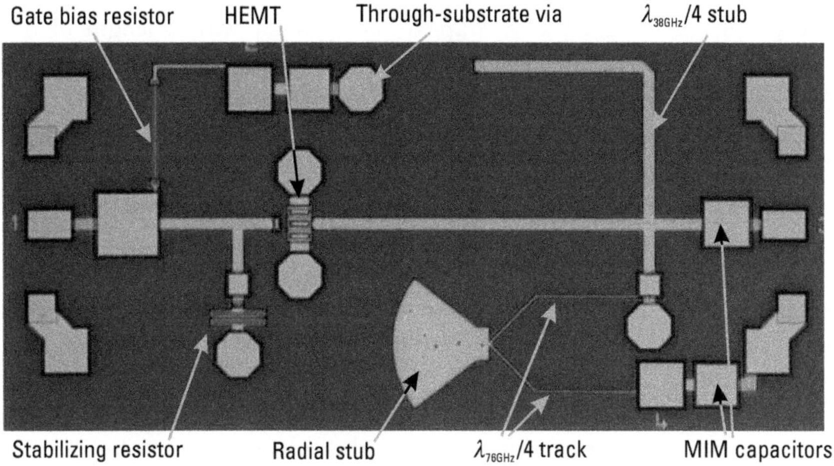

Figure 5.25 A 38-GHz–76-GHz multiplier MMIC using a radial stub to decouple the drain bias at high frequencies. (*Source:* Bookham Inc., 2006. All Rights Reserved.)

each other. This is the case when microstrip transmission lines are placed with separation distances that are equal or less than the substrate height. An example is the single-section microstrip coupled-line directional coupler shown in Figure 5.26, where power input at port 1 propagates to port 2 and some power is coupled to port 3 at a 90° phase difference to that at port 2. There should be zero or very little power output from port 4, so the ratio (in decibels) of this power to the input power is known as the isolation of the coupler. Another measure of the quality of a coupler is the directivity, which is equal to the isolation minus the coupling factor, or, in other words, how much better port 4 is isolated compared to the coupled power from port 3.

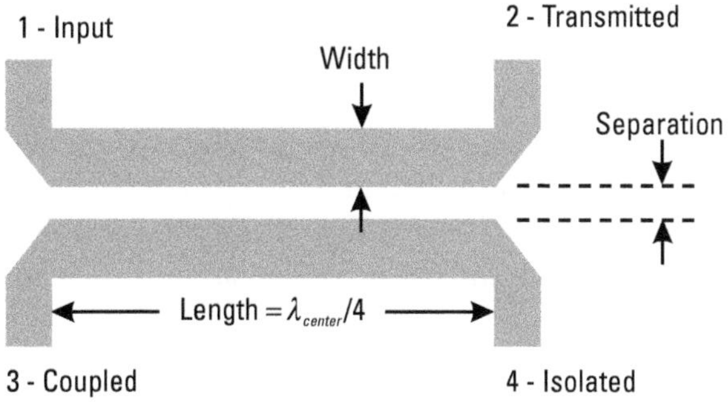

Figure 5.26 Single-section microstrip coupled-line directional coupler.

The coupler is designed as a coupled-line segment with a length of one quarter-wavelength at the center frequency. In simple terms, the closer the microstrip lines, the higher the coupling factor (coupling = $20\log(V_3/V_1)$ in decibels, where V_1 and V_3 are the voltages at ports 1 and 3, respectively) and the more power is transferred to the coupled port. The design of this type of coupler can be easily performed by tuning the width and separation of the $\lambda_{center}/4$ microstrip lines using the equivalent element in a standard CAD simulator, and detailed design rules are also available on the Web[3] and in the literature [11–13]. The 90° phase difference between the through and coupled ports means this is a quadrature coupler and is useful when designing balanced mixers.

When transmission lines are coupled in this manner, they can propagate in two modes: an even mode where the voltages on each line are the same, and an odd mode where the voltages on each line are of opposite polarity. The key to getting a broadband response from this type of coupler is to ensure that the propagation velocities of both modes are the same. This can be achieved using the so-called wiggly line coupler [14, 15], where the sawtooth edge of the coupler slows the odd mode to a propagation velocity much closer to the even mode. The multisection approach [16] also enables the designer to get much higher coupling factors, such as –3 dB rather than the typical values of –10 dB to –20 dB from the single-section approach. A similar broadband two-way splitter is the multisection-coupled microstrip directional coupler described by E. G. Cristal and L. Young [17].

5.2.3.1 Branch-Line Couplers

The branch-line coupler, shown in Figure 5.27, is constructed with lengths of microstrip that are one-quarter of the wavelength at the center frequency, two of which have the same characteristic impedance as the system (Z_0); the other two have an impedance of ($Z_0/\sqrt{2}$) [18, 19]. The coupling value of this style of coupler is usually fairly high, such as –3 dB, and the coupled and transmitted signals are 90° out of phase (quadrature), making this coupler what is known as a hybrid coupler. This is just one example of a branch-line coupler; the impedance of the series and shunt arms can be varied to produce different coupling ratios [20]. One feature of this coupler is that all the ports are connected at dc, which may be a consideration if used within a mixer circuit.

3. http://www.ee.bilkent.edu.tr/~microwave/programs/magnetic/dcoupler/dcoupler.htm, last access June 5, 2006.

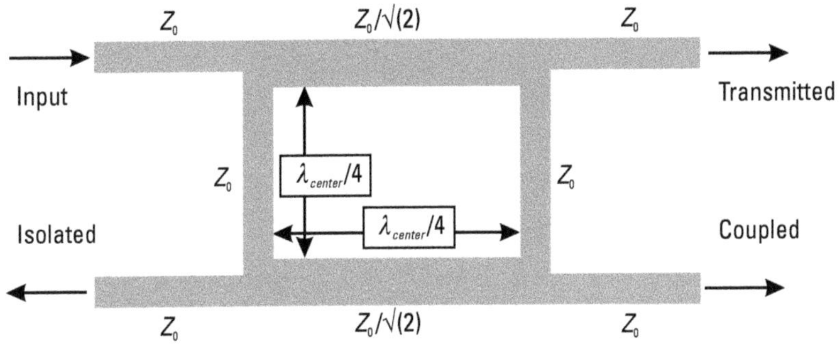

Figure 5.27 Branch-line coupler.

5.2.3.2 Rat-Race Couplers

The rat-race coupler, shown in Figure 5.28, is designed with microstrip transmission lines that are one-quarter and three-quarters of the center frequency wavelength, and with characteristic impedance of $Z_0/\sqrt{2}$. The input signal is split where it meets the circle of microstrip and combines with different phases at the other ports. The signals combine in phase at the transmitted port and are 90° different from the input port; they are also in phase at the coupled port and 270° different from the input port, with the result that the transmitted and coupled signals are now 180° out of phase with

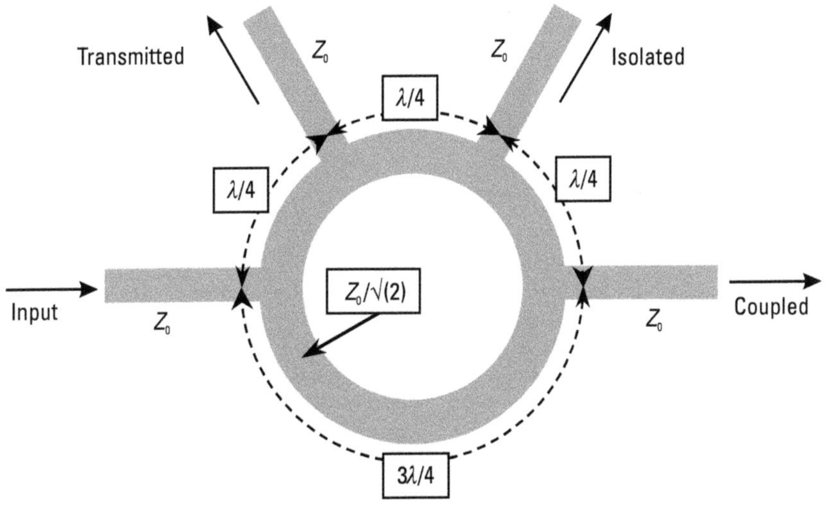

Figure 5.28 Rat-race coupler.

each other. This property is utilized in many mixer circuits. The signals combine out of phase at the isolated port so that power does not propagate from it. Further information on this style of circuit is available in the literature [21, 22].

5.2.3.3 Lange Couplers

The Lange coupler, shown in Figure 5.29, is essentially a 3-dB coupler, dividing power equally between two ports opposite the input, with one port being 90° out of phase with the other. The design principles are described in J. Lange's original paper [23], and the coupler has approximately an octave bandwidth. The track width and separation of the multiple coupled lines determine the impedance and coupling, while the length of the coupled lines determines the frequency response. The length of the coupled-line section is approximately one-quarter of the coupled-line wavelength at the lowest operating frequency and one half-wavelength at the highest operating frequency. This wide bandwidth is achieved because the combination of the bond-wires and multiple lines compensate for the odd- and even-mode dispersion.

Apart from their broad bandwidth (50% bandwidth defined as range/center frequency), their key feature is the fact that power reflected from the input ports is coupled into the isolating resistor and not back out of the input. In this way, two poorly matched devices may be combined in parallel between two back-to-back Lange couplers, as in Figure 5.30, and still exhibit excellent matches from the whole circuit. M. C. Tsai [24] has demonstrated even greater bandwidths of 2 to 15 GHz by developing tapered coupled-line structures based on the Lange structure. The main drawbacks to Lange couplers are that they are usually large compared to other MMIC components, as shown in Figure 5.31, where the Lange coupler is folded to take up the minimum space on the MMIC, so they are only practical at frequencies above 10 GHz. Their other limitation is that they only give a two-way power

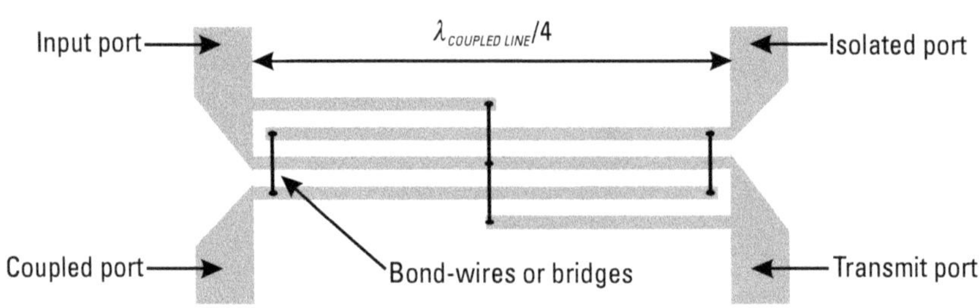

Figure 5.29 Lange coupler.

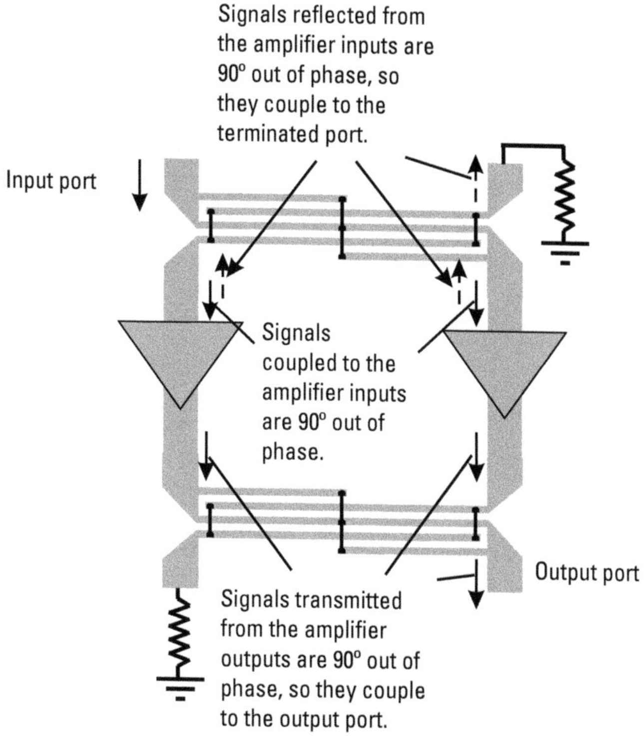

Figure 5.30 Two power amplifiers mounted between Lange couplers.

split, so they must be combined into a tree structure to give 4-, 8-, or 16-way power splits.

5.2.3.4 Wilkinson Splitters

The Wilkinson splitter is also a 3-dB coupler, dividing power equally between two ports opposite the input, with both ports being in phase with each other. These can be generally separated into two types, distributed and lumped-element Wilkinsons. The distributed Wilkinson is similar to the *N*-way power divider described in J. Wilkinson's original paper [25], with the exception that in MMICs the quarter-wave transmission lines and isolation resistors are now implemented as microstrip transmission lines and thin-film NiCr on a semiconductor substrate. The basic principles of the Wilkinson design are shown in Figure 5.32, where the microstrip lines have length $\lambda/4$ and impedance $\sqrt{2}Z_0$, and the resistors connecting to the common node have the same resistance as Z_0. These have been successfully

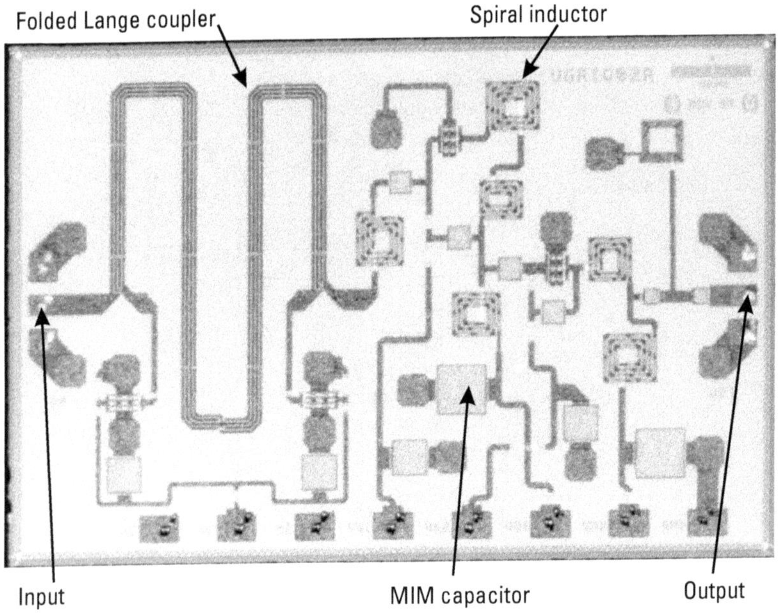

Figure 5.31 Folded Lange coupler used in a variable-gain amplifier. (Courtesy of Alcatel Alenia Space France.)

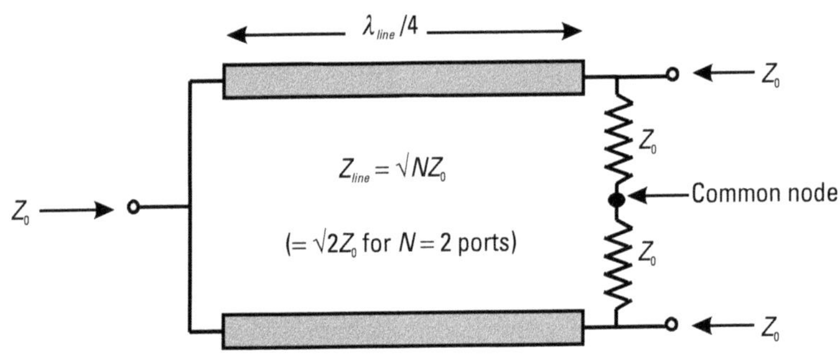

Figure 5.32 Wilkinson two-way splitter.

implemented, in coplanar waveguide (CPW) and microstrip, on MMICs at frequencies up to 110 GHz [26, 27].

Distributed Wilkinson combiners are low-loss components and, so, are used often as the final output combiner in power amplifiers. They also have good isolation, which means the power amplifier will exhibit graceful degradation in the unlikely event of failure of one out of the many combined devices. The bandwidth is also good (typically 30%), and the length of the quarter-wave microstrip lines enables good in-phase combining of devices, which are spaced far apart. This type of combiner, being based on transmission lines, is also very tolerant of MMIC process variations.

The primary disadvantage of Wilkinson combiners is that they are physically large, making them inappropriate at frequencies below 10 GHz, and above ($N = 2$)-way combining, it is not possible to implement a symmetric common node for the isolating resistors in a planar structure, as shown by the three-way Wilkinson splitter in Figure 5.33. The solution for combiners with more than two ports is to omit the isolating resistors. This is no longer the classical Wilkinson structure, and the loss of isolation between devices reduces the chances of graceful degradation and increases the risk of odd-mode oscillation. However, assuming all the output devices are fed in phase, this structure can be used successfully.

The lumped Wilkinson simply uses lumped-element capacitors and inductors to mimic the quarter-wave transmission lines (in a similar way to the equivalent circuit for a transmission line in Figure 2.20). Typical circuit topologies are shown in Figure 5.34, and a lumped Wilkinson used as an RF and local oscillator (LO) combiner in a mixer MMIC is shown in Figure 5.35. The advantage of the lumped Wilkinson is that it is very compact while still maintaining good bandwidth, isolation, and process tolerance. The disadvantages are that it is lossier than the distributed Wilkinsons and, so, tends to be used for signal splitting before the last gain stage of a power amplifier; also, it has less spatial separation of the split ports. Greater than two-way splitting is also more difficult to design in a planar environment.

Both distributed and lumped Wilkinson combiners are incorporated in the J-band power amplifier MMIC shown in Figure 5.36. The compact

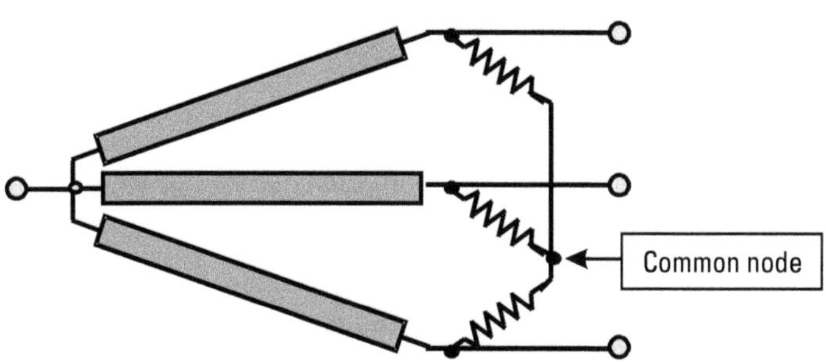

Figure 5.33 Three-way Wilkinson splitter with isolating resistors.

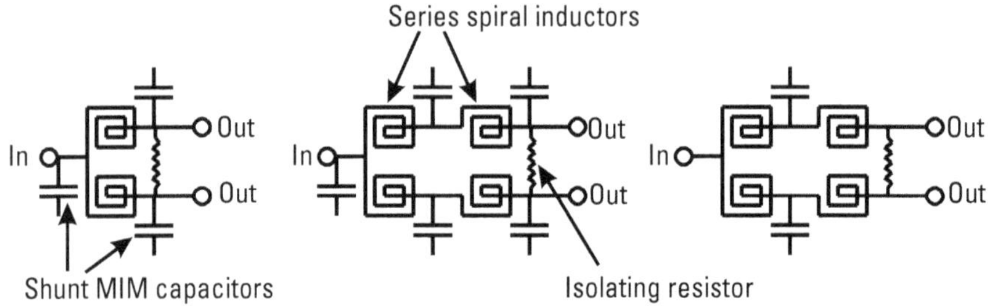

Figure 5.34 Topologies of lumped Wilkinson two-way splitters.

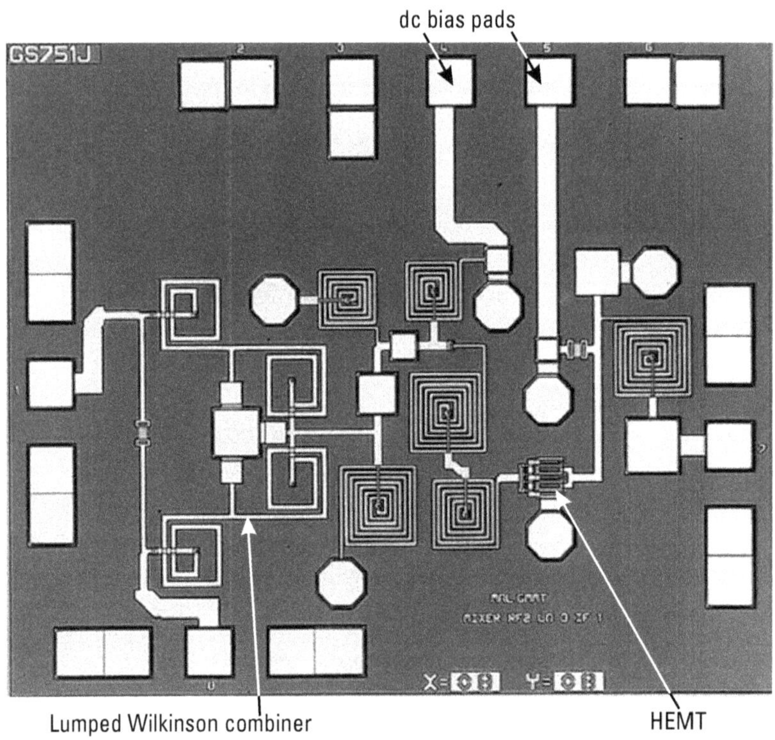

Figure 5.35 Single-ended mixer MMIC incorporating a lumped Wilkinson combiner. (Courtesy of TM R&D, Malaysia.)

lumped Wilkinson splitters can be seen between the first and second and the second and third gain stages. The larger distributed Wilkinson combiner is situated after the output matching of the last stage and brings the RF power

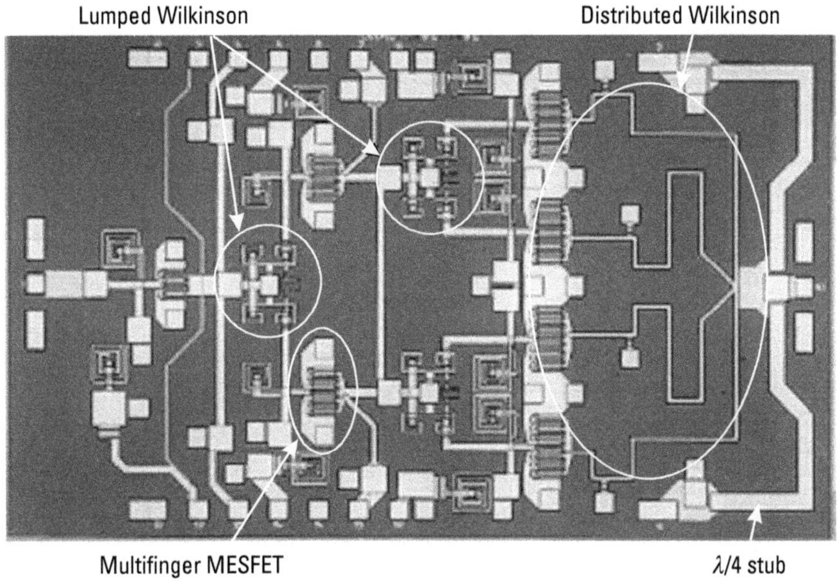

Figure 5.36 Power amplifier MMIC using distributed and lumped Wilkinson splitters and combiners. (*Source:* Bookham Inc., 2006. All Rights Reserved.)

together at the output RFOW pads. In this chip, the dc bias to the last stage is applied via the wider, quarter-wavelength, short circuit stubs and fed along the distributed Wilkinson combiner to the active devices.

5.2.3.5 Distributed Transmission-Line Combiners

Distributed transmission-line techniques involve using transmission lines of various widths to transform the low FET impedance at the output stage of a power amplifier into higher output load impedance. The general layout of such a combiner is shown in Figure 5.37, and an example from the literature can be found in [28]. One limiting factor is that the line loss restricts the minimum track width, hence, the maximum line impedance available to the designer. Two other disadvantages of this technique are that it is fairly narrowband (15%) and that external bias T's are required. Also, the structure tends to be quite large, so it is only really feasible at millimeter-wave frequencies, and care must be taken to ensure the odd-mode stability of the circuit.

5.2.3.6 Traveling-Wave Combiners

These circuits are described by A. G. Bert and D. Kaminsky [29] and consist of a set of unequal power splitters connected in series. For example, a

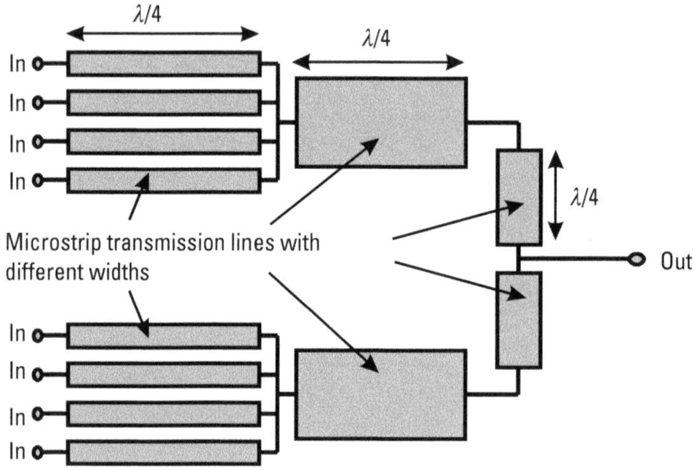

Figure 5.37 Distributed transmission-line combiner.

four-way splitter would be constructed from a 6-dB coupler, followed by a 4.8-dB coupler, followed by a 3-dB coupler, as in Figure 5.38. The 6-dB coupler takes one-quarter of the power, the next coupler takes one-third of the remaining power, and the final coupler splits the remaining power between the last two ports. The result is that one-quarter of the incident power is coupled to each of the four output ports.

Two examples in the literature [30, 31] use an off-chip design for compact combining of several power MMICs within one package. This design technique is further discussed by K. J. Russel [32] and could be applied on chip, but probably only at millimeter-wave frequencies. An advantage of this series connection means that an extra port can be added to the design at the input simply by adding another power splitter with a $10\log_{10}(N + 1)$ coupling coefficient (where N is now the total number of splitters). The only

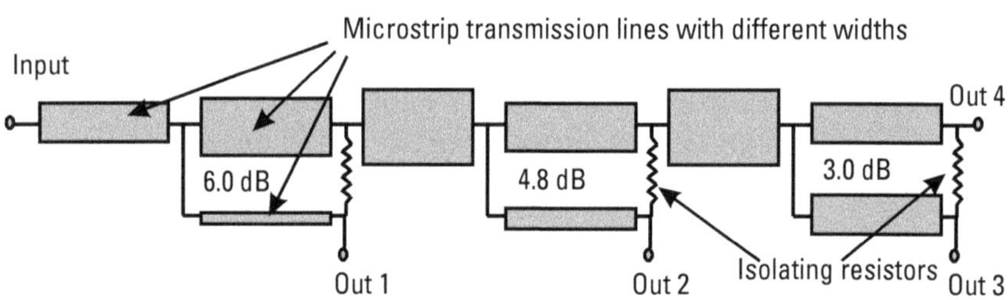

Figure 5.38 Traveling wave four-way splitter.

limit to the number of ports is the ability to accurately make the individual power splitters with high coupling coefficients.

The benefits of this type of combiner include a broad bandwidth (50%), good isolation, and low loss, which helps maintain the efficiency of the overall amplifier. On the minus side, this combiner requires prematched devices and a high-resolution process capable of reproducing the fine features of the couplers.

5.2.3.7 Tapered Transmission-Line Splitters

Tapered transmission lines can be used to form an *N*-way splitter or combiner, such as the Dolph-Chebychev technique used by W. Yau and J. M. Schellenberg [33] and N. Nagai, E. Matkawa, and K. Ono [34], shown in Figure 5.39. The design methodology starts with a Dolph-Chebychev tapered transmission line, which transforms the load impedance (50Ω) into a much lower impedance ($50/N\Omega$). The curve of the tapered transmission line determines the in-band return loss, and the length sets the low end of the frequency band. The broad part of the transmission line is then segmented into sections of nonuniform coupled transmission lines with a network of isolating resistors between the adjacent lines. General conditions for achieving equal phase and amplitude division between the ports are set out in [35] and, in summary, state that the capacitance between any line to its adjacent lines must be equal, from each line to ground must be the same, and to

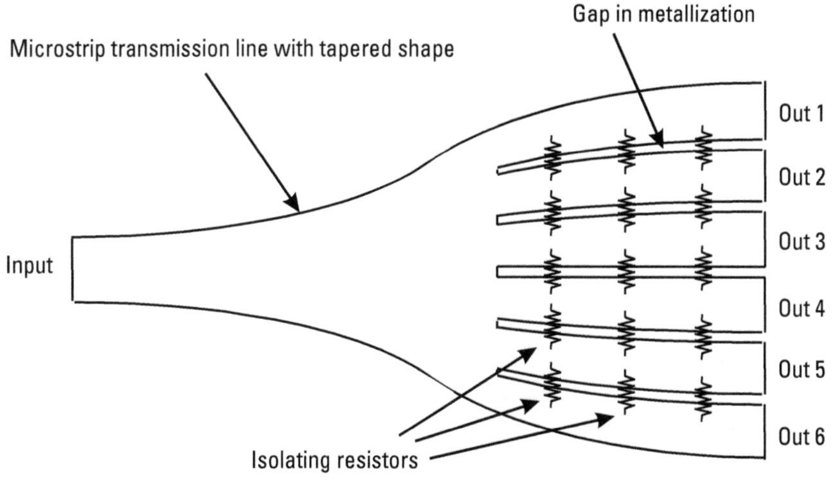

Figure 5.39 Tapered transmission-line six-way splitter.

nonadjacent lines must be negligible. The circuit is then analyzed in terms of its odd and even modes to derive the values of the resistors. Any subsequent imbalance between the ports, caused by asymmetry in the capacitance between the coupled lines, can be compensated for by adjusting the isolating resistor values.

The main advantage of this type of structure is that it is very broadband (100%) and that a planar *N*-way split is obtained while maintaining good port isolation. The disadvantage is that, because of its size, it is still mainly an off-chip technique, and this is likely to limit it to millimeter-wave applications. Also, the devices need further matching circuits, and the design technique is not trivial.

5.2.4 Baluns

A balun is a device that joins a balanced line (one that has two conductors with equal currents in opposite directions, such as a twisted pair of wires or slotline) to an unbalanced line (one that has just one conductor and a ground, such as microstrip or coax). One pair of the balun terminals is balanced (i.e., the currents are equal in magnitude and opposite in phase), while the other pair of terminals is unbalanced (i.e., one side is connected to electrical ground and the other carries the signal). Baluns in MMICs are sometimes used to feed antennas, but their primary role is to provide balanced signals to help cancel unwanted mixer products (see Section 5.5).

The most common way to make a balun in MMICs is to use an equal power splitter followed by different phase shifters to create the two signals with the opposite phase, as shown in Figure 5.40. One example [36, 37] is to use a Lange coupler with 90° coupled-line phase shifters [38], and another is to use a Wilkinson splitter followed by low-pass and high-pass filter phase shifters.

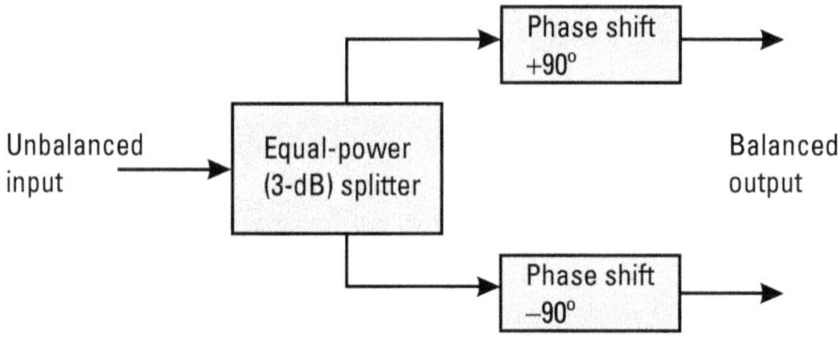

Figure 5.40 Balun implementation.

5.2.4.1 Marchand Balun

The Marchand balun [39], shown schematically in Figure 5.41, can be designed to work on the planar surface of an MMIC using one-quarter-wavelength sections of coupled lines [40–42]. Figure 5.42 shows Marchand baluns being used at two different frequencies in a quad-FET ring mixer MMIC [43].

5.2.5 Filters

The resistive loss of MMIC lumped components, such as spiral inductors, is low but significant enough to reduce the Q-factor to values of around 10 to 30. This means that filters designed with spiral inductors and MIM capacitors will have similarly low Q-factors, and their frequency selectivity will be poor. The main practical methods for producing filters on MMIC with higher Q-factors is to use either distributed components as in coupled-line filters or to use transistor gain to counteract the loss, as in active filters.

5.2.5.1 Coupled-Line Filters

The coupled-line filter consists of quarter-wavelength sections of coupled lines connected together in series [7, 44]. This type of filter can achieve a maximally flat response with 20% bandwidth and an equal ripple response with 30% bandwidth. The coupling of the sections is a function of their separation distance compared to the substrate height and increases as the separation is reduced. Full design information is already available in the literature [45, 46].

5.2.5.2 Active Filters

Active filters are designed using transistors to produce some signal gain within the filter to compensate for, or cancel out altogether, the resistive

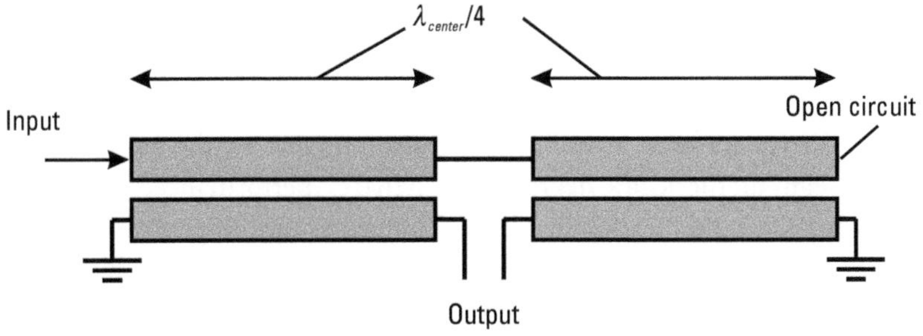

Figure 5.41 Marchand balun in planar form.

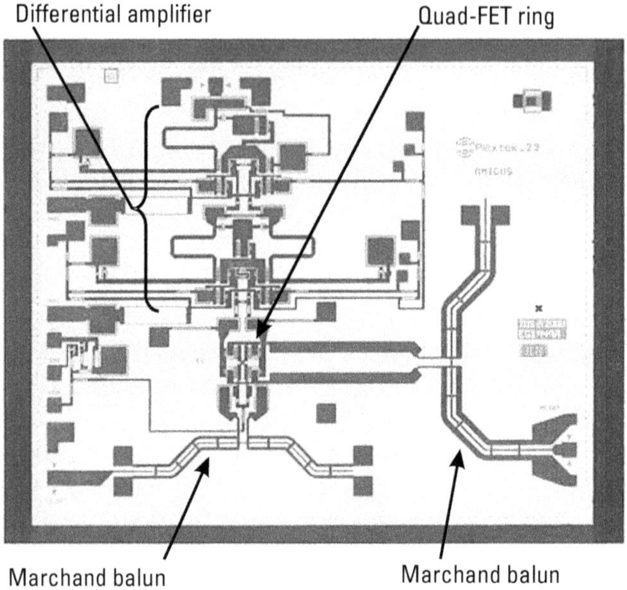

Figure 5.42 A pHEMT quad-FET ring mixer incorporating two different Marchand baluns and a differential amplifier. (Courtesy of Plextek Ltd. and QinetiQ [43].)

losses within the passive elements. They can also be designed to be tunable. There are two main filter techniques, distributed microstrip, or inductor-capacitor (LC), filters [47, 48], and lower frequency resistor-capacitor (RC) filters [49, 50]. RC filters are easier to implement in an active form, and a number of them have been designed using gyrator circuits [51, 52]. The main problem with active filters is ensuring they are stable against temperature variations because the gain of the transistors is a function of temperature.

5.2.6 Antennas

When MMICs are used at millimeter-wave frequencies, the chip-to-system interface can dramatically degrade the MMIC performance, especially if bond-wires are used. Variability in the bonding and chip positioning can produce performance uncertainties that are unacceptable for the system application. To overcome this problem, designers have investigated placing antennas on the MMIC and using free-space or quasioptic transmission to interface with the overall system.

The easiest antennas to create on MMICs are half-wavelength dipoles and patch antennas, as shown in Figure 5.43. An example single patch antenna on GaAs with dimensions 840 μm × 617 μm yielded a measured return loss of –9.4 dB with 2.6-dBi gain at 65 GHz [53].

Patch antennas have also been designed for silicon ICs [54] and can be fed with both microstrip [55] and CPW [56].

Unfortunately, a large amount of the propagating radiation is coupled into the substrate, meaning that only around 40% efficiency is achieved. It has also been shown that microstrip antennas on high-dielectric-constant MMIC substrates, such as GaAs and silicon, suffer from very narrow bandwidths (0.5%) and poor radiation patterns due to the triggering of unwanted surface waves [57].

5.3 Amplifiers

Of all the types of MMIC circuits designed, the most common by far is the amplifier, and this comes as no surprise because an MMIC is made primarily by a transistor fabrication process with the additional capability of making passive components as well. For this reason the first "whole chip" (as opposed to a subcircuit) design type considered is the amplifier, and a large part of this chapter is given over to MMIC amplifier design techniques. Also, because the whole chip design is now being looked at, the first section discusses the initial considerations that are required for all MMIC design, in particular the effects and consequences of the chip-to-substrate/package/system interface.

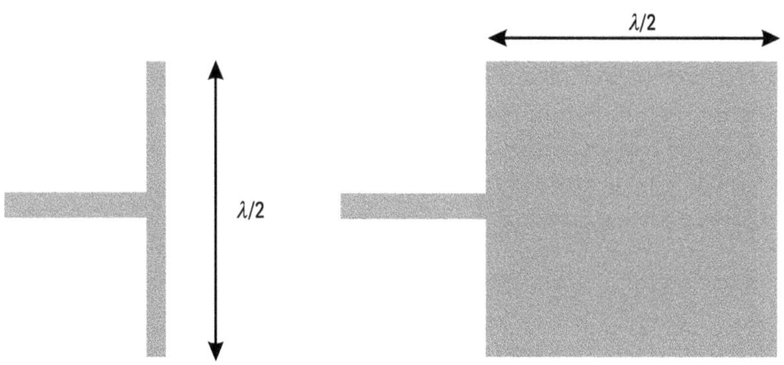

Figure 5.43 Dipole and patch antennas.

5.3.1 Initial Considerations

When considering the design of an MMIC, whatever the circuit type, there are a few general points that the designer must note in relation to the desired specification. The first concerns the RF path through the chip and connecting to the system it must work in. It is important to remember that although the system impedance may be 50Ω, there is almost certainly going to be an interface between the chip and the system that will not be 50Ω but will have parasitic impedances associated with it. The simplest interface will either be just a bond-wire or a flip-chip solder joint, but there may also be a package wall to go through and a further connection between the package and the system. The second concerns the dc path through the chip, where the bias voltages and currents are fed into the chip to power the transistors and control tunable elements. The designer must know how to layout the correct set of dc pads so that the chip will perform correctly not only when mounted into the system but also when tested RF-on-wafer (RFOW).

5.3.1.1 Chip-to-System Interface

The interface between an MMIC and the system it is embedded in has parasitic impedances that affect the chip performance, in particular the input and output matches. A simple system may be a module made from an alumina substrate with the bare MMIC mounted on the same ground plane within a hole in the substrate. The MMIC will then be connected to the 50Ω transmission lines on the alumina using bond-wires, so each RF port will have the type of interface shown in Figure 5.44.

The system requirement will be for the MMIC to have a certain minimum return loss in the plane of the system transmission line so that any parasitic impedance caused by the bond-wire and the RFOW pads must be taken into account in the MMIC design. Therefore, the bond-wire and RFOW pads must be modeled and included in the MMIC simulation when designing the input- and output-matching circuits. The characteristics of the RFOW signal pad are fairly well characterized and do not vary. Their electrical model may be provided by the foundry, or alternatively it can be modeled as a parallel plate capacitor to the substrate ground, as given in (5.5), where ξ and d are the dielectric constant and the height of the substrate, and a is the area of the RFOW signal pad. As an example, a 120-μm square pad on a 100-μm GaAs ($\xi_r = 12.95$) substrate will have a capacitance of about 16.5 fF.

$$c = \xi\, a/d = \xi_0 \xi_r\, a/d \tag{5.5}$$

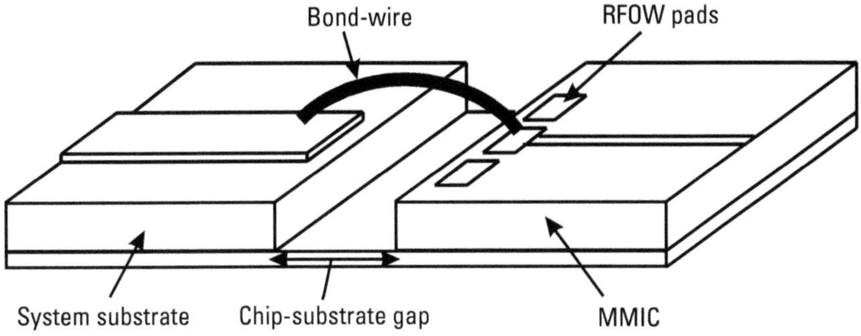

Figure 5.44 Chip-to-system interface.

The bond-wire inductance is dependent on the length of the bond-wire (l) and the wire diameter (d), according to (5.6) (assuming that the wire is made from a nonferromagnetic material such as gold) [58]. This is approximately 0.73 nH/mm for 50-μm,-diameter gold wire.

$$L = 2l[2.303 \log(4l/d) - 0.75 + (d/2l)] \quad (5.6)$$

This length is a variable that is a function of the chip-substrate gap and the bond loop height, which also tends to vary according to which operator is bonding the wire. In practice, the bond-wire length varies between 400 μm and 600 μm, which, for a 50-μm-diameter gold wire, gives a parasitic inductance of 0.22 to 0.38 nH. The impact of this on the port matches can be seen in Figure 5.45, where the magnitude of the reactance is plotted against frequency for these two inductance values. At frequencies of 40 GHz, the reactance of the bond-wire is greater than 50Ω, so unless techniques are utilized to ameliorate this effect [59], it is important that the MMIC matching circuit be designed to include this parasitic impedance.

The variable nature of this parasitic inductance due to assembly variations, together with modeling errors and process variations, means that the designer cannot ensure a perfect input and output match. Matches (return loss) of less than −10 dB (VSWR of 1.9) are the minimum level that is usually acceptable in a system and are readily achievable. Matches of less than −15 dB (VSWR of 1.4) are desirable for most systems and are achievable over a narrow band or at frequencies lower than 10 GHz. Matches less than −20 dB (VSWR of 1.2) are not easily achieved even at frequencies lower than 1 GHz.

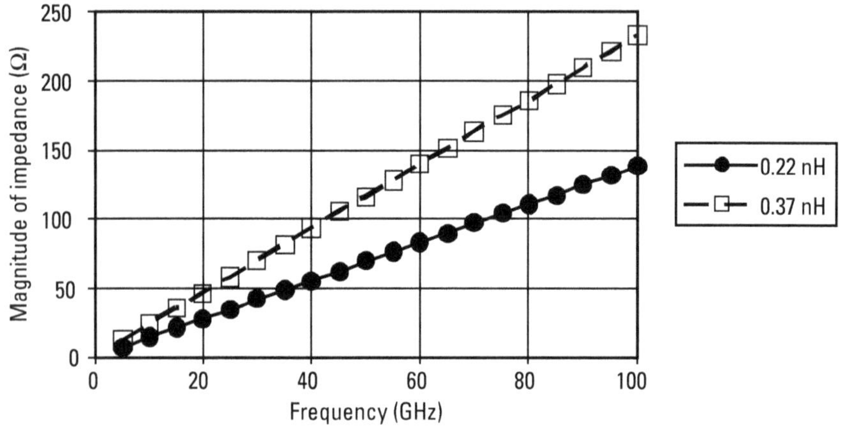

Figure 5.45 Impedance variation with bond-wire inductance.

5.3.1.2 The dc Connections

The classic layout for an MMIC is shown in Figure 5.46, where the RF signal is input via the RFOW pads on the west (left) side of the chip and output via the RFOW pads on the east (right) side of the chip, and the dc pads are on the north (top) and south (bottom) sides of the chip.

During RFOW testing, RF probes are brought into contact with the chip from the east and west sides of the probe station using micromanipulators, and the dc pads are contacted using probe cards attached to micromanipulators on the north and south sides of the probe station. Each side of the chip can only have either RFOW pads or dc pads and not both because there is only room to bring one or the other to each side of the chip.

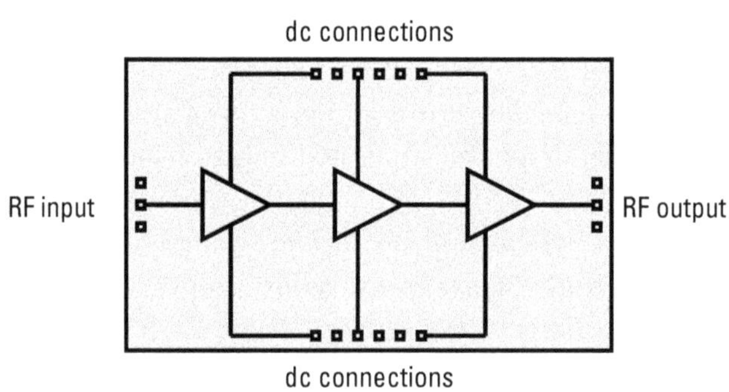

Figure 5.46 Typical MMIC layout.

The dc probe cards are constructed from a printed circuit board (PCB)–style board with probe pins soldered to its surface (shown in Figure 5.47), which normally have been aligned to the dc pads using a dummy wafer.

Foundries prefer the dc pads to be at the sides of the chip and laid out in a common pattern so that the same dc probe cards can be used to measure many different MMIC designs. For example, one foundry recommends using 120-μm square pads on a pitch of 240 μm and has a stock of dc probe cards with two to eight pins at the same pitch. If the MMIC is to be mounted into the system using flip-chip, then the dc and RF pads can be anywhere on the chip, but custom dc probe cards will need to be made especially for that chip design in order for it to be tested on wafer.

The dc pads must be decoupled from the RF signal on the MMIC to prevent the RF signal from propagating along the dc bias wires to the power supplies as this often causes unwanted feedback and can make the chip oscillate at an arbitrary frequency. If the chip is oscillating, this can be observed by probing the bias lines with an oscilloscope, and if the pattern on the

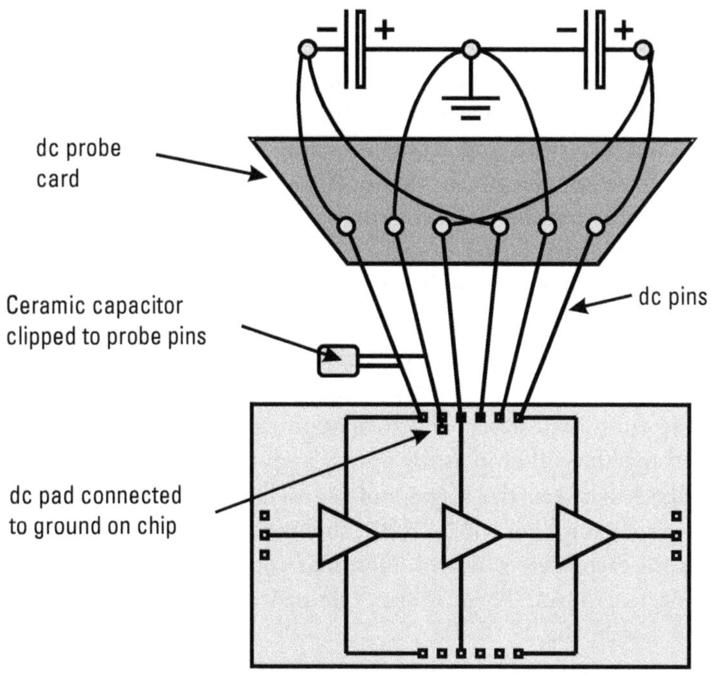

Figure 5.47 A dc probe card connecting the dc bias to the chip during RFOW measurements.

oscilloscope changes when the biasing wires are touched or moved, then poor decoupling is allowing feedback along the bias wires. To prevent this, it is good practice to have each dc pad coupled to ground using a large MIM capacitor (typically 5 to 10 pF) and a through-substrate via. If the chip is to operate at low frequencies (<1 GHz), this value of on-chip capacitor is not large enough to decouple the RF to ground, and the chip can still oscillate in the megahertz frequency range. When this is the case, off-chip MIM decoupling capacitors (which can have much higher values by utilizing higher dielectric materials) will need to be mounted next to the chip, and the dc bias fed across these grounded capacitors. Chips with a lot of low-frequency (<10 MHz) gain are also difficult to measure on wafer because these large off-chip capacitors are not yet mounted next to the chip. To overcome this, leaded ceramic or tantalum capacitors can be mounted on the dc probe card or even clipped to the probe pins. It is difficult for the designer to ensure there will be no low-frequency oscillation during the RFOW test of a MMIC design, so it is wise to take every possible precaution during the design stage. One valuable precaution is to place extra dc pads between each bias voltage pad and to connect these extra pads to ground using a through-substrate via. Then, if there are low-frequency oscillations during RFOW testing, a large capacitor can be clipped to the bias pad pin and the grounded pad pin, creating a very short and low-impedance path to the chip ground, as shown in Figure 5.47. If an on-chip grounded pad is not included on the MMIC, the capacitor must be connected between the bias pad pin and the wafer ground, which is the metal chuck that the wafer is sitting on. The RF path to ground in this case will be at least the length of the wafer diameter (when probing the center of the wafer) and so inductive that it is not a low-impedance path to ground. Another precaution is to use separate dc pads for each stage of an amplifier, even if they are using the same voltage. This is because when they are connected together on-chip, there is nothing that can be done to improve the decoupling from each other, and oscillation cannot be stopped. If they are connected together off-chip at the power supply, there is more scope to prevent feedback between the stages and fix oscillation problems. An example 60-GHz power amplifier MMIC [60], shown in Figure 5.48, has separate dc bias pads for each stage gate and drain bias voltage, and each has capacitive decoupling to ground. There is also a dc pad connected to ground using a through-substrate via.

In terms of current carrying capacity, each dc pin is rated up to handle 0.5A, so if a stage of a power amplifier requires more current, then multiple pads and pins must be used.

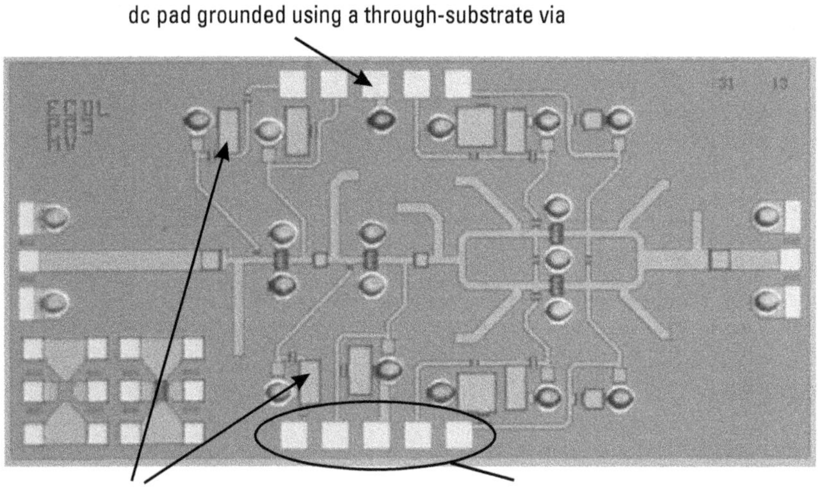

Figure 5.48 A 60-GHz power amplifier MMIC showing decoupled and earthed dc bias pads. (With kind permission of Springer Science and Business Media [60].)

5.3.1.3 Gain and Stability

When designing an MMIC amplifier, it is useful to understand the amount of gain that it is reasonable to achieve for the whole chip and the amount of gain per amplification stage (per transistor). The maximum possible gain per MMIC chip is limited only by the need to be stable and not oscillate. For the overall chip, this is achieved by keeping the chip gain lower than any feedback paths from the output RFOW pads back to the input of the chip (Figure 5.49). This feedback is referred to as the chip reverse isolation and is expressed in decibels ($-B$ dB). This means that the chip gain G (also expressed in decibels) must be less than B for stable operation.

There are many possible feedback paths, the most obvious being the simulated reverse isolation (S_{12}) of the fully laid out MMIC. There is also a path through the dc bias tracks, probe needles, cables, and power supplies due to the poor RF decoupling discussed in Section 5.3.1.2. If, when mounted in the system, the MMIC sits within a metal enclosure or package, then cavity waveguide modes [62] can also feed power back to the chip input. If there is poor grounding to the module base plate, this can lead to common-ground feedback, or parallel plate modes [63] can propagate under the chip. There is also the possibility that during RFOW measurements, there will be radiation

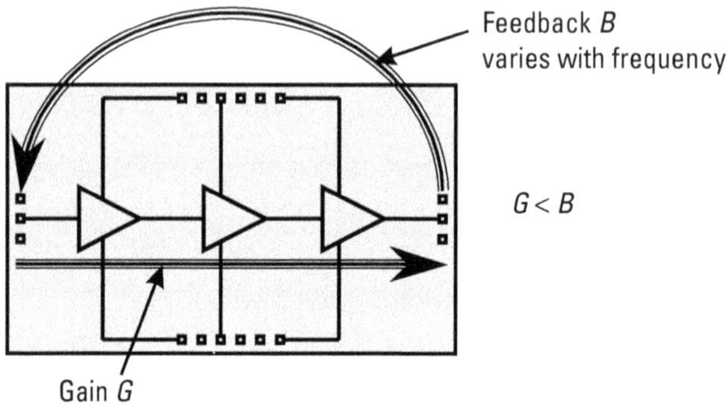

Figure 5.49 Schematic diagram of an MMIC.

from the output RFOW probe to the input RFOW probe. All of these paths make accurate assessment of the total feedback a difficult task, but, in general, the feedback is inversely proportional to the frequency. For example, chips at 2 GHz can exhibit isolation of –80 dB, while chips at 40 GHz tend to exhibit isolation of only –30 dB.

The gain-per-amplification stage needs to be as high as possible to cut down on the number of stages required and to maximize the amplifier efficiency. Choosing the transistor with the highest gain is not necessarily the correct solution because these devices may be difficult to stabilize over the whole frequency range. For example, HEMT devices have good gain at millimeter-wave frequencies but also have very high gain below 1 GHz, so the stability needs to be checked over a wide range of frequencies. As a rule of thumb, a good target figure for the gain of an individual stage is 10 dB, and the minimum practical gain per stage should be around 5 or 6 dB.

5.3.2 Small-Signal Amplifiers

In the design of small-signal amplifiers, the assumption is made that the RF signal levels are low enough that the active devices behave in a completely linear manner. When this is the case, the amplifier can be designed using standard *s*-parameter techniques, and the simulation tools can quickly calculate the unique performance solution of the particular circuit. Small-signal amplifier design usually covers all MMIC amplifier design, except specifically power amplifiers, and the sections here include low-noise amplifiers (LNAs), distributed amplifiers, distributed cascode amplifiers, and amplifiers designed for low power consumption.

5.3.2.1 Low-Noise Amplifiers

Low-noise amplifiers are required in receiver systems to increase the amplitude of the very low-level signals from the antenna without adding too much noise. The subsection below discussing noise theory shows that the first amplifiers in any system dominate the overall system noise performance, and they must be designed to have a low noise figure and high signal gain. Techniques such as series inductive feedback are described, which allow the amplifiers to have good input match, as well as a very low noise figure, and other issues, such as stability and off-chip components, are also covered.

Noise Theory

At any temperature above absolute zero, the thermal motion of charge carriers within a resistor (or any conductor) will induce a randomly fluctuating voltage, and this effect is known as resistor, thermal, or Johnson-Nyquist noise [64, 65]. The thermal noise of a resistor is equal to $V_t = \sqrt{(4\mathrm{k}TBR)}$, where V_t is the rms noise voltage, k is Boltzmann's constant (1.38×10^{-23} joules/kelvin), T is temperature (kelvin), B is the bandwidth over which the noise is measured, and R is the resistance in ohms.

If a current is passing through the resistor, there is also a contribution to the total noise, called shot noise. This consists of random fluctuations of the electric current caused by the fact that the current is carried by discrete charges (electrons). The strength of this noise increases with the growing magnitude of the average current flowing through the conductor.

When analyzing the noise in a transistor, the thermal and shot noise produced at the output is referred back to an equivalent noise at the input of the transistor. Figure 5.50 shows an amplifier stage with gain G_A and noise referred to its input of N_A. An RF signal is applied at the input with a signal-to-noise ratio of S_I/N_I, and at the output of the amplifier, the resulting signal-to-noise ratio is $G_A\ S_I/G_A(N_I + N_A)$, so the degradation of the signal-to-noise ratio is given by the ratio in (5.7). The noise factor, when expressed in decibels, is known as the noise figure [(5.8)].

$$(S_O/N_O)/(S_I/N_I) = (N_I + N_A)/N_I = \text{total noise/source noise} = \text{noise factor} \tag{5.7}$$

$$\text{Noise figure (dB)} = 10 \times \log_{10}(\text{noise factor}) \tag{5.8}$$

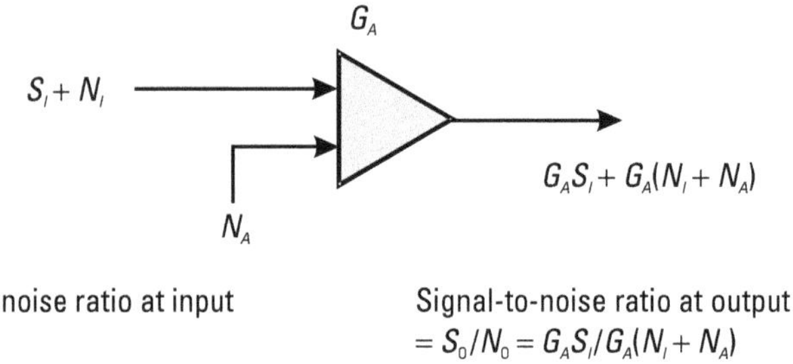

Figure 5.50 Degradation of the signal-to-noise ratio due to the noise contribution from an amplifier.

As seen in Section 5.1.1, there is an optimum reflection coefficient that, when presented to the input of a transistor, minimizes the noise figure, known as Gamma opt (Γ_{opt}) [3]. When the transistor is presented with Γ_{opt}, the noise figure is known as F_{min}, and gain is known as the associated gain (G_{Ass}).

In a multistage amplifier such as the three-stage amplifier in Figure 5.51, the overall noise figure can be calculated from the individual noise figures and associated gains using the work of H. T. Friis [66, 67], as shown in (5.9).

$$F_{total} = F_1 + \frac{F_2 - 1}{G_1} + \frac{F_3 - 1}{G_1 G_2} + \ldots \tag{5.9}$$

It can be seen from (5.9) that the noise figure of the first stage is dominant and that the noise figure of subsequent stages is reduced by the gain of the preceding stages.

Therefore, to get the minimum noise figure from a multistage amplifier, the first stage must be designed to have the lowest possible noise and the highest possible gain. The second stage should be designed with reasonably low

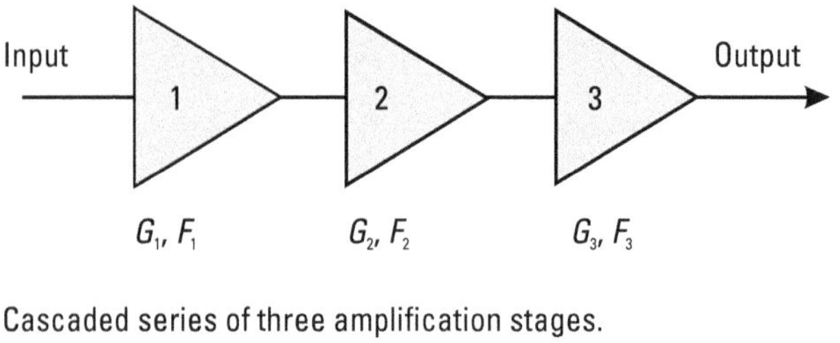

Figure 5.51 Cascaded series of three amplification stages.

noise figure, but stages after that contribute an insignificant amount to the total noise figure.

Noise Characteristics of a Transistor

Chapter 2 showed that HEMTs are the best technology choice for making low-noise amplifiers, and this is because the electrons are held within a quantum well, away from the lattice atoms, which reduces their shot noise contribution. Figure 2.5 also showed that the bias point is typically around $I_{DSS}/2$, which is a compromise point for low noise and reasonable gain. Figure 5.52 shows the variation of the noise figure and associated gain versus frequency for three different sized HEMT devices; it can be seen that the minimum noise figure increases with increasing frequency, and the associated gain drops. The smallest device, the 2 × 50 μm HEMT, has the lowest noise figure and highest associated gain over the whole frequency range, so it would appear to be the ideal device for the first stage of a low-noise amplifier. This is often the case, but if the device is too small, it has very high input impedance at low frequencies and requires a Gamma opt, which is very close to an open circuit, as shown in Figure 5.53. Increasing the device size slightly can make a big difference in the ease of designing a matching circuit for minimum noise figure over the whole band.

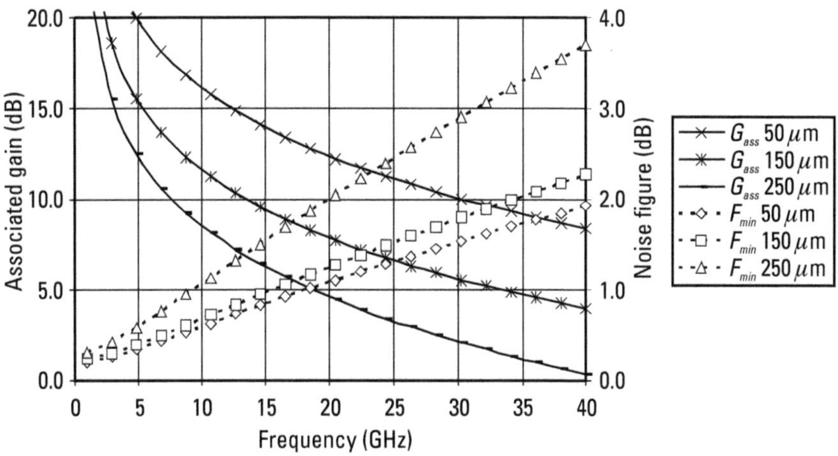

Figure 5.52 Minimum noise figure and associated gain of 2 × 50 μm, 2 × 150 μm, and 2 × 250 μm HEMTs versus frequency. (Simulated with Agilent ADS and the PH25 Design Kit, courtesy of UMS.)

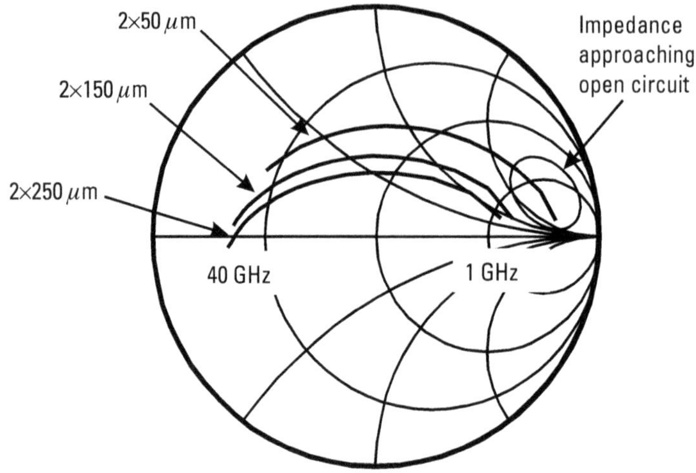

Figure 5.53 Gamma opt for 2 × 50 μm, 2 × 150 μm, and 2 × 250 μm HEMTs versus frequency. (Simulated with Agilent ADS and the PH25 Design Kit, courtesy of UMS.)

Input Matching and Series Inductive Feedback

The specification for a low-noise amplifier will generally require a low noise figure and a good input match, but the conjugate of S_{11} (required for a good match) is seldom the same as Γ_{opt} (required for low noise figure), so these cannot be achieved simultaneously. This is observed in Figure 5.54, where S_{11}^* and Γ_{opt} are plotted for a 2 × 75 μm HEMT over the frequency range 10 to 30 GHz.

However, if a small amount of inductance is placed between the HEMT source contact and ground, this increases the real part of the HEMT impedance and brings S_{11}^* and Γ_{opt} much closer, as in Figure 5.55, where 0.15-nH inductance is used. This effect, known as series inductive feedback [68–70], also decreases the gain slightly and improves the stability factor.

Stabilizing a Low-Noise Amplifier

MMIC low-noise amplifiers tend to use HEMT devices to achieve the requirements for high gain and low noise at microwave and millimeter-wave frequencies, and these devices also have very high gain in the megahertz and low gigahertz frequency ranges. This makes amplifier stages designed with these devices very unstable at low frequencies, so techniques are needed to reduce the low-frequency gain without affecting the in-band millimeter-wave

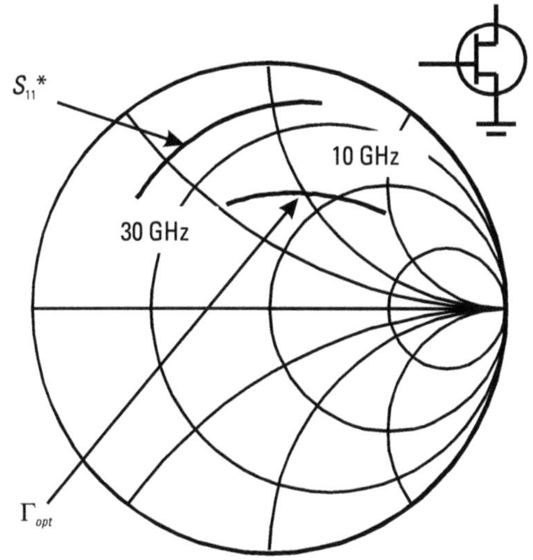

Figure 5.54 Conjugate of S_{11} and Gamma opt for a 2 × 75 μm HEMT from 10 to 30 GHz. (Simulated with Agilent ADS and the PH25 Design Kit, courtesy of UMS.)

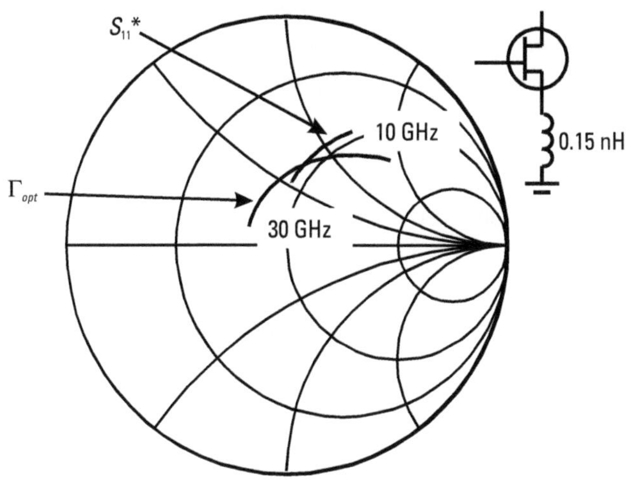

Figure 5.55 Conjugate of S_{11} and Gamma opt for a 2 × 75 μm HEMT from 10 to 30 GHz with 0.15-nH series inductive feedback. (Simulated with Agilent ADS and the PH25 Design Kit, courtesy of UMS.)

performance. A common amplifier design technique is to load the gate and drain of the transistor with resistors that limit the low-frequency gain, but a resistor connected directly to the gate adds noise to the circuit and degrades the noise figure of the overall amplifier. To avoid this degradation in a low-noise amplifier, the resistors should be decoupled in some way from the transistor. The best way to do this is to design the input-matching network to include a shunt inductive element and then place the gate stabilization resistor at the grounded end of the inductor [71]. At low frequencies, the inductive matching element has a low reactance, so the resistor is effective at reducing the transistor gain, while at high frequencies, the inductive element has a high reactance and does not reduce the gain or allow the resistor thermal noise to degrade the noise figure. Figure 5.56 shows typical matching networks for a low-noise amplifier, including the inductor decoupled gate stabilization resistor. An example 60-GHz three-stage LNA MMIC [60] is shown in Figure 5.57, where series inductive feedback has been implemented on the first stage, and reactively decoupled stabilizing resistors are used on all three stages.

Off-Chip Matching

The spiral inductors fabricated by the MMIC process have significant resistive losses, so when they are used in the input-matching network of a low-noise amplifier, they directly contribute to the noise figure of the LNA. Because of this, it is occasionally appropriate to use off-chip components within the matching circuit [72]. This technique tends to be used only at low gigahertz or RF frequencies, where bond-wire connections to off-chip components are not too reactive, and the commercial applications demand the minimum noise possible. At these lower frequencies, it is possible to purchase

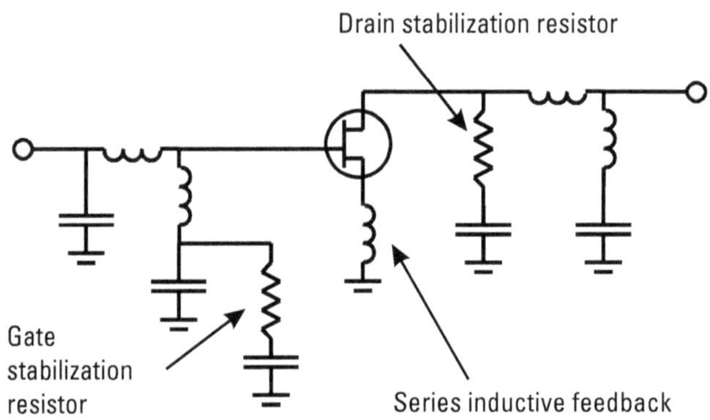

Figure 5.56 Typical low-noise amplifier matching networks showing an inductor decoupled gate stabilization resistor.

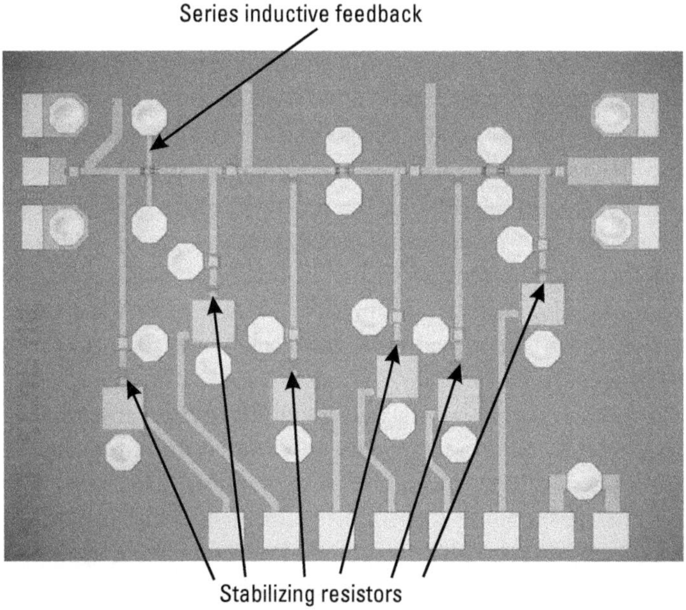

Figure 5.57 A 60-GHz three-stage low-noise amplifier MMIC. (With kind permission of Springer Science and Business Media [61].)

microminiature air-core inductors with less resistive losses than MMIC spiral inductors. Incidentally, MMIC power amplifiers operating at these low frequencies also tend to use off-chip output matching because the commercial pressure for output power and efficiency prevents the use of lossier MMIC spiral inductors [73].

5.3.2.2 Distributed Amplifiers

Distributed amplifiers [74, 75], as their name suggests, are not designed as single transistor gain stages with lumped-element matching networks but are designed with their gain distributed over a larger physical distance on the surface of the chip. The effect of this is that they have much larger bandwidths than lumped-impedance-matched amplifiers and are useful for such applications as electronic warfare, digital optical-fiber communications (SDH), and measurement instruments. Distributed amplifiers can achieve very broadband flat gain with good input and output matches over several octaves of frequency; in fact, bandwidths can be as high as 110 GHz [76].

They are designed with an input line made from series inductors and shunt capacitors that make up an artificial transmission line, and this is terminated in a matched resistor, as shown in Figure 5.58. The series inductors are made from narrow transmission lines or spiral inductors, and the shunt capacitors are the input capacitance of the transistors. With careful choice of the transistor size and design of the series inductors, the impedance of the artificial transmission line can be matched to the system impedance (normally 50Ω), and the terminating resistor is made to have the same value of resistance. In this way, the input signal travels along the artificial transmission line and into the resistor with very little mismatch, giving rise to the alternative name for this style of amplifier, the traveling-wave amplifier. The input transmission line provides the extremely wideband input match of the distributed amplifier.

The voltage of the signal across the input of the transistors is amplified by the transconductance and appears on a similar output artificial transmission line connected to the output of the transistors. The signal at each transistor output travels along the output transmission line in both directions, and with the appropriate phase response of the connecting inductors, the signals can be designed to add constructively in the direction of the output. Any signals traveling in the opposite direction are terminated in the resistor.

The frequency response of the distributed amplifier is dominated by the attenuation on the input line, and the gain of the distributed amplifier is dominated by the attenuation on the drain line [77]. The design procedure starts by considering the input and output capacitances of the available transistors, which for a HEMT device are the gate-source capacitance (C_{gs}) and the drain-source capacitance (C_{ds}), respectively. The characteristic impedance of the artificial transmission lines is given in (5.10), so when this is matched to 50Ω, the unit inductance per transistor (L_g, L_d) can be calculated.

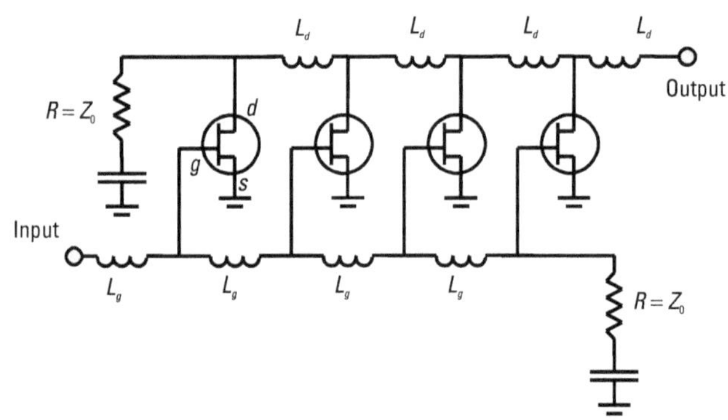

Figure 5.58 HEMT distributed amplifier.

$$\text{Characteristic impedance } Z_0 = \sqrt{(L_g/C_{gs})} = \sqrt{(L_d/C_{ds})} \quad (5.10)$$

With the unit inductance per transistor calculated, the cutoff frequency for each artificial transmission line can be determined from (5.11) and (5.12).

$$f_{cg} = 1/\pi\sqrt{(L_g C_{gs})} \quad (5.11)$$

$$f_{cd} = 1/\pi\sqrt{(L_d C_{ds})} \quad (5.12)$$

The loss or attenuation along the artificial transmission line is dominated by the parasitic resistance of the transistor, which is R_i on the input side of a HEMT and R_{ds} on the output side. Therefore, the RC cutoff frequencies of the input and output transmission lines are given in (5.13) and (5.14).

$$f_{RCg} = 2\pi/(R_i C_{gs}) = \frac{1}{2\pi(R_1 C_{gs})} \quad (5.13)$$

$$f_{RCd} = 2\pi\sqrt{(R_{ds} C_{ds})} = \frac{1}{2\pi(R_{ds} C_{ds})} \quad (5.14)$$

The gain of a distributed amplifier increases as the number of transistors along the transmission lines increases, but the gain does not increase monotonically and eventually additional transistor do not add to the overall gain [78]. This is because the extra attenuation of the transmission line exceeds the extra gain of the additional transistor. The optimum number of transistors in a distributed amplifier can be derived from (5.15) [77], where A_g and A_d are the attenuations per section in the NEPER of the gate and drain transmission lines (The NEPER is the natural log of a ratio of two amplitudes and is equivalent to 8.686 dB).

$$N_{opt} = \ln(A_d/A_g)/(A_d - A_g) \quad (5.15)$$

The attenuations per section, together with the bandwidth and low-frequency gain of the distributed amplifier, can then be derived from the transmission-line cutoff frequencies and the RC cutoff frequencies [77]. The designer will find that because of the constraints of (5.10) (i.e., the characteristic impedances are the same but C_{gs} and C_{ds} are somewhat different), the cutoff frequencies of the two artificial transmission lines are not the same. The phase velocity on this type of transmission line is a well-known function of the line cutoff frequency, so the phase velocities will be different, and the output signals will be unlikely to add up in phase at the output. This can be compensated for by adding inductors between the drain contact of the HEMT and the output artificial transmission line, as shown schematically in Figure 5.59 and in the photo of the dual-channel distributed amplifier in Figure 5.60.

Successful distributed amplifiers have been designed by this method, but, often, because HEMTs are not unilateral and there are uncertainties in the value of the HEMT model parameters, such as C_{gs}, R_i, and the like, the resulting design does not meet the specification and requires optimization of the circuit. In industry, experienced MMIC distributed amplifier designers often take a more trial-and-error approach to meeting the specification. In practice, the designer needs to select a transistor that will allow the input transmission-line cutoff frequency (f_{cg}) to be 50% higher than the highest frequency in the required specification. This choice will then determine the starting value for the transmission-line inductors, assuming a 50Ω system. The designer should then construct three-, four-, and five-stage distributed amplifiers in a microwave simulator using the selected transistors,

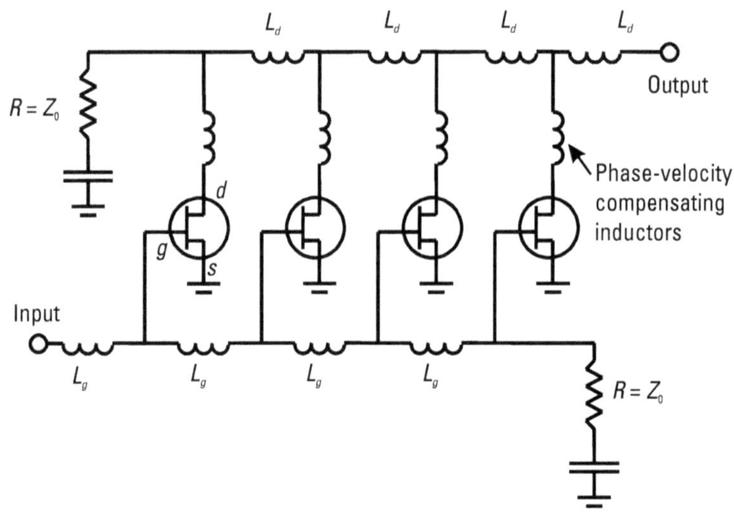

Figure 5.59 HEMT distributed amplifier with inductive phase-velocity compensation.

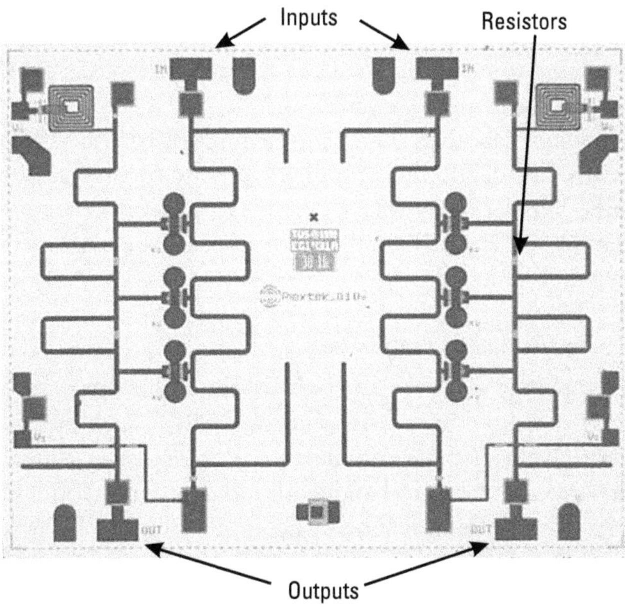

Figure 5.60 Dual 2–18-GHz distributed amplifier MMIC. (Courtesy of Plextek Ltd. and QinetiQ [79].)

transmission-line inductor, and phase-velocity compensation inductors initially set to 0. It is then necessary to optimize the three-, four-, and five-stage amplifiers to the required gain, bandwidth, and match, allowing all of the inductors to be independently varied. The most promisingly sized amplifier should then be taken forward into a more detailed design process.

As all the transistors are connected to the gate and drain transmission lines, it is common for them to be biased through them. On the gate side, this is easy because the gates draw negligible current and can be biased through the 50Ω terminating resistor, and as long as this resistor is coupled to ground with a large enough MIM capacitor, it has no effect on the distributed amplifier circuit. It is more difficult on the drain side as significant current must be supplied to the drain transmission line; thus, biasing through a resistor would be inefficient. Also, the bandwidth of the amplifier is usually very wide, so a tuned circuit would not be suitable. Common practice is to use a combination of different sizes of spiral inductors to provide enough decoupling of the RF from the dc bias over the whole bandwidth. The element nearest the RF-live transmission line is typically a small spiral inductor that reactively blocks the high frequencies, followed by a larger spiral inductor to maintain the reactance at lower frequencies, with a resistor shunting the large spiral to broaden the frequency response (lower the Q-factor) of the drain decoupling circuit.

The designs described here are manufactured with MESFET or HEMT devices, but distributed amplifiers can also be easily designed using HBT and other bipolar transistors.

An example CPW distributed amplifier MMIC operating from 5 to 55 GHz is shown in Figure 5.61.

5.3.2.3 Cascode Distributed Amplifiers

The cascode topology was invented to solve the Miller effect [81] in triode amplifiers and later used in transistor circuits [82]. The Miller effect is the apparent multiplying of the feedback capacitance (C_{gd}) in common-source FETs, which affects the gain-bandwidth product of the transistor. The cascode topology is the combination of a common-source FET followed by a common-gate FET, as shown in Figure 5.62. The common-gate FET stops the Miller effect in the common-source FET from impacting the output of the circuit and greatly reduces the feedback. The resulting combined pair of transistors has much higher output impedance, which reduces the loading on the drain line of distributed amplifiers. A distributed amplifier using cascode pairs of FETs is shown in Figure 5.63, and many examples are found in the literature [83, 84].

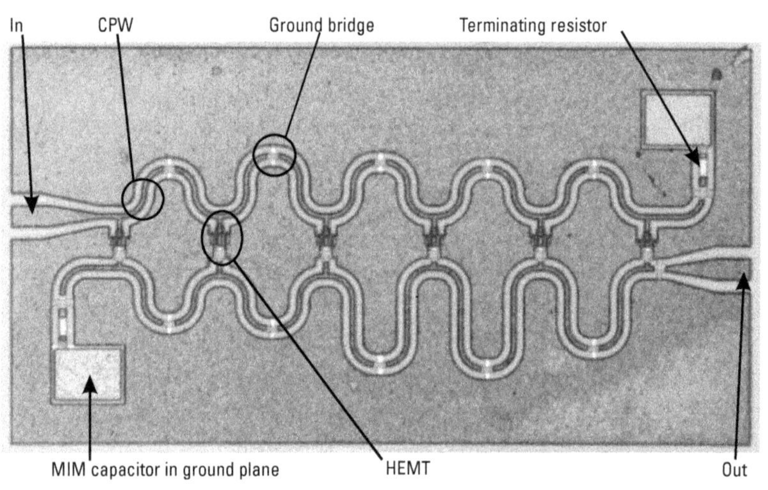

Figure 5.61 Photograph of a 5–55-GHz CPW distributed amplifier MMIC. (Courtesy of Katholieke Universiteit, Leuven/IMEC [80].)

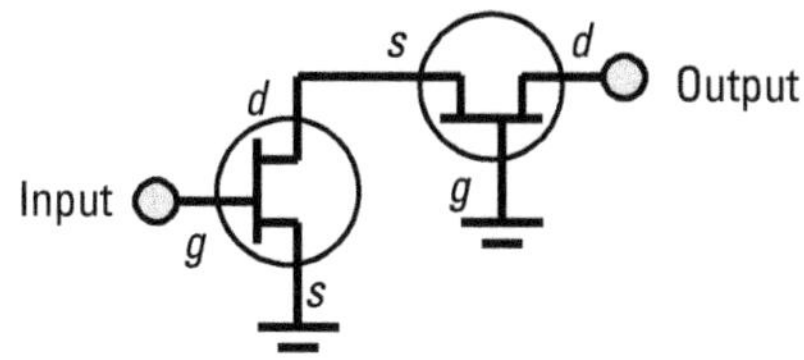

Figure 5.62 Cascode topology of common-source and common-gate FETs.

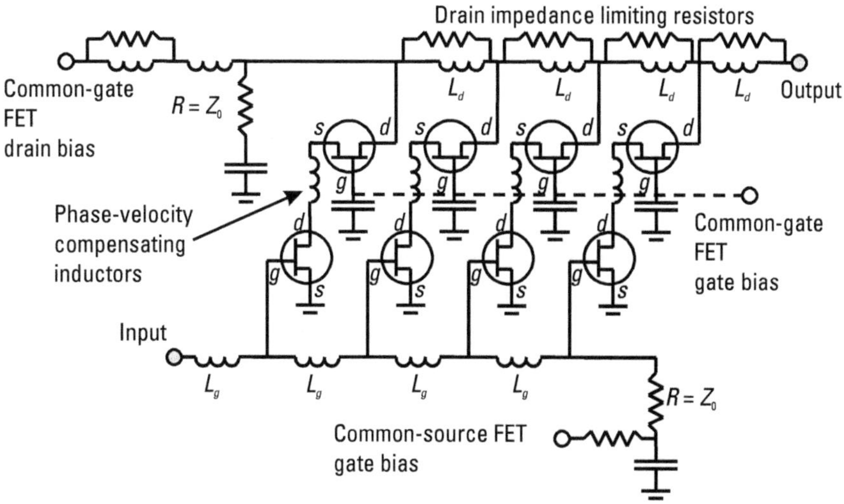

Figure 5.63 Cascode FET distributed amplifier.

Another result of the removed Miller effect is that the transistor pair has higher gain than a single common-source FET, and the gain does not roll off as quickly versus frequency.

The bias voltage to the common-gate transistor can be varied to change the gain of the pair without affecting the input match because the impedance of the common-source FET is not altered. This is utilized to make voltage-controlled MMIC amplifiers as the voltage-control feature does not upset the impedance presented to the system. The phase-velocity difference between the gate and drain transmission lines is not as great with a cascode topology, but inductive compensation is still used and commonly placed between the drain of the common-source FET and the source of the common-gate FET, as shown in Figure 5.63.

The cascode distributed amplifier can suffer from instability at the high end of its frequency band, especially if the drain transmission-line cutoff frequency is not much higher than the top of the frequency band. This is due to the drain transmission-line impedance's rising as it approaches its cutoff frequency. This can be solved by placing resistors across the drain line inductors, which limits the maximum impedance to which each section of the drain line can rise at high frequencies. These resistors are evident in the dual-channel common-source distributed amplifier in Figure 5.60.

5.3.2.4 Low-Power-Consuming Amplifiers

One of the largest and fastest-growing markets for MMICs is the mobile telephone industry, where consumers are demanding lighter and smaller handsets with longer battery life. This means that MMICs used for the RF power stage in the handset must operate from lower voltages and consume less current.

Foundries tend to offer special processes for low power consumption; for example, low-current processes can be offered that utilize shallow, highly doped, active regions to reduce the current density and give a low threshold voltage. Also, low operating voltage processes, such as HEMT processes, are available, giving good gain at 2.7V, which is typically the value from a handset battery.

Other system applications require a minimal number of supply voltages to make implementation easier, particularly when they have large numbers of MMIC to bias, as in phased array radar applications. Alternative circuit design topologies can be used, together with a low-power process, to reduce the overall power consumption of an MMIC or to reduce the total number of dc bias supplies required. Two such examples are self-bias and stack-biasing techniques.

Self-bias

FET self-bias is used to remove the need for a negative bias voltage supply to the gate of the device and allow the FET to be biased at any percentage of I_{dss} using a single bias supply on the drain. This is achieved by placing a resistor in the dc path from the source to ground, then dc coupling the gate to ground, as shown in Figure 5.64. The drain-source current flowing through the resistor raises the voltage at the source to a positive voltage, making V_{gs} a negative voltage. This technique can be extended to a diode-regulated self-bias circuit, which also suppresses the FET drain-current variations due to threshold voltage nonuniformities [85].

The drawbacks of self-bias are, first, that the bias current is now fixed and cannot be adjusted externally without affecting the drain voltage and,

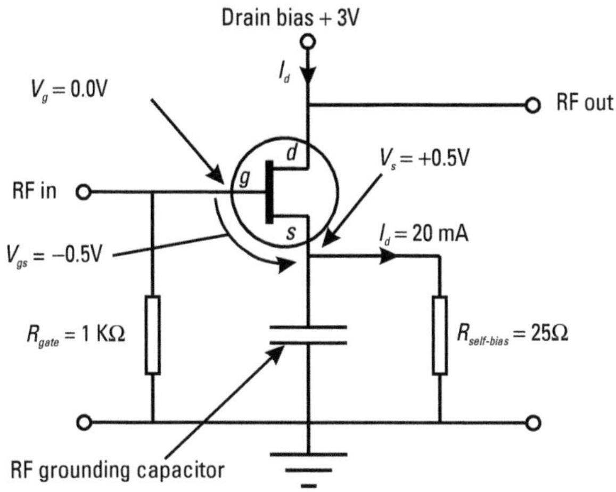

Figure 5.64 Self-bias of an FET.

second, that the fixed bias point is a strong function of the self-bias resistor value, and as this varies across the wafer or from wafer to wafer, the bias current will vary accordingly. Therefore, tight control must be placed on this resistor if the self-bias topology is to be used for product design. An MMIC with self-bias is shown in Figure 5.65.

Stack Bias

Stack bias is used to reduce the overall current consumption of a multistage amplifier by reusing the same dc current in each stage. The advantage of stack bias is that only a single drain supply is required; the drawback is that a higher supply voltage is needed. The dc current path through a stack-biased FET circuit is shown in Figure 5.66, and the RF path is shown in Figure 5.67.

This arrangement is also sensitive to resistor variation across the wafer or from wafer to wafer, so this must be tightly controlled. Great care must also be taken to ensure that the RF signal is not fed back to the first stage along the stack-biased dc path as this may well cause the gain stage to oscillate. A photograph of an MMIC using a stack-biased arrangement is shown in Figure 5.68.

5.3.3 Power Amplifiers

MMIC power amplifier design is much more complicated than small-signal amplifier design because the larger voltages and currents within the circuit

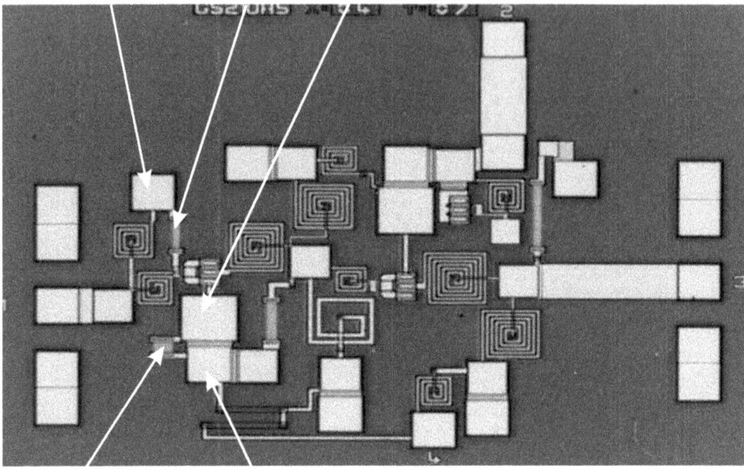

Figure 5.65 Photograph of self-biased MMIC. (*Source:* Bookham Inc., 2006. All Rights Reserved.)

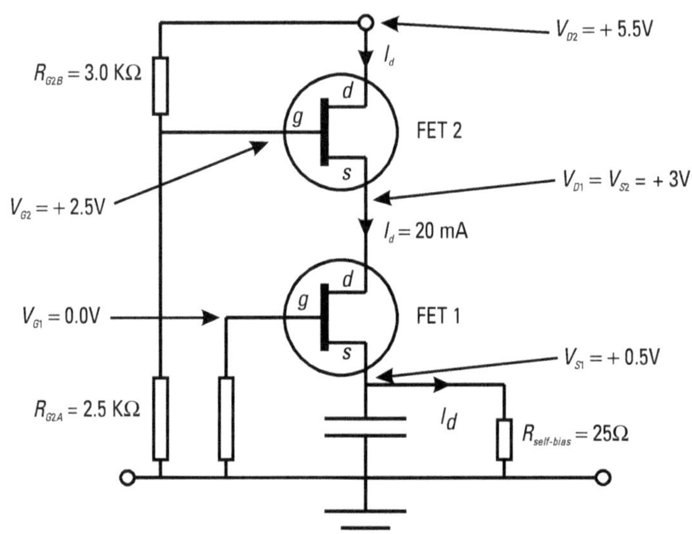

Figure 5.66 The dc current path through a stack-biased FET circuit.

can cause the MMIC components to behave nonlinearly. Simulation of a circuit with nonlinear components no longer has a unique solution and requires

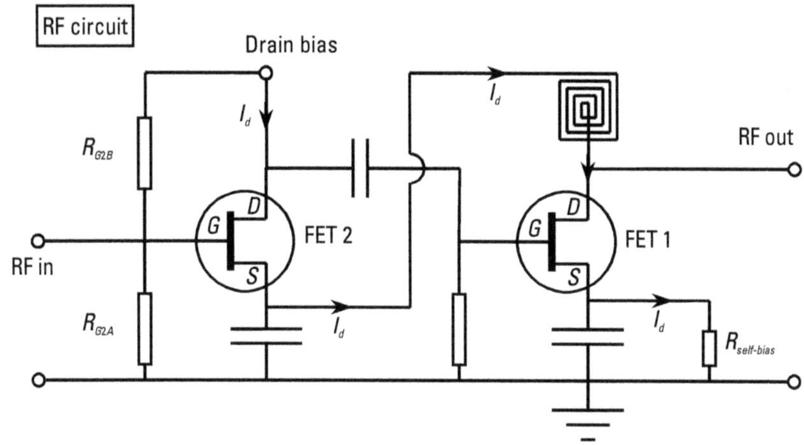

Figure 5.67 RF path through a stack-biased FET circuit.

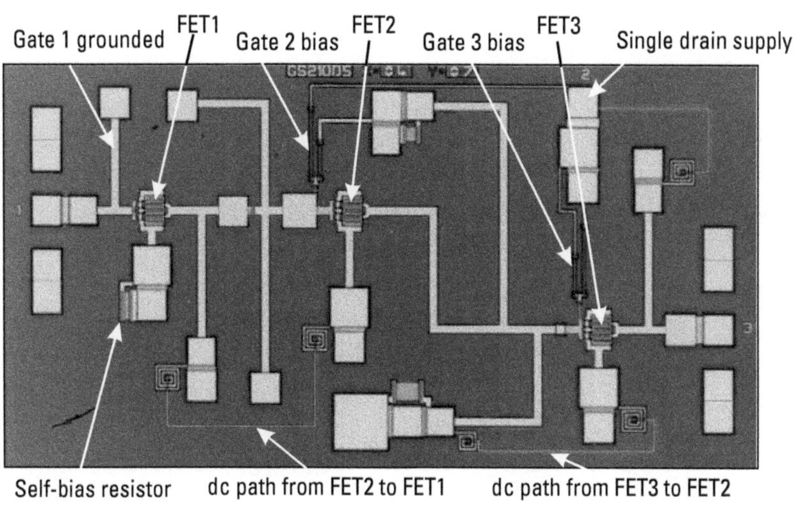

Figure 5.68 Photograph of a stack-biased MMIC. (*Source:* Bookham Inc., 2006. All Rights Reserved.)

iterative techniques, such as harmonic balance to predict the performance characteristics. This section starts by describing these large-signal effects on the active devices and goes on to describe one practical design methodology that takes account of them. A very useful reference book on power amplifier design has been published by S. C. Cripps [86].

5.3.3.1 Large-Signal Effects on Active Devices

It is helpful to start by defining what is meant by linear and nonlinear devices and what assumptions are made when referring to small-signal and large-signal operation:

- A *linear device* has properties (e.g., resistance, transconductance) that are independent of the voltage or current applied to the device.
- A *nonlinear device* has properties (e.g., resistance, transconductance) that are a function of the voltage or current applied to the device.
- Under *small-signal* operation, it is assumed that the voltage and current variation applied to the device is small enough that the device is completely linear.
- Under *large-signal* operation, it is assumed that the voltage or current variation applied to the device is large enough that the device may become nonlinear.

An example of how a metal semiconductor field effect transistor (MESFET) operates linearly under small-signal operating conditions is illustrated in Figure 5.69, where the load line is plotted at the center of the *I*/*V* (current/voltage) characteristics of the device. Under small-signal operating conditions, the variation of the gate voltage caused by the applied signal (ΔV_G) is small, and the resulting drain current variation (ΔI_D) is small; hence, the final drain voltage variation (ΔV_D) across the load resistor (R_L) is also small. Within this central area of the MESFET *I*/*V* characteristics, the

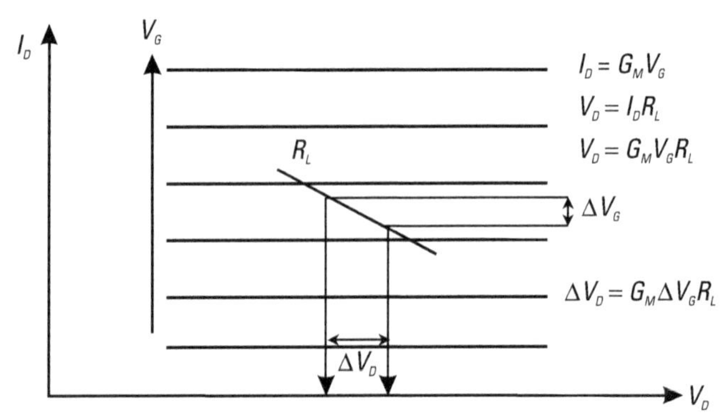

Figure 5.69 Load line plotted at the center of a MESFET *I*/*V* characteristic.

transconductance ($G_M = I_D/V_G$) is constant, so the relationship in (5.16) between the signal voltage applied to the gate and the output drain voltage is entirely linear.

$$\Delta V_D = G_M \Delta V_G R_L \tag{5.16}$$

The nonlinear behavior of the MESFET under large-signal operating conditions becomes apparent when considering the full *I*/*V* characteristics as shown in Figure 5.70. In the central or saturation region, the transconductance is primarily constant, as discussed above. However, in the so-called linear region, where the drain voltage is very low, the MESFET channel is not completely saturated, so the transconductance varies with both the gate and drain voltages. Similarly, in the breakdown region, where the drain voltage is high, the transconductance also varies with both the gate and drain voltages as the semiconductor begins to break down.

Large-signal variation on the gate of the MESFET causes large excursions of the drain voltage and current along the load line, until they reach the linear (or knee) and breakdown regions, where the transconductance is no longer constant, as shown in Figure 5.71. In this case, the transconductance is a function of both the gate and drain voltage ($G_M(V_G, V_D)$), and the overall expression relating the drain voltage to the gate voltage in (5.17) is nonlinear.

$$\Delta V_D = G_M (V_G, V_D) R_L \tag{5.17}$$

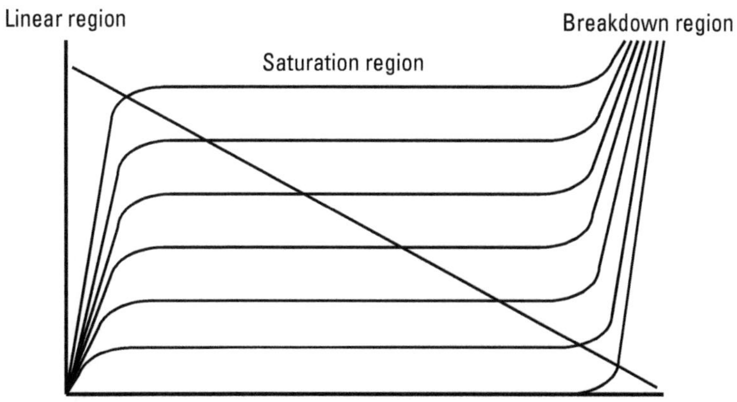

Figure 5.70 Complete plot of a MESFET *I*/*V* characteristic.

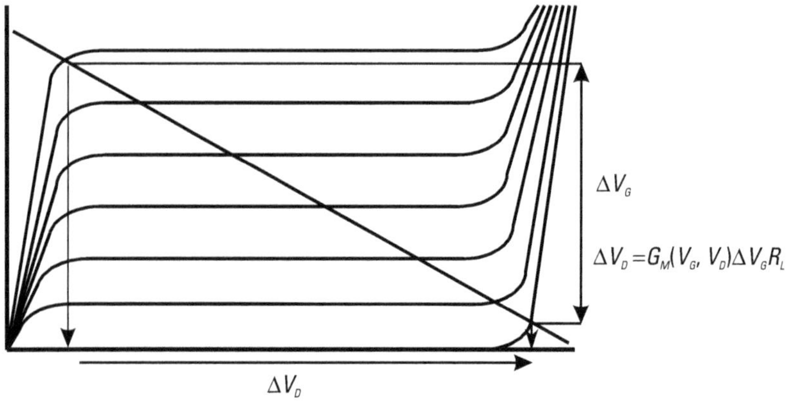

Figure 5.71 Large-signal swing along the load line of a MESFET.

The Nature of Nonlinearity

It has been seen that the MESFET exhibits nonlinear characteristics under large-signal operating conditions, but the form or nature of these nonlinearities has not been discussed. In general, nonlinearities in devices can be divided into two categories, namely, strong and weak nonlinearities. The distinction is important to understand because different design techniques and simulation tools are appropriate for the efficient design and analysis of each type.

Devices in the strong nonlinearity category are those that are mainly nonlinear in their behavior. These devices are utilized for their mainly nonlinear characteristics in components such as detectors, mixers, multipliers, and oscillators. The output spectrum from such a device typically contains many harmonics of the signal frequency at quite high levels. Commonly, the levels of these harmonics are sufficiently high to produce clearly visible distortion in the output waveform.

Devices in the weak nonlinearity category are those that are mainly linear in their behavior. These devices are used for amplifiers, where their small nonlinearities are seen as a perturbation on top of their primarily linear characteristics. The harmonic levels in such devices are low, but high enough to cause concern in certain high-linearity applications, such as wireless communications. The distortion produced by these low-level harmonics is not always apparent on the output waveform. These weak nonlinearities need to be understood, analyzed, and accounted for during conventional power amplifier design.

Analysis of Nonlinear Elements

When a voltage is applied across a linear resistor, as in Figure 5.72, the current flowing through the resistor (I) is equal to the voltage (V) divided by the resistance (R), according to Ohm's law.[4] If the voltage is a sinusoidal signal with amplitude A and angular frequency ω [$V = A\cos(\omega t)$], then the expression for the current is given by (5.18). Hence, the current in the linear resistor has the same angular frequency as the applied voltage and amplitude directly proportional to the amplitude of the applied voltage.

$$I = A/R\cos(\omega t) \tag{5.18}$$

Contrast that with the case of the nonlinear resistor in Figure 5.73, whose resistance (R) is an arbitrary function of the applied voltage (V). Whatever shapes the curve takes, the current (I) can be expressed as a polynomial expansion of the applied voltage as given in (5.19).

$$I = 1/R_1 V + 1/R_2 V^2 + 1/R_3 V^3 + 1/R_4 V^4 + 1/R_5 V^5 + \ldots \tag{5.19}$$

If the voltage is a sinusoidal signal with amplitude A and angular frequency ω[($V = A\cos(\omega t)$], it may be substituted into (5.19) to give a new expression for the current, as shown by (5.20).

Figure 5.72 Current in a linear resistor as a function of the applied voltage.

4. Ohm, G. S. (1787–1854) formulated the relationships among voltage, current, and resistance as follows: The current in a circuit is directly proportional to the applied voltage and inversely proportional to the resistance of the circuit.

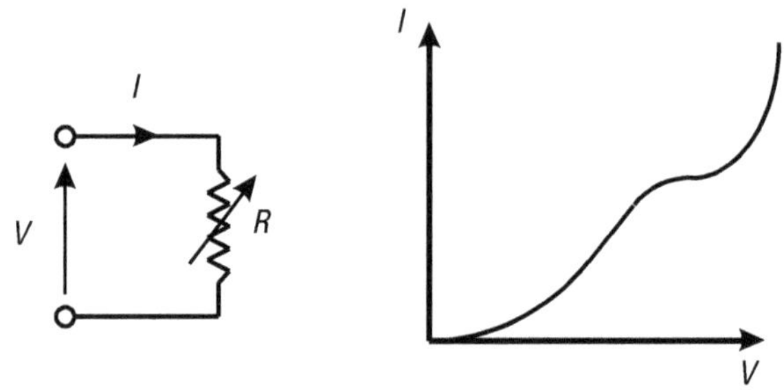

$$I = 1/R_1 V + 1/R_2 V^2 + 1/R_3 V^3 + 1/R_4 V^4 + 1/R_5 V^5$$

Figure 5.73 Current in a nonlinear resistor as a function of the applied voltage.

$$I(t) = A/R_1 \cos(\omega t) + A/R_2 \cos^2(\omega t) + A/R_3 \cos^3(\omega t) + \\ A/R_4 \cos^4 + A/R_5 \cos^5(\omega t) + \ldots \tag{5.20}$$

Using De Moivre's theorem [87], we can express the $\cos^N$ terms as the cosines of multiple angles [cos ($N\omega t$)], examples of which are given in (5.21) to (5.23).

$$\cos^2(\omega t) = 1/2\left(1 + \cos(2\omega t)\right) \tag{5.21}$$

$$\cos^3(\omega t) = 1/4\left(3\cos(\omega t) + \cos(3\omega t)\right) \tag{5.22}$$

$$\cos^4(\omega t) = 1/8\left(3 + 4\cos(2\omega t) + \cos(4\omega t)\right) \tag{5.23}$$

Consequently, it can be seen in (5.24) that the current (I) now consists of many different frequency components (ω, 2ω, 3ω, and so forth), whose amplitudes are complex functions of the original signal amplitude (A).

$$\begin{aligned} I(t) = &\left\{A^2/2R_2 + 2A^4/8R_4 + 10A^6/32R_6\right\} \\ &+\left\{A/R_1 + 3A^3/4R_3 + 10A^5/16R_5 + \ldots\right\}\cos(\omega t) \\ &+\left\{A^2/2R_2 + 4A^4/8R_4 + 15A^6/32R_6 + \ldots\right\}\cos(2\omega t) \\ &+\left\{A^3/4R_3 + 5A^5/16R_5 + \ldots\right\}\cos(3\omega t) \\ &+\ldots \end{aligned} \tag{5.24}$$

Observations of this expression for the current show some general points. First, a dc term is generated (by the even-order terms A^2, A^4, and so forth), and there are frequency components at the original (fundamental) signal frequency (ω) and at harmonics of the fundamental frequency (2ω, 3ω, and so forth). Second, the odd-order harmonics are generated by the odd-order terms (A, A^3, and so forth), and the even-order harmonics are generated by the even-order terms (A^2, A^4, and so forth). Third, the amplitude of the harmonic components are higher-order functions of the amplitude [e.g., $A^2/2R_2 \cos(2\omega t)$] than the fundamental [$A/R_1 \cos(\omega t)$]. The result is that as the input-signal amplitude is increased, more of the power goes into the harmonics, and less power is at the original signal frequency.

For a MESFET with a nonlinear transconductance, this has the following effects. There is less power in the fundamental as the input power increases correspond to compression of the device gain; the dc component produced can change the dc bias point of the device; the increasing power at the harmonic frequencies distorts the output waveform; and the generation of power at other frequencies can have serious consequences for the system performance. For these reasons, a more complex design methodology is required for effective design of power amplifiers.

5.3.3.2 Power Amplifier Design Methodology

Bearing in mind the nonlinear effects on active devices, the design methodology for power amplifier design can be broken down into three main sections: architecture design, small-signal design, and large-signal optimization.

Architecture design is the process of determining the basic layout of the circuit, such as the number and size of the active devices required, working from the specification through to a rough design. This should result in a design plan capable of meeting the gain and power specifications.

Small-signal design is the fast and simple way of designing the matching and biasing networks to achieve the basic bandwidth, gain, and input match. This is acceptable even for a power amplifier because the transistor is primarily a linear device over the majority of the load line.

Large-signal optimization is the optimization of the output-matching circuits for the best performance under large-signal operating conditions. This involves tweaking the output-matching circuit using the minimum amount possible of time-consuming nonlinear analysis to take account of the "weak" nonlinearities encountered at the extreme ends of the load line.

Architecture Design

This section describes the initial processes involved with power amplifier design: taking the requirement specification and deriving from it the basic layout and structure of the power amplifier. This involves deciding which MMIC process to use, what size of unit cell device, how many gain stages, and what power splitter and combiner techniques, as well as which bias point and efficiency mode to operate at. At the end of this stage, the designer will be able to draw a power budget diagram for the amplifier and check that it is capable of meeting the specification.

A typical requirement specification for a power amplifier will include the following:

- Frequency bandwidth;
- Transmission gain (S_{21});
- Output power at 1-dB gain compression (P_{1dB});
- Saturated output power (P_{SAT}, usually reached between P_{2dB} and P_{3dB});
- Power-added efficiency (PAE), defined as $(P_{OUT} - P_{IN})/P_{DC}$;
- Input match (S_{11}, output match is normally omitted);
- Operating temperature range;
- Linearity (expressed as a third-order intercept point referred to the output: P_{TOI}).

MMIC power process selection. The choice of the MMIC process is primarily determined by the operating *frequency* and *efficiency* required. At RF and low microwave frequencies (<10 GHz), MESFET and heterojunction bipolar transistor (HBT) processes can be used, with HBTs typically being the more efficient. At higher microwave and millimeter-wave frequencies (>10 GHz), high-electron-mobility transistor (HEMT) processes are usually required.

In each technology, it is possible to have a process optimized for power applications, so this may also influence the choice of process. These power processes are generally defined in terms of watts of output power per unit size of the active device. For MESFET and HEMT devices, the current flows under the gate and is scaled up by increasing the width of the gate, so these processes are defined in terms of the output power per millimeter width of the device gate (W/mm). In HBT devices, the current flows down through the emitter contact and is scaled by increasing the length of the emitter, so these processes are defined in terms of the output power per millimeter length of the device emitter (W/mm). Foundries may quote the output power per millimeter length of a process in terms of the saturated or 1-dB-compression output power of a device and at a Class A or B bias point, so care must be taken to clarify the process capability.

Once the process is selected, the output power per millimeter length is known, and the *output power* specification for the amplifier defines the total size of the devices required in the output stage of the design (total width of the gate in millimeters for HEMT and MESFET and total length of emitter in millimeters for HBT). For example, a 5W amplifier designed on a 1W/mm HEMT process would require at least 5 mm of total gate width in the output stage of the amplifier. Bear in mind that the output-matching circuit of an MMIC power amplifier may exhibit 0.5 to 1 dB of loss, so it is a good rule of thumb to increase the specified output power by 1 dB when making this initial calculation.

Optimum unit device size for the output stage. The total size of the output stage device has just been defined, so the question becomes whether we can have just one device of this size or if we need to use several smaller devices. To decide this, we need to look at the general rules applying to scaling of MMIC active devices:

- First, as the device size increases, the output power increases.
- Second, as the device size increases, the gain decreases.

To understand the trade-off between these effects, we need to look more closely at the basic device characteristics and how they scale with increasing device size.

The output power (P_{out}) is proportional to the device size because it is a function of the drain voltage (V_{drain}), the drain current (I_{drain}), and the efficiency (η). The drain voltage is limited by the breakdown voltage, which is itself a function of the substrate material and the device layout and is fixed for each process. The drain current is a function of the device structure, channel doping density, and so forth, and is proportional to the width of the gate. The efficiency is a function of the device type (i.e., HBT or HEMT) and is fixed for each particular process. Therefore, the output power is a constant multiplied by the total width of the device, as given in (5.25), so it is proportional to the device size.

$$P_{out} = \eta \, V_{drain} \, I_{drain} = \text{constant} \times \text{gate width} = X\,\text{W/mm} \qquad (5.25)$$

Conversely, the device gain is inversely proportional to the device size because the increased parasitics associated with the larger device degrade the device gain. These large FET parasitics include the gate-source capacitance, the source inductance, the phase errors along the gate, the phase errors between the gate fingers, and the thermal effects.

Large power FETs are constructed by increasing the width of the individual gate fingers and using multiple gates, as shown by the FET in Figure 5.74, which has six gate fingers, each of which is 120μm wide. The gate-source capacitance is the capacitance between the gate feed and the source feed metallization and increases with the number of gate fingers being fed. This increased capacitance lowers the input impedance of the device and causes the device gain to roll-off more quickly with frequency.

The source inductance is the inductance produced by the current path from all the channels along the source feed metal and down through the through-GaAs vias to the back-face ground plane. As the number of gate fingers increases, the increasing distance between the central gate fingers and the source vias at the side of the device produces significantly more inductance.

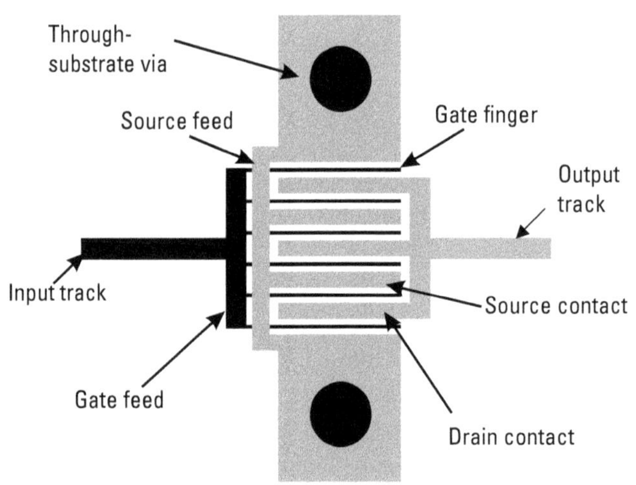

Figure 5.74 Schematic diagram of a large power FET device.

Inductance in the source path of a transistor has the effect of producing negative feedback and directly reduces the transistor gain.

Phase errors along the individual gate fingers are produced when their dimensions approach a significant fraction of the wavelength of the signal traveling down them. When this happens, the current flowing under the gate near the gate feed metal may be slightly out of phase with the current flowing under the far end of the gate. If the currents from all the devices gates do not add up in phase, the effective transconductance is reduced and the device gain is decreased.

Similarly, phase errors between gate fingers are produced when the device becomes large because the distance between the input signal track and the central gate fingers is different from that between the input signal track and the outer gate fingers. Again, the out-of-phase current components reduce the device's effective transconductance and lead to decreased device gain.

Large FET devices also suffer from thermal effects because the transconductance is an inverse function of the channel temperature. Each FET channel produces heat because the dc bias power applied to the device is not completely converted to RF output power (i.e., the efficiency is not 100%). The heat produced raises the temperature of the channel until an equilibrium is reached with the surrounding environment. If many gates are placed close together, as in the case of a large power FET, the channel temperature can be very high, and the resulting semiconductor transconductance is reduced, leading to lower device gain. An example power amplifier MMIC using six large transistors is shown in Figure 5.75.

Therefore, all these parasitic effects of large power FETs add up to the other general rule that as the device size increases, the device gain decreases. In other words, the device gain is inversely proportional to the device size.

Bear in mind that the influence of these parasitic effects is also a function of frequency. At frequencies below 1 GHz, the phase errors become insignificant, and only the thermal effects need to be considered. At these frequencies, two or more smaller output devices may be used instead of one large device for thermal-management purposes. At frequencies between 1 and 30 GHz, all the parasitic effects are significant and will influence the design. At frequencies above 30 GHz, device size tends to be limited by frequency response, so the thermal-management problems become less important.

We can now choose the optimum unit device size for our output stage on the understanding that there is a trade-off between output power and

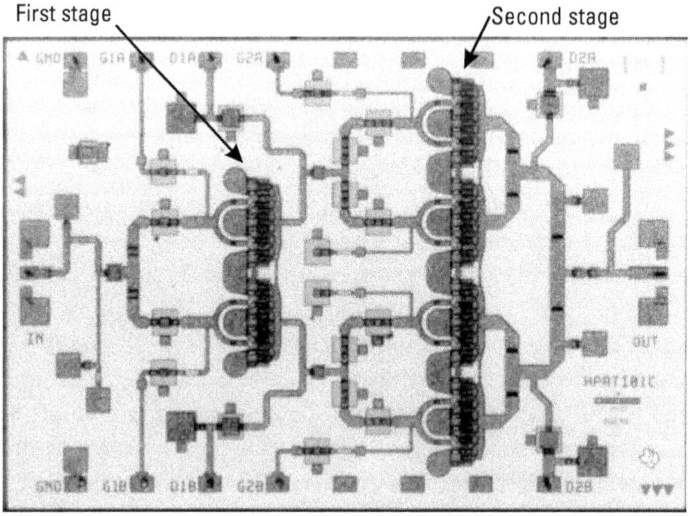

Figure 5.75 High-power amplifier MMIC using six large transistors. (Courtesy of Alcatel Alenia Space France.)

device gain as the device size increases. The optimum device is one that gives the maximum amount of power and still has useable signal gain over the specified frequency range. A good rule of thumb is to select the largest device that still exhibits a G_{max} of 10 dB at the high end of the specified frequency range, as indicated in Figure 5.76. This typically results in an amplification

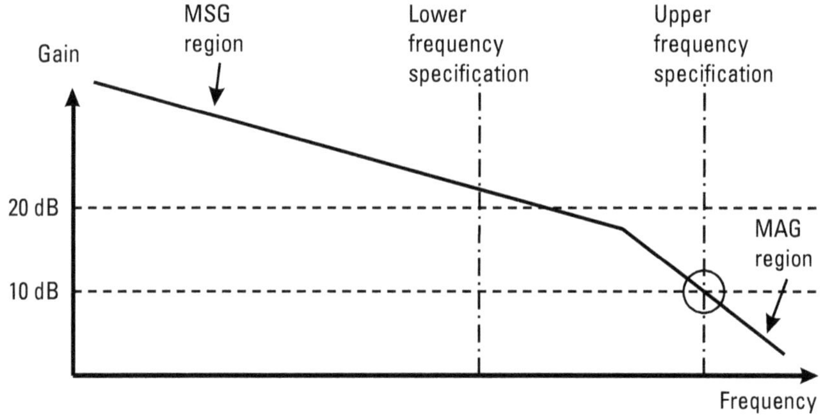

Figure 5.76 Plot of maximum gain versus frequency (G_{max}) for an optimum-sized output device for a power MMIC.

stage with enough gain to make an efficient output power stage. The relationship between the stage gain and stage efficiency is described in the next section.

The selected unit device size has a known and fixed output power level, so if the specification requires more power than a single device can provide, more than one unit device must be used in parallel in the output stage. When this is the case, power splitters and power combiners must be used to divide the input signal between the output devices and recombine it when it has been amplified by the output device.

Number of gain stages. The gain per stage in an amplifier is typically around 10 dB, but higher gain per stage is desirable for two reasons. First, the higher the gain per stage, the fewer the stages required, and the design complexity can be significantly reduced. Second, the overall amplifier efficiency is improved if fewer higher-gain stages are used.

It is worth, at this moment, to consider the definition of, and the factors influencing, power amplifier efficiency. The PAE is a measure of how efficiently the design converts the dc bias power into additional power at the RF frequency. This is defined in (5.26), where P_{in} is the input RF power, P_{out} is the output RF power, and P_{dc} is the power applied as the dc bias.

$$\text{Power added efficiency } (PAE) = (P_{out} - P_{in})/P_{dc} \tag{5.26}$$

The output RF power (P_{out}) is the input RF power (P_{in}) multiplied by the power gain (G) of the amplification stage, as given in (5.27).

$$\text{RF output power } (P_{out}) = GP_{in} \tag{5.27}$$

High power gain in an amplification stage improves efficiency because, at the limit of infinite stage gain, P_{in} tends to 0, and the power-added efficiency reduces to simply the output RF power divided by the dc bias power, as shown in (5.28).

$$PAE(\text{when } G \text{ is very high, } P_{in} \to 0) = P_{out}/P_{dc} \tag{5.28}$$

The expression in (5.28) is also sometimes referred to as the drain efficiency of an amplification stage as it relates how efficiently the dc drain bias is converted to RF output power from the drain port of a single-stage FET amplifier.

The reason that fewer gain stages improves the overall amplifier efficiency can be understood by considering the two-stage amplifier shown schematically in Figure 5.77. We know from the above discussion that the output power (P_{out}) is defined by the size of the output gain stage and that the efficiency of this output stage increases as it exhibits more power gain. Thus, peak efficiency is achieved when all the gain is in the last stage; in other words, there is only one gain stage and the PAE is the output power (P_{out}) divided by only the output stage dc bias (P_{dc2}), given by (5.29). Any additional gain stages before the output gain stage increase the overall amplifier gain but do not increase the output power capacity of the output stage. These additional input gain stages require their own dc bias power and increase the overall dc power consumed by the amplifier, as indicated by (5.30), and decrease the overall amplifier efficiency. The gain specification for a power amplifier normally requires more than one stage of gain, so the ideal efficiency cannot always be obtained; in general, however, the number of gain stages should be minimized to maximize the overall efficiency.

$$PAE = P_{out} / P_{dc2} \tag{5.29}$$

$$P_{dc} = P_{dc1} + P_{dc2} \tag{5.30}$$

The next issue that the designer of a multistage power amplifier must be aware of is the relationship between the efficiency of the amplifier stage and the compression of the stage gain. Figure 5.78 shows the power-compression curve for a power amplifier. At low input-power levels, the amplifier is linear, and the output power increase is directly proportional to

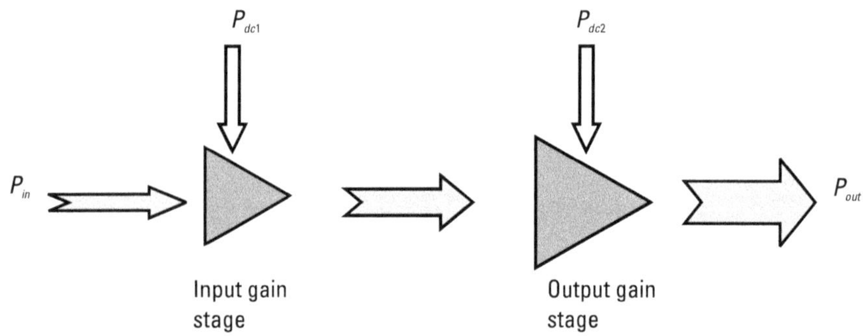

Figure 5.77 Schematic diagram of a two-stage amplifier.

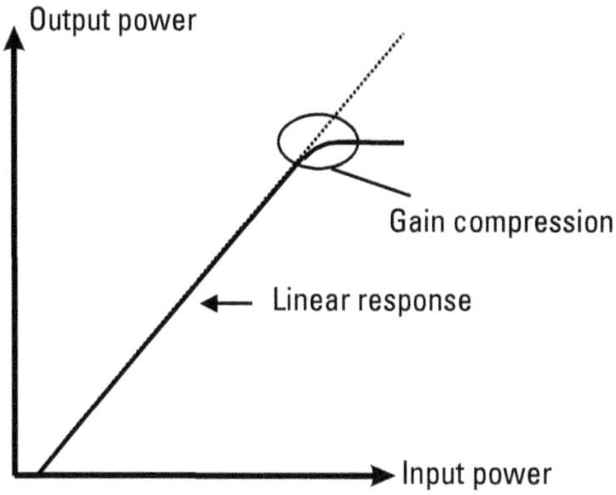

Figure 5.78 Plot of output power versus input power for a power amplifier.

the input power, according to the device small-signal gain, which is represented by the gradient.

As the input power reaches a certain point, the voltage and current swing along the load line reaches the "knee" and "breakdown" regions, and the behavior of the device transitions from purely linear to nonlinear. As discussed in Section 5.3.3.1, some power at the fundamental RF frequency is transformed to power at other harmonic frequencies, and the device gain at the fundamental frequency is reduced or compressed. This corresponds to the point in Figure 5.78 at which the gain deviates from the linear response. The output power at the point on the curve where the gain of the overall amplifier is reduced by 1 dB is known as the P_{1dB} compression point. The maximum power output from the amplifier is know as the saturated output power (P_{SAT}) and occurs typically at between 2-dB and 3-dB gain compression.

Power amplifiers are commonly specified to have high efficiency at the P_{1dB} compression point. The question can arise as to which stages should be operating within their linear region and which should be operating with gain compression.

The preceding paragraphs have shown that the best overall amplifier efficiency is dominated by the efficiency of the output stage. The peak efficiency in the output stage occurs when, for a given dc supply to that stage, this stage is producing the most power, in other words, when the input power to this stage is high enough that the gain is compressed by 1 dB. Thus, output-stage efficiency increases with the level of power it is handling until it reaches its P_{1dB} compression point.

The overall gain compression is the sum of the compression of each stage, so if earlier gain stages compress, then the overall amplifier reaches the P_{1dB} compression point before the output stage is fully compressed by 1 dB. If the large output stage is not operating at its own P_{1dB} compression point, then some of the dc power in the output stage is wasted, and the PAE of the overall amplifier at P_{1dB} is reduced. Therefore, for peak efficiency, the power amplifier should be designed to exhibit all of the gain compression in the output stage and have all the preceding input stages operating within their linear region.

The choice of the device size for the input stages is usually a trade-off between several factors. High gain per stage is an advantage, so a small device might seem preferable. However, for peak efficiency at the P_{1dB} compression point, all the gain compression must be in the last stage, so the preceding stages must not be so small, or they will start to compress. The preceding stages must be able to supply the input power to the last stage while it is operating within its own linear region (i.e., preferably with output powers 3 dB below its own P_{1dB} compression point).

Taking all this into account, the gain requirement will define the number of gain stages necessary.

Power splitting and combining. Having decided on the number of devices in each amplification stage, as well as the number of stages, we need to determine what technique to employ to split and recombine the RF/microwave power between them. This section describes a range of techniques that are currently used on, or could be applied to, power amplifier MMICs. There are numerous techniques for splitting and combining power off chip using alumina hybrid structures, but these are referred to only if the technique has the potential for monolithic implementation in the future. Only passive splitters and combiners are covered as active structures are inefficient and suitable only for small-signal applications. In general, the same structures are used for splitters and combiners, so the two components are considered separately only where appropriate.

The most important attributes of all these types of combiners are insertion loss, physical size, frequency bandwidth, bias compatibility, and effect on odd-mode stability. Equal phase of the signal at the split ports is also important, but good symmetry of the splitter can ensure this, and reproducible symmetry can be readily achieved with a photolithographic process. If layout constraints do not allow perfect symmetry, K. B. Niclas [88] has recommended keeping phase deviations between ports below ±10°. The signal loss through the combiner is particularly important after the last gain stage, and this must be minimized to maintain good efficiency in the amplifier. Size must also be kept to a minimum as the cost of the chip is proportional to its size, although some combiners make use of long transmission lines to maintain equal phase from devices with large spatial separation. The maximum bandwidth of the combiner is important, and also whether it includes, or requires separate, device matching. Power amplifiers have demanding dc bias requirements, and it is an advantage if the bias can be fed via the combiner to the devices, eliminating the requirement for a separate bias T. Finally, combining structures can be large, approaching a significant proportion of the signal wavelength, so their influence on odd-mode stability can be important. In this section, the performance advantages offered by each type of combiner in each of these areas are compared, and trade-offs between the combiner types are analyzed.

The types of combiner discussed below are parallel devices, synthesized transformer networks, parallel matching networks, and bus-bar combiners. Other passive networks are often used, such as Wilkinson combiners, Lange couplers, traveling-wave combiners, distributed transmission-line combiners, and Dolph-Chebychev tapered-line combiners, all of which were discussed in previous sections. An example power amplifier using Wilkinson splitters and combiners is shown in Figure 5.79.

Parallel Devices

These are simply two or more devices connected up in parallel by transmission lines, as shown in Figure 5.80. This approach is only really useful at low RF frequencies (<1 GHz), where the phase difference between the device feeds is a sufficiently small fraction of the transmission-line wavelength. This technique is mainly applied in order to separate devices physically for thermal management within the MMIC. The main disadvantage of this approach is that the resulting impedance at the common node is very low and hard to match over a broad frequency range.

Synthesized Transformer Networks

The synthesized transformer network [89] consists of a network of lumped and distributed components similar to that shown in Figure 5.81. This technique uses filter synthesis programs to design a circuit with a band-pass response, sections of which are split symmetrically in order to form the required number of ports. Network transformations are applied to the prototype design to get a circuit that is then translated into a microstrip and planar

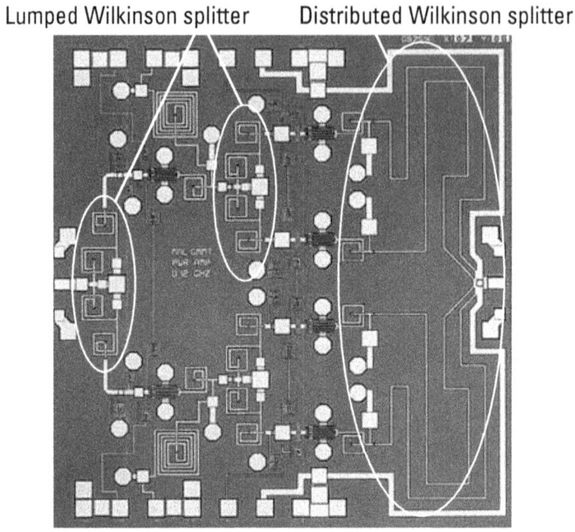

Figure 5.79 Power amplifier MMIC using lumped Wilkinson splitters on the input and a distributed Wilkinson combiner on the output. (Courtesy of TM R&D, Malaysia.)

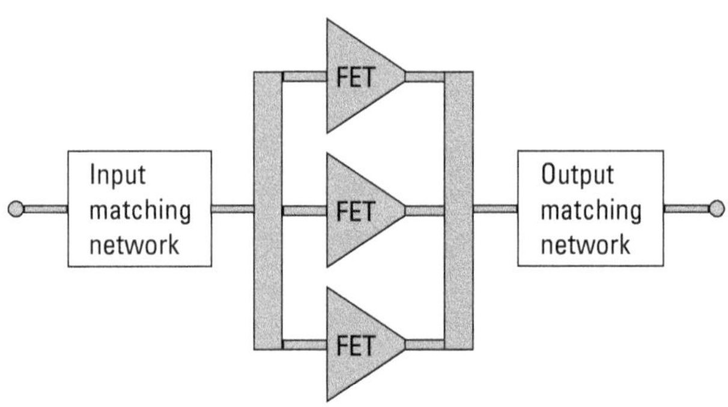

Figure 5.80 Parallel devices.

capacitor layout. This method can achieve a flat response over a very broad bandwidth (100%) and require little further device matching, although the design methodology is fairly complex.

Parallel Matching Networks

Parallel matching networks consist of lumped and distributed elements, like synthesized transformer networks, but are realized by a different method.

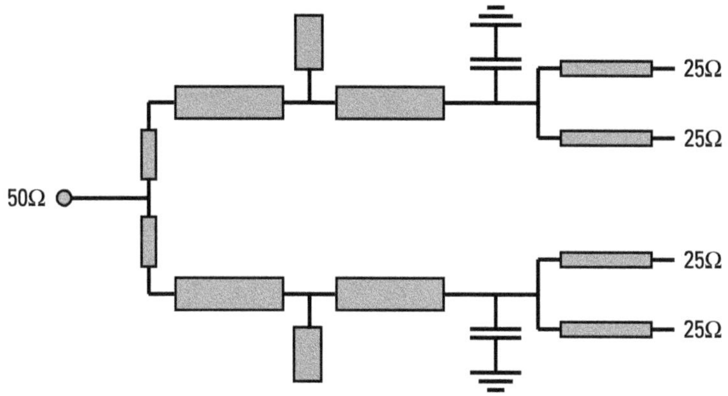

Figure 5.81 Synthesized transformer network.

The design procedure starts by considering the various matching configurations that will produce a good match to an individual device over the required frequency range. If there are to be four devices in the output stage, then each matching network should match the devices to 4 × 50Ω (i.e., 200Ω) as they will eventually be combined in parallel. It is normally possible to find input and output-matching networks for large power devices that use series inductance and shunt capacitance elements, like those shown in Figure 5.82. At this stage, it is also useful to choose a network with shunt inductance elements close to the active device through which the dc bias can be supplied.

Two of these networks can be placed side-by-side, as in Figure 5.83, and their common nodes can be connected together with "ideal" wires, as in Figure 5.84. In reality, the series inductance elements can be formed from narrow microstrip transmission lines, where their physical length can be used to bring the matching elements to common nodes.

The matching elements that are now in parallel, shown by the dotted lines in Figure 5.85, can be transformed into single elements by standard

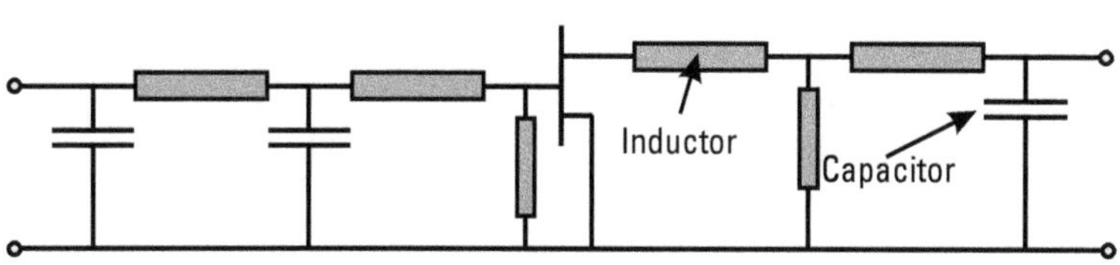

Figure 5.82 Typical impedance matching network.

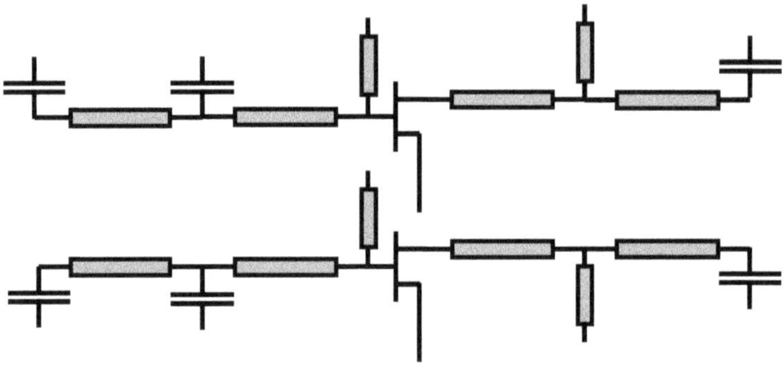

Figure 5.83 Two devices with matching networks in parallel.

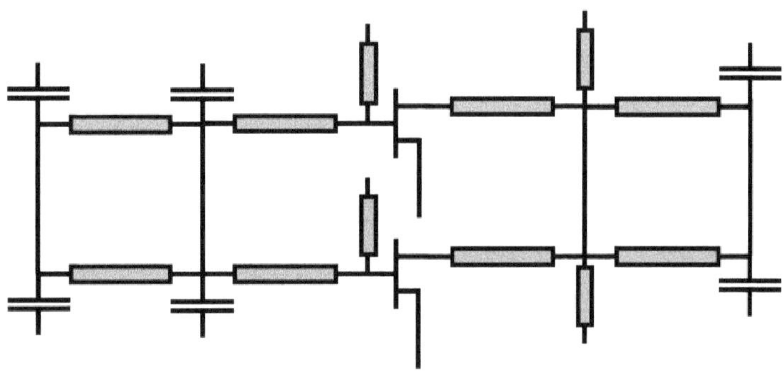

Figure 5.84 Matching networks connected together at common nodes.

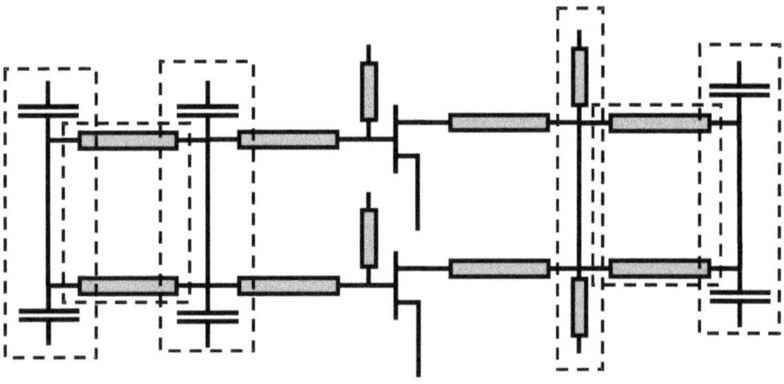

Figure 5.85 Parallel matching elements that can be combined into equivalent single elements.

network transformations to leave the circuit shown in Figure 5.86. This circuit now consists of two active devices with single input and output ports and must be reoptimized to the 100Ω (2 × 50Ω) match condition using a linear circuit simulator. In the same way, two of the circuits in Figure 5.86 can be brought together in parallel to create the matching network for four devices, as shown in Figure 5.87, and the input and output matches will now be close to 50Ω.

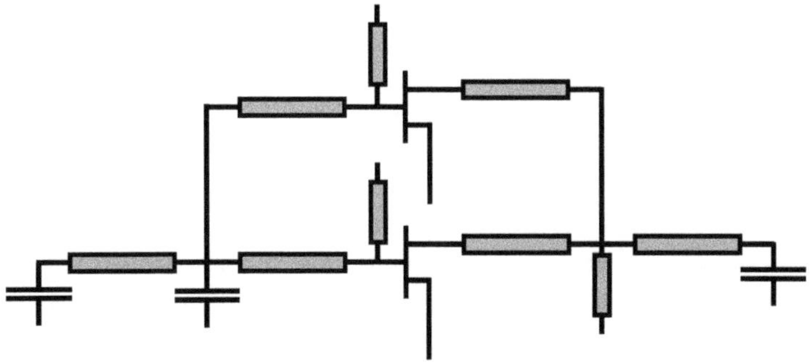

Figure 5.86 Matching network for two devices with single input and output ports.

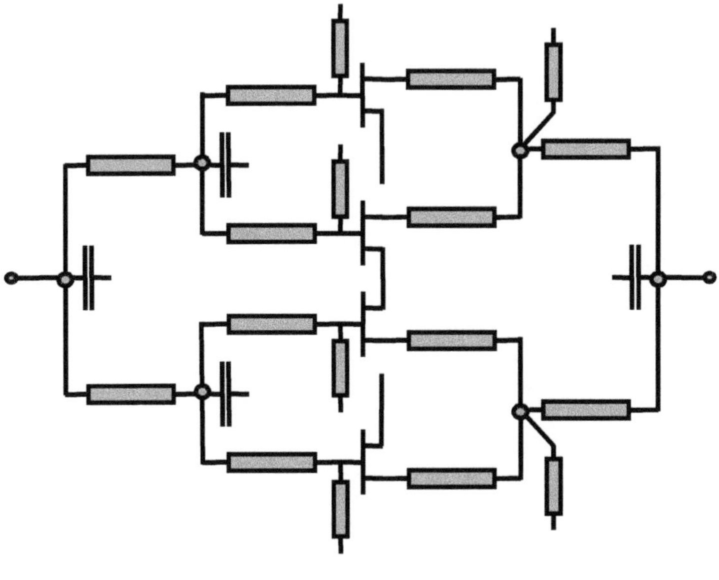

Figure 5.87 Parallel matching network for four devices.

This arrangement is very compact as the matching, biasing, and combining is performed by the one network and is also less lossy than using separate matching and combining. The main drawback is that it has only a moderate bandwidth (typically 20%). An example of a K-band power amplifier designed using this method is shown in Figure 5.88. The thin microstrip transmission lines can be seen splitting the power between device inputs. Where a low series inductance is required, as for the output matching of the last stage, the lines are broadened to facilitate the physical connection with the minimum of inductance.

Bus-Bar Combiners

In this type of combiner, shown schematically in Figure 5.89, all the outputs of the last stage devices are connected together by a wide bus-bar, which feeds dc current to all the devices. Shunt matching elements are placed at symmetric points on the bus-bar, and RF power is tapped off at other symmetric points along the bus-bar. The RF power at these different points is combined together by microstrip lines and shunt capacitors, which act as series matching elements, as well as the RF power combiner.

The effect of the bus-bar on the RF is negligible, assuming that all the output devices are fed in phase and the distance between device outputs is a

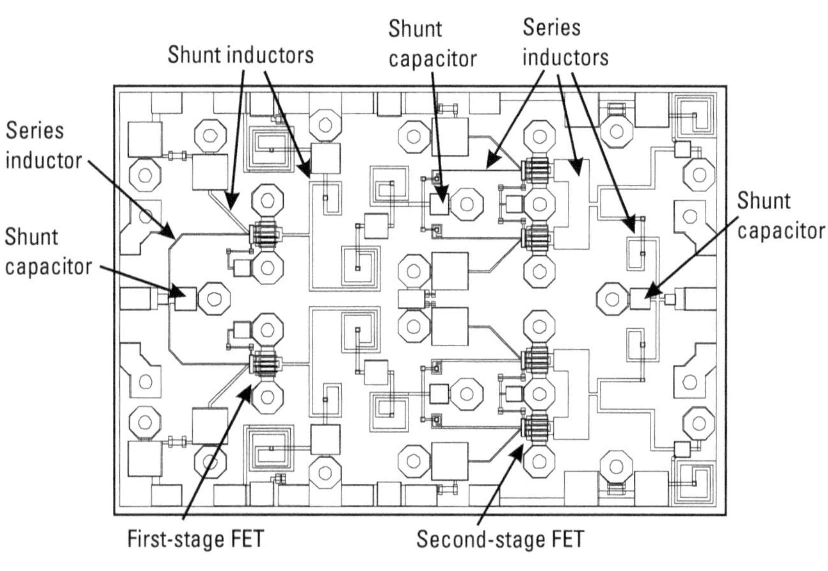

Figure 5.88 K-band power amplifier designed using parallel matching networks. (*Source:* Bookham Inc., 2006. All Rights Reserved.)

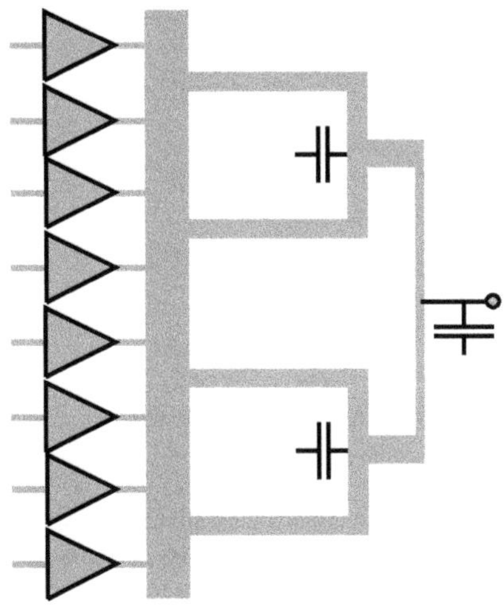

Figure 5.89 Bus-bar combiner.

small enough fraction of the signal wavelength. When this is the case, no RF current flows between the devices, and the point on the bus-bar midway between devices can be seen as a virtual open circuit. The design of such a combiner starts with building the output stage as a set of symmetric subcells, with two devices in each cell and shunt and series matching elements, as shown in Figure 5.90. The subcells may then be connected in parallel at the virtual open circuit points to form larger subcells or macrocells, as in Figure 5.91. At this stage, the two matching components where the subcells joined can be transformed into their equivalent single component.

Two macrocells can be joined in the same way, and the series elements can be combined with shunt capacitors to form the RF combiner, shown in Figure 5.89.

One bus-bar combiner from an HBT MMIC has been described by M. McCullagh et al. [90]. An example of a large FET bus-bar combiner can be seen after the last stage of the MMIC detailed in [91], and this is shown in

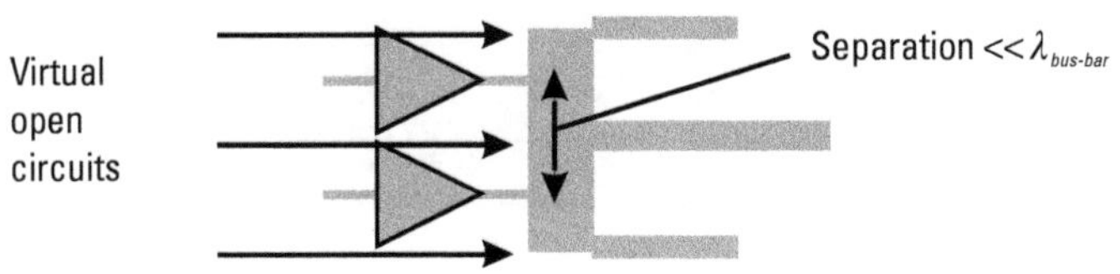

Figure 5.90 Bus-bar symmetric subcell with two devices.

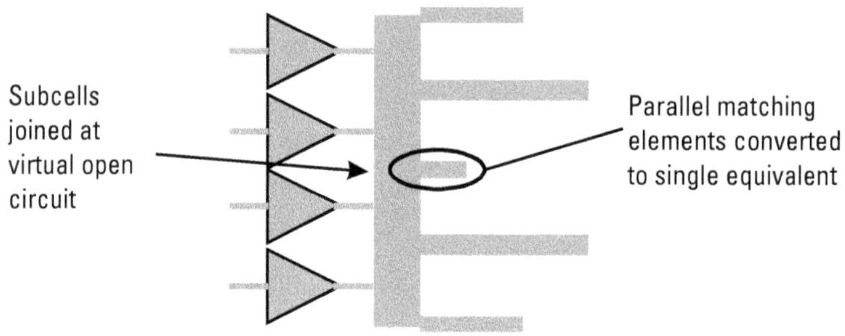

Figure 5.91 Bus-bar macrocell made from two subcells.

Figure 5.92. This design breaks the twelve last stage devices into four subcells consisting of three devices each. The 100-μm-wide bus-bar can be seen running the full width of the chip after the row of last stage devices. This sort of bus-bar connection can also be used at the input side of the gain stages, and the lower dc current levels at this point mean that narrower tracks can be used.

The most obvious advantage of this type of combiner is the ease of providing the high dc current to all the output devices from either side of the chip. Furthermore, this direct bus-bar connection between devices usually eliminates all odd-mode instability [92]. These structures tend to be quite

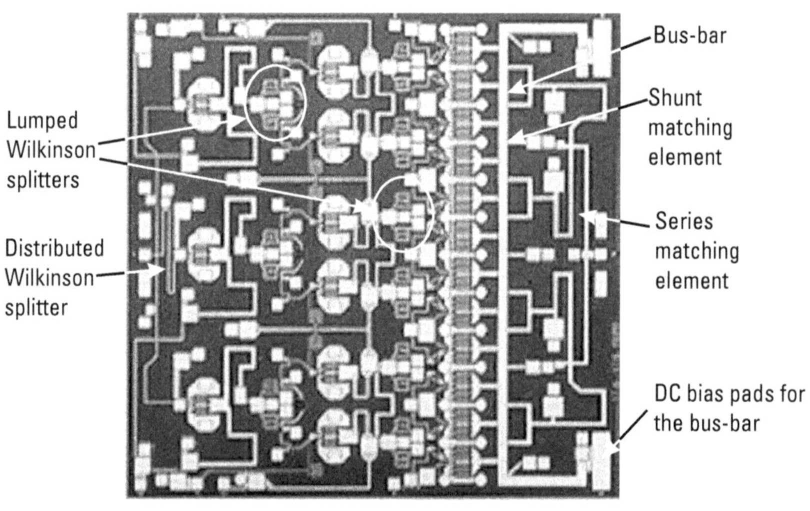

Figure 5.92 X-band 5W power amplifier using a bus-bar combiner after the output stage. (*Source:* Bookham Inc., 2006. All Rights Reserved.)

compact and to exhibit moderate bandwidths (20%). The disadvantages are that 2.5D or full 3D electromagnetic (EM) simulation is required to achieve full design confidence in the combiner, and shunt inductance matching is needed on the ends of the bus-bar for easy bias feeding and decoupling.

In summary, the most commonly used MMIC power splitters and combiners are the Wilkinson, bus-bar, and parallel matching networks, and this is likely to remain the case for the foreseeable future because these are the easiest to design and optimize using the standard simulators available today. Some new and interesting coupling techniques are still being developed, such as the three-way planar divider described by J. Lim and S. Eom [93], and these are likely to provide useful alternatives for the power amplifier designer to use over the next few years.

The Lange couplers, distributed transmission-line combiners, and traveling-wave combiners, while capable of monolithic implementation at millimeter-wave frequencies, are still generally used in an off-chip hybrid arrangement. As the demand for volume products offering high output power in the millimeter-wave frequency range increases, these techniques are likely to be developed further and implemented on a wider scale.

The most interesting broadband techniques are the synthesized matching networks and the Dolph-Chebychev tapered transmission-line approaches. As with any very broadband design technique, these methods require a more careful understanding of the circuit synthesis software, over and above the use of standard simulators.

Power Budget

Once the size of the devices, the number of gain stages, and the splitting/combining techniques have been chosen, a power-budget diagram for the power amplifier can be drawn up. This is a map of the RF power levels at every point throughout the power amplifier architecture, and an example for a two-stage power amplifier is shown in Figure 5.93.

The following general assumptions can be made when drawing the power-budget diagram:

- The gain of the device is its G_{max} value at the highest specification frequency.
- The gain of last stage is the same $G_{max} - 1$ dB.
- There is a 0.5-dB loss in each matching circuit and splitter or combiner (if constructed separately).

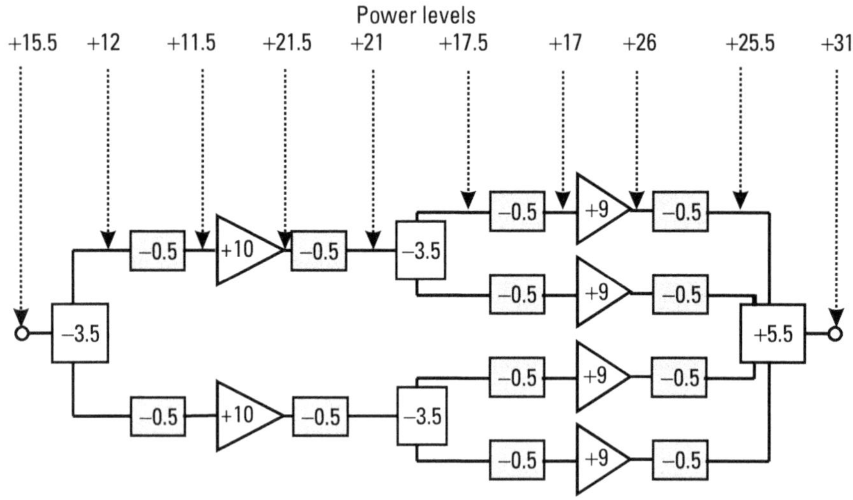

Figure 5.93 Power-budget diagram for a two-stage power amplifier where the gain and loss are in decibels and the power levels are in dBm.

The power-level values are then added by working back from the required output power. The overall chip gain and the power levels at the active devices can then be checked against the specification and foundry recommendations throughout the circuit. For example, using the power-budget diagram in Figure 5.93, the overall chip gain at P_{1dB} is equal to the output power minus the input power (31 − 15.5), which equals +15.5 dB. The diagram also indicates that the output-stage devices must be able to produce +26 dBm[5] with only 1 dB of gain compression and that the input-stage devices must be able to output +21.5 dBm with negligible gain compression (i.e., to operate within their linear region, or 3 dB below their specified P_{1dB}).

If the overall chip gain is too low, or the active devices cannot produce the required power with an acceptable amount of gain compression, the architecture needs to be changed. In this way, power-budget diagrams are a useful tool to examine the trade-offs between different power amplifier architectures.

5. dBm expresses power levels referenced to 1 mW power: $(\text{dBm}) = 10 \log \dfrac{\text{power(mW)}}{1 \text{ mW}}$.

Small-Signal Design

Once the architecture of the power amplifier is defined, the matching circuits for the active devices must be designed and/or incorporated with the splitters and combiners and dc bias networks. As discussed at the beginning of this chapter, most power amplifiers can be viewed as linear amplifiers with the output-stage matching adjusted for the weak nonlinearities at the P_{1dB} compression point, so the matching circuits are initially designed by linear small-signal techniques. This is a fast and simple way of designing the matching and biasing networks to achieve the basic specification of bandwidth, gain, and input match.

This section assumes the design uses a Class A bias point, but the bias point and matching circuits can be designed specifically for operation in higher-efficiency modes. The Class A bias point is the standard operating bias point for a power MESFET and is at 50% of the drain-source saturation current (I_{DSS}), with the drain voltage midway between the knee and breakdown, as shown in Figure 5.94. Class A bias is usually the bias point at which the power rating of the process (W/mm) is quoted by the foundry. Under Class A operation, drain current flows during the full period of the fundamental frequency cycle and is sinusoidal, making this a very linear operating point. The theoretical maximum efficiency for this class is 50%, so if higher efficiency is required, then other operating modes must be considered. Some of the trade-offs between the different modes are given in Table 5.1, and further information can be found in Cripps [86].

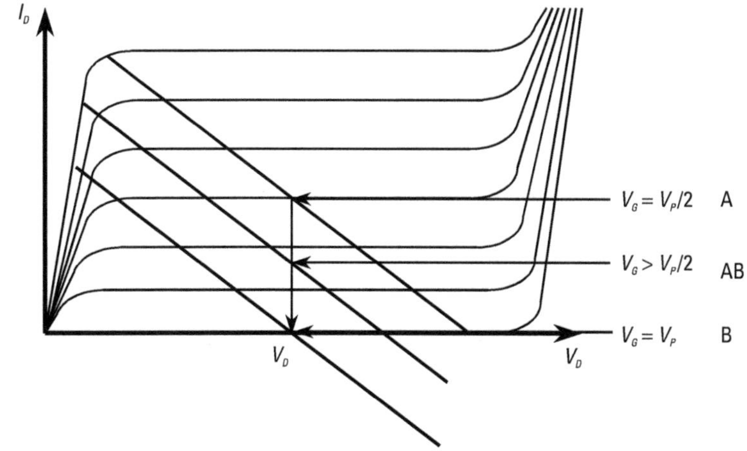

Figure 5.94 Class A, AB, and B bias points.

The starting point for the small-signal design is to obtain the *s*-parameters of the unit cell active device at the selected bias operating point. This data may be available as a linear model within the foundry-supported CAD simulation tools, or it can be extracted from a nonlinear model that is biased at the desired operating point. It may also be obtained by measuring the *s*-parameters of a real device biased at the desired operating point.

With the *s*-parameters of the selected unit cells within the CAD simulation tool, the designer can explore possible matching circuits for the unit cell and the number of poles needed to achieve the required bandwidth. Multipole matching can achieve wider band responses, but too many poles can lead to a matching circuit with unacceptable loss, which is especially important after the last stage.

Combine the matched devices and power splitters or combiners according to the architecture design, and check that they can be physically laid out together. At this stage in the design, start to consider ways of compressing the layout to reduce the chip size.

Ensure that the matching circuit or power combiner allows dc bias to be fed to the active devices. In general, the biggest problem to overcome is

Table 5.1
Characteristics of High-Efficiency Modes

Class	Gate Bias	Maximum Efficiency (%)	Gain Reduction (dB)	Load Required	Harmonic Levels
A	$V_{Pinch\text{-}off}/2$	50	0	Resistive	0
AB	Near $V_{Pinch\text{-}off}$	60	Minimal	Resistive	Low
B	$V_{Pinch\text{-}off}$	78	6	Reactive	Medium
C	$< V_{Pinch\text{-}off}$	90	>6	Reactive + shorted harmonics	High
Saturated A	$V_{Pinch\text{-}off}/2$	80	Minimal	Resistive	Low
D	$V_{Pinch\text{-}off}$	100	6	Reactive + finite for odd harmonics	High
E	$V_{Pinch\text{-}off}$	100	6	Capacitive + resonator	High
F	$V_{Pinch\text{-}off}$	85	6	Reactive + finite for third harmonic	High

feeding the high current levels to the drain of the output stage. It is necessary that the dc bias be fed in at points where the effect of the dc probes is adequately decoupled from the RF response. Note that MMIC power amplifiers destined for phased array applications should be designed to be fed wholly from either side of the chip, as this simplifies their integration into the array modules.

The designer must also take care that all components are rated to handle the dc and RF power levels that may be passing through them. Again, the biggest problem is the high dc current level to the drain of the output stage. The microstrip track must be wide enough to handle the current without overheating or degrading due to effects such as electromigration. A typical safe rating is 10 mA/mm of track width, but this tends to vary between foundries. High RF power levels are usually less of a problem. With standard separation of components, high-impedance tracks can safely handle the RF voltage associated with over 10W of power, and even low-impedance tracks can handle the current induced by over 50W of RF power. However, more care is required when the same track handles high dc and RF power at the same time.

The result at the end of this stage of the design method is a fully laid out amplifier that

- Meets the gain and match requirements of the specification and small-signal operating conditions;
- Allows the high levels of dc bias to be fed to the active devices with sufficient decoupling from the RF response;
- Has an architecture with the potential to meet the power performance specification;
- Does not meet the power performance requirement under large-signal operating conditions.

Large-Signal Optimization

The final stage of power amplifier design is the optimization of the output-matching circuits for the best performance under large-signal operating conditions. This involves tweaking the output-matching circuit using the minimum of time-consuming nonlinear analysis to take account of the "weak" nonlinearities encountered at the extreme ends of the load line.

Under large-signal operation, for example at 1-dB gain compression, certain of the active device parameters become nonlinear. These parameters, shown in Figure 5.95, are mainly the gate-to-source capacitance C_{gs}, the gate-to-drain capacitance C_{gd}, the transconductance G_m, and the output resistance R_{ds}.

From the discussion in Section 5.3.2.1, it can be seen that RF signals are produced by the compressing active device at all the harmonics of the incident signal. The harmonic signals interact with the linear matching circuits, and a proportion is reflected back to the active device, which in turn produces more harmonics. Large-signal simulation of nonlinear circuits is known as harmonic balance because the simulator must adjust the power levels in all the harmonics until the signals transmitted and reflected from the linear matching circuits balance with the signals transmitted and reflected from the nonlinear device. This is an iterative process and continues until the errors between the signals drop below a predetermined level.

The small-signal design, when analyzed with nonlinear models under large-signal conditions, will generally exhibit a gradual reduction in gain with increasing output power; this is known as soft compression and is shown in Figure 5.96. The output power at P_{1dB} is typically ≈2 dB less than that achievable with the optimum output match. To achieve the peak output power performance, the output-matching circuit on the last stage must be optimized for nonlinear large-signal operation.

Optimization of the amplifier for nonlinear large-signal operation means analyzing the best load line and presenting this to the output of the last stage device over the specified frequency range. Optimum load line R_L, shown in Figure 5.97, allows maximum power transfer from the device current

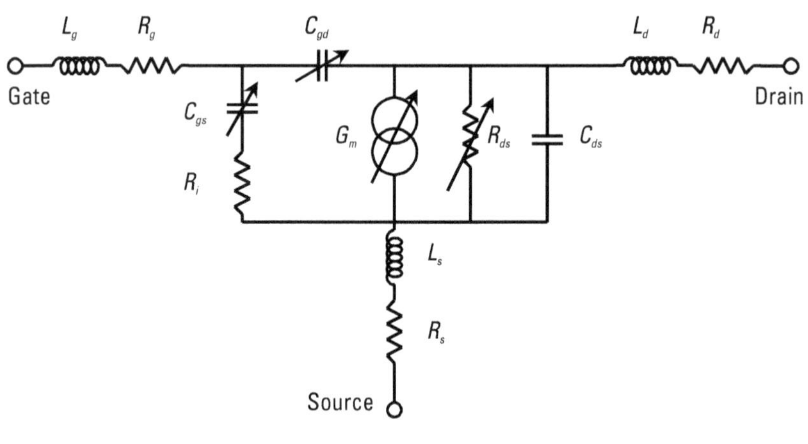

Figure 5.95 Model elements that become nonlinear for large signals.

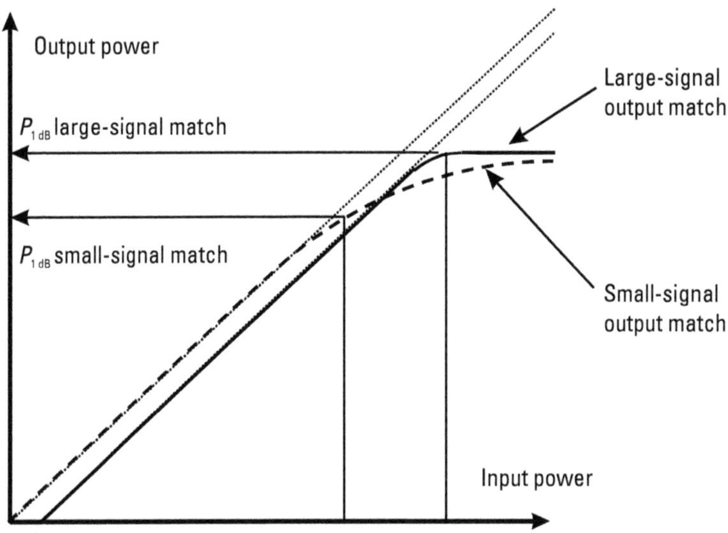

Figure 5.96 Power-compression curves showing soft and hard characteristics.

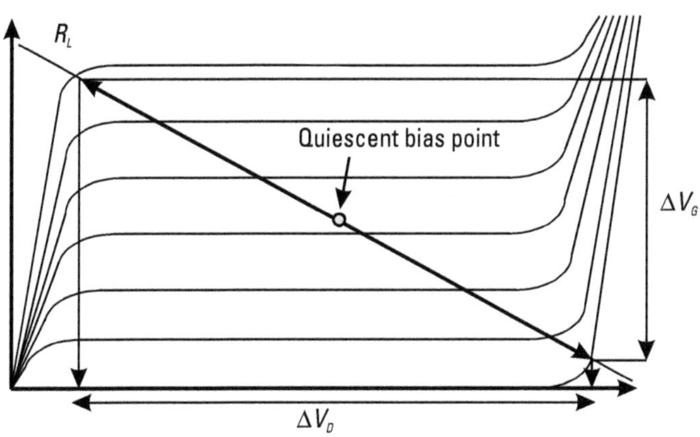

Figure 5.97 Optimum load line for large-signal operation.

source into the load. The methods that can be used to determine the optimum load line include

- Plotting the load line on the intrinsic dc parameters;
- Load-pull measurements;
- Large-signal simulation using a nonlinear model of the active device.

Plotting the Load Line on the Intrinsic dc Parameters

This technique is commonly known as the Cripps technique [94]. The current source is separated from the other FET output parameters and parasitics, as shown in Figure 5.98, and the load line is calculated from the dc parameters. The resulting purely real load resistance is presented to the current source by the matching network and through the FET output parasitics.

The load line is calculated from the dc *I*/*V* characteristics shown in Figure 5.99 and (5.31).

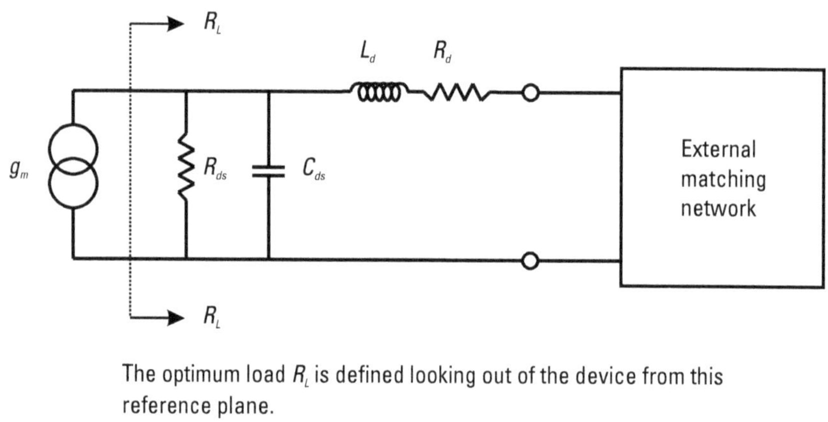

Figure 5.98 Optimum load reference plane.

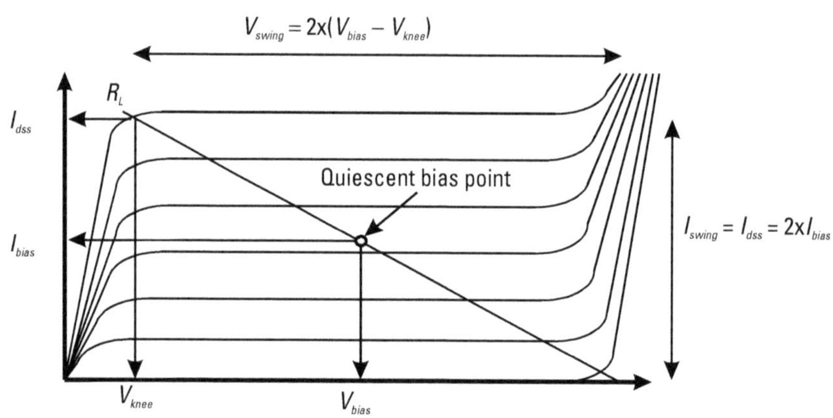

Figure 5.99 *I*/*V* characteristics showing voltage and current swing and bias point.

$$R_L = V_{swing} / I_{swing} = (V_{bias} - V_{knee}) / I_{bias} \tag{5.31}$$

The resulting purely real load resistance is presented to the current source by the matching network and through the FET output parasitics. The device output capacitance has been "absorbed" into the external matching network with the result that the optimum impedance is frequency independent. This technique requires knowledge of the device parasitic elements and does not take account of source inductance or resistance, so it is less valid at high frequencies. It does derive a frequency-independent load, so it can be used for broadband circuits. Further information can be found in [86, 94].

Load-Pull Measurements

This technique involves measurements of actual devices at their operating bias point. Mechanical or electronic tuning elements present the full range of impedances to the output of the device. The power performance, at each output impedance, is measured in terms of P_{1dB}, P_{sat}, and PAE. Each parameter can then be plotted on the Smith chart, with contours connecting points of equal value and the optimum impedance found from the center of the power contours. The circles of constant P_{1dB} shown in Figure 5.100 are examples of the sort of data obtained by load-pull measurements. These are circles of impedance, which, when presented to the device output, will give the same circuit P_{1dB} performance.

When the optimum impedance at the tuner has been found from the Smith chart, this impedance must be transformed to the required impedance at the device output reference plane. This means de-embedding the effects of the coax cable, RFOW probes, RFOW pads, and track on the substrate, as shown in Figure 5.101. The matching network must then be designed to present the optimum impedance to the output reference plane of the device, as in Figure 5.102.

This technique is dependent on the accurate transforming of the tuner impedance to the actual impedance presented to the device, and this is often the hardest part. The range of impedances is dependent on the loss from the tuner to the device, but this may be overcome by the more complicated technique of active load pull. The procedure must also be repeated over the whole range of operating frequencies. The advantage of this technique is that it is based on measurements of real devices; thus, it is reliable and requires no prior knowledge of the device parasitics.

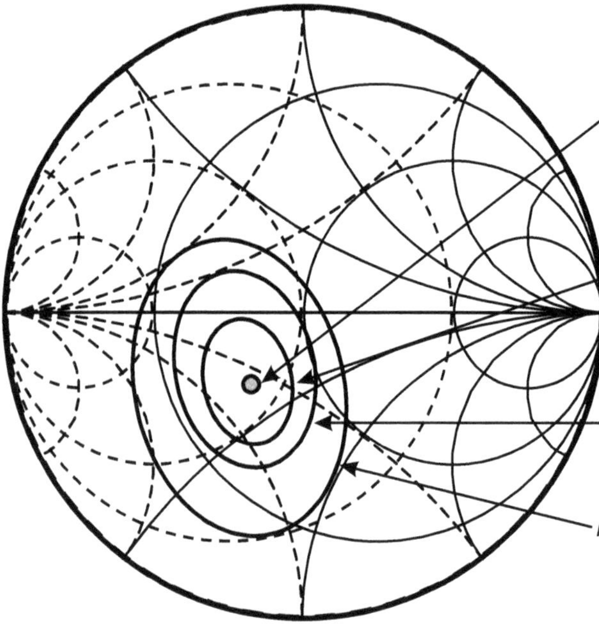

Figure 5.100 Smith chart showing circles of constant P_{1dB}.

Large-Signal Simulation Using a Nonlinear Model of the Active Device

This technique requires a nonlinear circuit model of the active device, fitted over all possible values of voltage and current, and installed in a nonlinear simulator. It is not a trivial task to develop a nonlinear model for an active device, so the range of nonlinear device models supplied by the foundry is usually limited. Assuming a nonlinear model is available for the chosen unit device size, the small-signal circuit design can then be simulated under large-signal conditions and its performance found.

Using the simulator "optimizer" function does not always work as well for nonlinear simulation as it does for linear simulation, so the chances of finding a false minima in the error function is high. A practical way of finding the optimal load impedance with nonlinear models is to simulate load-pull measurements. This can derive the optimum impedance, and the matching network can be designed to present this impedance to the output of the active device.

The procedure is, first, to replace the small-signal *s*-parameter representation of the active devices with the nonlinear models, as in Figure 5.103,

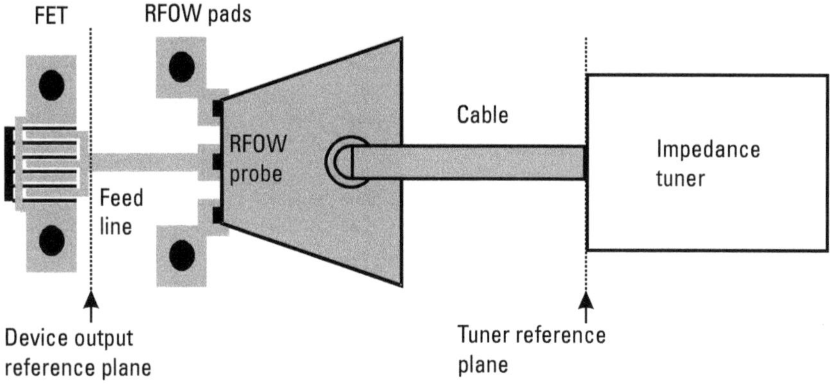

Figure 5.101 Diagram of components between the impedance tuner and the device reference planes.

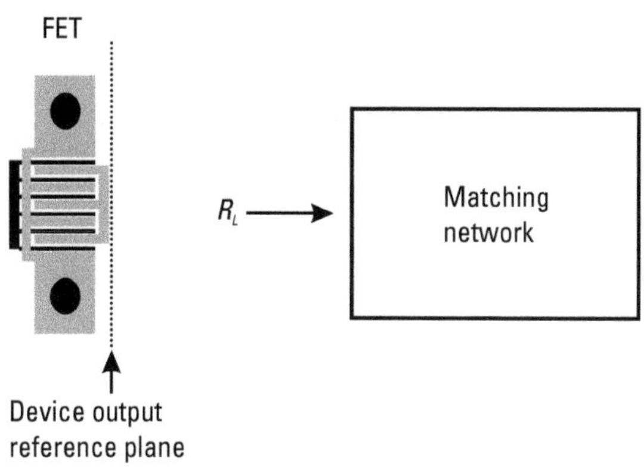

Figure 5.102 Device reference plane for output-matching network.

and to ensure that they are dc-biased correctly. Second, simulate the large-signal performance of the small-signal circuit design, calculating the P_{1dB}, P_{sat}, and PAE. This gives a benchmark of the performance of the amplifier design with small-signal matching.

Disconnect the output-matching circuit from the last stage, and measure the impedance currently presented to the output of the active device by the output-matching network, as indicated by Figure 5.104.

Next, use linear optimization of the output-matching circuit to change the impedance by a set amount. For example, the variation could be ±10% of the real or imaginary part of the impedance. Then, reconnect the

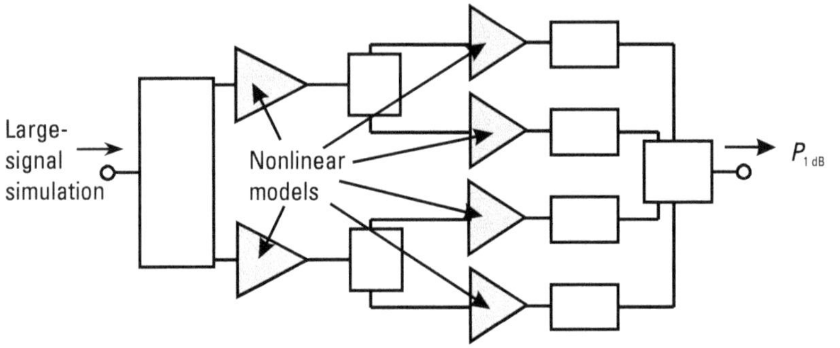

Figure 5.103 Large-signal simulation of the circuit design using nonlinear models.

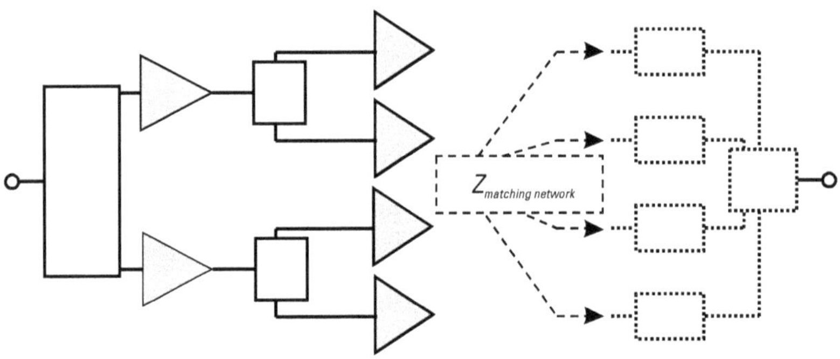

Figure 5.104 Measuring the impedance presented by the output-matching network.

output-matching circuit to the last stage and simulate the large-signal performance of the modified circuit design, shown in Figure 5.105. If the large-signal performance characteristics have improved or degraded, this will indicate the direction in which to make the variation of the impedance presented by the output-matching network. Continue to change the impedance presented by the output-matching circuit iteratively until the best power performance is obtained.

If the power compression curves (output power versus input power) are plotted for the amplifier with both the small-signal and large-signal output-matching circuit, they are likely to resemble those shown in Figure 5.96. Note that the small-signal gain of the overall amplifier is likely to be reduced by ~1 dB because the output match of the last stage has been changed away from the small-signal, or best gain, value. The gain-compression curve with the large-signal match should have a harder characteristic, with P_{1dB} only

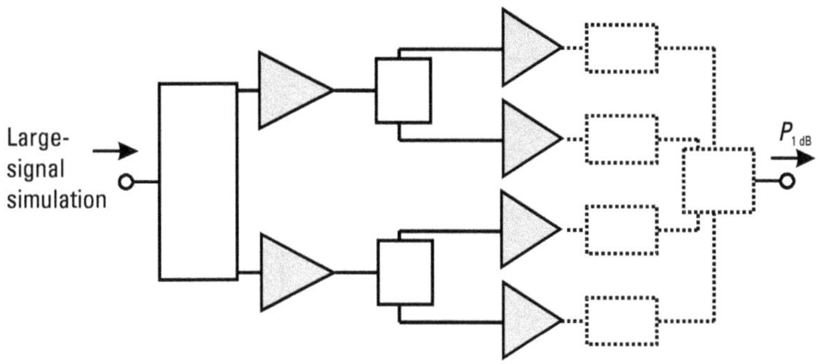

Figure 5.105 Large-signal simulation of the circuit with modified output-matching network.

about 1 to 2 dB below P_{sat}. The small-signal output match will be poor; however, under large-signal conditions, the match should be good.

This technique is critically dependent on the quality of the nonlinear model (i.e., how well the model describes the device behavior at all the points along the load line). The main advantage of this technique is that it fully simulates the amplifier large-signal performance after the circuit has been optimized for the large-signal match.

Thermal Considerations

A power amplifier which produces 2W of RF power and is 40% efficient will require 5W of dc power and, consequently, dissipates 3W of power as heat. The device has finite thermal impedance, so this heat generation raises the device above ambient temperature. Active device parameters are functions of their temperature; moderate increases in temperature decrease device gain and power, and large increases in temperature compromise the device's reliability. For these reasons, the temperature of the active devices must be managed carefully. A technique for determining the junction temperature of HBTs under high self-heating operating conditions can be found in [95].

The source of the dissipated heat is the conduction channel underneath the gate in an FET and the area underneath the emitter finger in a bipolar transistor. The thermal conduction of semiconductors, such as GaAs, is itself a function of local temperature, so full 3D finite-difference analyses are required to model the temperature increases. A schematic diagram of the heat flow from a long, thin heat source on the surface of a typical MMIC is shown in Figure 5.106.

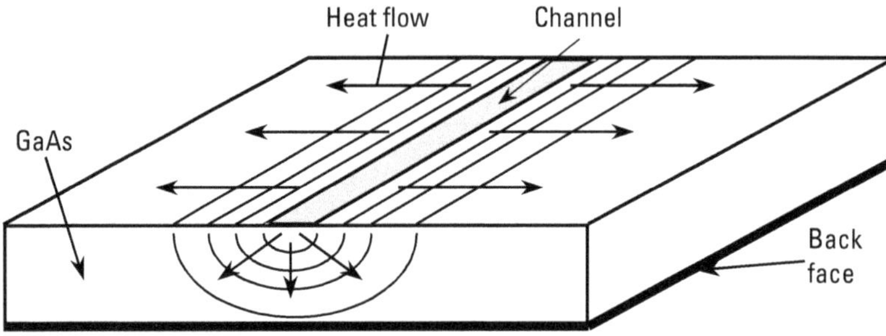

Figure 5.106 Heat flow from a heat source on the surface of an MMIC.

GaAs thermal conduction is worse at high temperatures, so the thermal gradient in Figure 5.106 is steepest close to the channel and shallow far from the channel. As the substrate is very wide (typically > 2,000 μm) compared to the 100-μm thickness, heat is best sunk out from the back face, where it may be mounted on a metal base, which can act as a good heat-sink. Sideways heat flow becomes important when channels or emitters are closer to each other than the back face, as shown in Figure 5.107. This leads to mutual heating and higher channel temperatures than for a single channel.

To deal with these issues, various thermal-management solutions are employed. These include device separation, pulsed operation, substrate thinning, integral heat-sinks, and thermal shunts.

Device operating temperatures can be reduced by removing the mutual heating between devices. The mutual heating between adjacent devices, or between adjacent channels in a multifinger device, can be minimized by physically separating them on the surface of the chip. Care must be taken

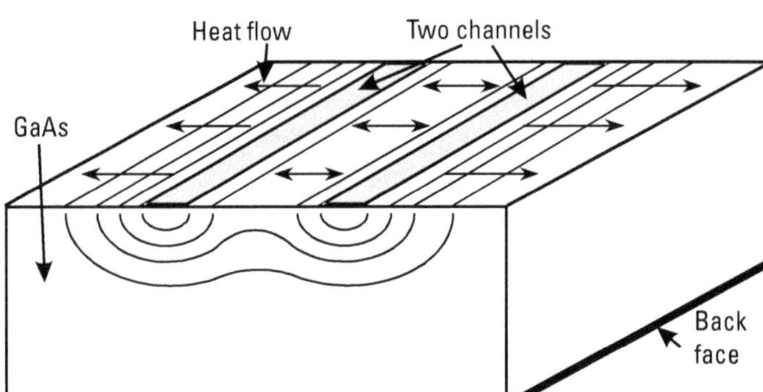

Figure 5.107 Mutual heating of closely spaced heat sources.

because physical separation can cause phase errors between the devices and reduce the RF performance.

If the system can operate under pulsed conditions, with the dc bias only being switched on during a short transmit pulse, the average power dissipation can be greatly reduced, resulting in lower channel temperatures. However, local heating of the device channel can still be significant unless the pulse length is very short and comparable with the local thermal time constants.

The closer the back face can be brought to the channel, the better the heat sinking. The thermal gradient is steepest near the channel, so reducing the thickness from 100 μm to 50 μm lowers the channel temperature by a greater amount than reducing the thickness from 150 μm to 100 μm. However, wafer and chip handling become very difficult at thicknesses below 100 μm, which leads to an associated drop in the yield of circuits after test and wafer sawing. Alternatively, the back face can be effectively brought closer using integral heat-sinks (as is shown schematically in Figure 5.108), which give the thermal benefits of a 50-μm substrate together with the yield benefits of a 100-μm substrate. This integral heat-sink technique, also known as "bath tubs," is a difficult process and not offered by many foundries.

The heat generated by the active devices can be managed using thermal shunts, as is shown schematically in Figure 5.109. These are high thermal-conductivity materials placed on the top surface of the chip, which help to spread the heat away quickly from the device's active area. They are constructed from plated-up gold metallization or sometimes diamond films. Multifinger HBT devices in power amplifiers often suffer from thermal runaway, which is where one emitter finger can end up conducting all the current [96]. This is because in a GaAs HBT, the built-in potential (V_{be}) reduces as

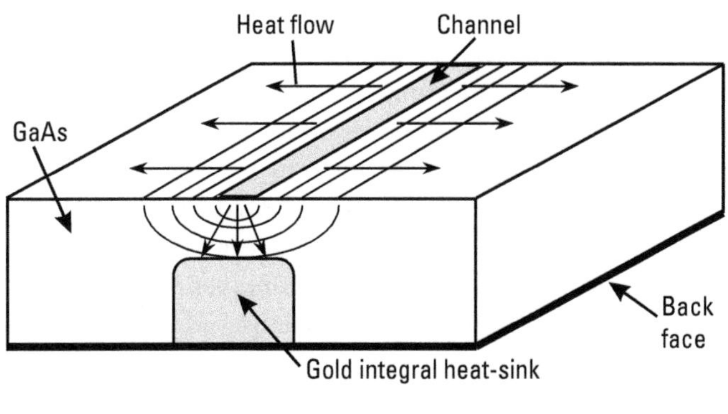

Figure 5.108 Integral heat-sink beneath the surface heat source.

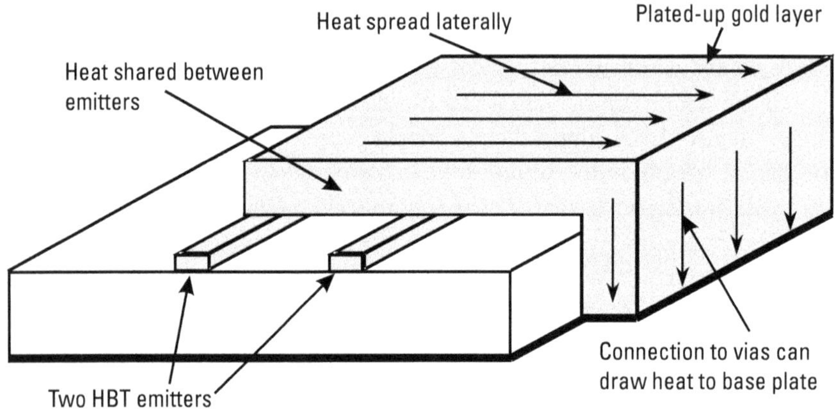

Figure 5.109 Schematic diagram of thermal shunts.

the base-emitter junction temperature increases; so, if one finger gets slightly hotter than another, its built-in potential drops compared to the other fingers, and it starts drawing more current. More current leads to more self-heating and a higher junction temperature, so the effects combine, and one finger ends up very hot and conducting all the current. Placing resistors in series with the emitters, known as ballasting [97, 98], can compensate for this effect but does reduce the gain of the device. Thermal shunts are used to equalize the temperatures in these multifinger devices, preventing thermal runaway in individual fingers without sacrificing device gain.

5.4 Oscillators

Oscillators convert dc power to ac power and are the source of RF, microwave, and millimeter-wave signals that are transmitted, received, and manipulated by radar and communication systems. They provide the ac signal at the specified frequency, which is then amplified, modulated, and transmitted through an antenna; they also provide the signal known as the local oscillator (LO), which is used to down-convert the received signals to lower frequencies, where they can be decoded easily. The important characteristics of oscillators are their frequency, frequency stability, frequency tuning and ability to lock the frequency to a reference frequency, amplitude, quality of oscillation, and phase noise.

MMIC oscillators can be made over the full RF, microwave, and millimeter-wave frequency bands, in fact, wherever the MMIC transistors have gain. They tend to be limited to low output-power levels (a few milliwatts) due to thermal and breakdown characteristics and usually require buffer amplifiers and power amplifiers at their output. These oscillators are often designed as voltage-controlled oscillators (VCOs), which allows their oscillation frequency to be tuned by an external control voltage. This ability also enables the oscillation frequency to be locked to an external reference signal so that its characteristics are improved. Many of the MMIC oscillator's characteristics are dominated by the type of resonator that is used; this is described in more detail in the following sections.

5.4.1 Oscillation Principles

The principles for oscillation in a circuit are well known [99, 100] and require that the total signal gain around the circuit be unity and that the total phase shift be 360°. This seems simple; however, it is actually a nonlinear problem that cannot be solved directly, but must be solved iteratively. This is because at oscillator startup, the active device needs to have more gain than the losses in the circuit for random noise fluctuations to initiate oscillation. As the oscillation stabilizes, the active device will have changed its operating point until the sum of the gain and losses around the circuit equal unity. Likewise, stable oscillation will only occur when the total phase shift is 360° around the circuit, which means that the sum of the reactances around the circuit must equal 0.

The actual practice of MMIC oscillator designers is somewhat different. In the microwave regime, it is common to look at the oscillator circuit in three parts, the resonator, the gain section, and the load, as shown in Figure 5.110. The resonator is needed to set the frequency of oscillation and satisfy the phase condition of oscillation, and in its simplest form, this could be a series inductance and capacitance, which at its resonant frequency of $f_{RES} = 1/2\pi\sqrt{LC}$ exhibits zero net phase shift. The gain section is the source of the power that drives the oscillations and is usually a transistor with positive feedback. This section is also known as the negative resistance section as typically its value must be 20% greater than the total of the load resistance and any parasitic resistance in the resonator for an oscillation to be continuous and satisfy the oscillation gain condition [101]. The load is simply the impedance of the system or following amplifiers that the oscillator delivers its power into.

Simulation of the oscillator can now be broken down into separate elements. The resonator components can be simulated using standard linear

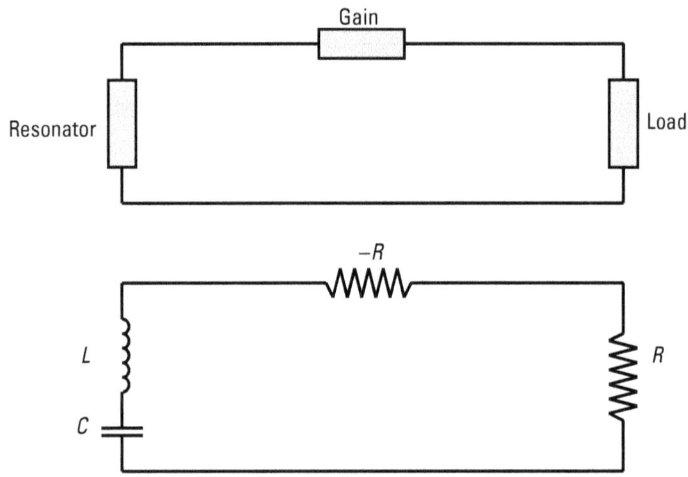

Figure 5.110 The three parts making up a microwave oscillator circuit.

simulation tools to check that their phase response crosses zero at the correct frequency. Note that the reactance of the negative resistance circuit must be taken into account because if it is slightly capacitive, the resonator will need to be slightly inductive to give zero-total reactance, and the oscillator will not oscillate where the resonator only crosses zero but where it looks slightly inductive. The transistor with feedback is also simulated with a linear *s*-parameter simulation tool, and the designer ensures that the output reflection (S_{22}) is significantly greater than unity at the required frequency. These two simpler analyses will normally ensure that the oscillator will generate an ac signal at the correct frequency. The disadvantage of this technique is that neither the output level nor the phase noise is actually simulated. The output power is estimated from the dc voltage and current supplied to the gain section with an assumption of 30% efficiency.

5.4.1.1 Resonators

MMIC on-chip resonators are almost exclusively limited to series and shunt combinations of capacitors and inductors (the only exception being FBAR-like devices, which are not yet available as foundry components), and so the Q-factor of these resonators is relatively low. This means that the resonator does not hold the oscillator very well to a fixed frequency against variations such as temperature, and the resulting phase noise is poor. This is acceptable in some applications, such as radar systems, where amplitude modulation is used.

The solution for systems that require temperature stability and low phase noise is to use an off-chip resonator. The most common example of this is the dielectric resonator oscillator (DRO), which uses a high dielectric, hence high Q-factor, puck as the resonator [102–104]. The dielectric puck is placed near a microstrip line that is connected by a bond-wire to the negative resistance circuit on the chip, as shown in Figure 5.111. The precise resonant frequency of the dielectric puck can be tuned slightly by adjusting the position of a metal disc suspended from the top of the enclosure. DROs are difficult and costly to assemble because they are sensitive to the metal cavity they are mounted in and often require positioning by hand.

Another off-chip resonator that is sometimes used is the yttrium iron garnet (YIG) resonator [105, 106].

5.4.1.2 Negative Resistance Circuits

The negative resistance section on an MMIC is typically formed either from a common-source MESFET or pHEMT with a capacitor between the source contact and ground to provide the positive feedback or from a common-base HBT with an inductor between the base contact and ground to provide the positive feedback, as shown in Figure 5.112. For an FET, the resonator is connected to the gate contact, and the output is extracted from the drain contact. When an HBT is used, the resonator is connected to the emitter contact, and the output is extracted from the collector contact. The circuit topologies shown in Figure 5.112 are just examples, and numerous different topologies can be derived from a three-terminal device, each with its own particular

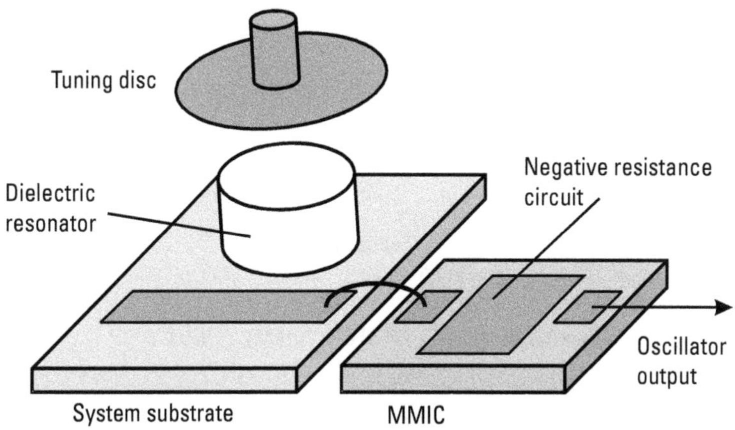

Figure 5.111 External dielectric resonator oscillator.

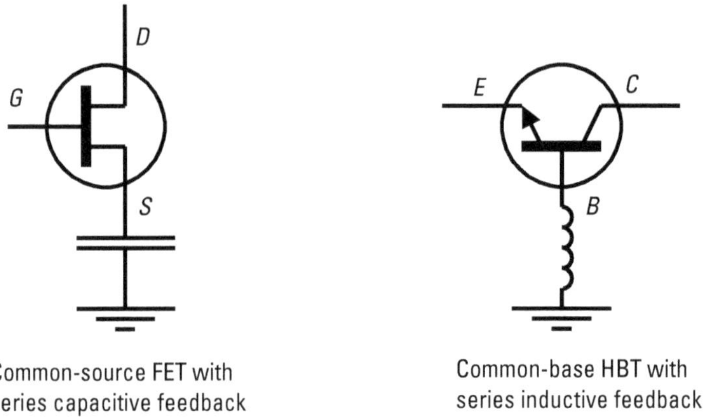

Figure 5.112 Negative resistance circuits.

advantages. For example, the common-base HBT topology tends to offer a wide-tuning-range characteristic, but it can be very sensitive to the value of the feedback inductance, while common-emitter and common-collector configurations using capacitive feedback are narrower band but are often less sensitive to process variations.

There are trade-offs between the different types of transistors, which must be considered when choosing one or the other for a particular application. The performance parameters, such as maximum frequency of oscillation, phase noise, and cost, are shown in Table 5.2, which shows that HBT oscillators tend to have the lowest phase noise, while HEMT oscillators operate at the highest frequencies. MMIC oscillators designed with various technologies are available in the literature [107–110].

5.4.1.3 Voltage-Controlled Oscillators

Voltage-controlled oscillators (VCOs) can be tuned to different oscillation frequencies by varying one or more bias voltages connected to the chip [111,

Table 5.2
Oscillator Performance Parameters Associated with the MMIC Transistor Technology

Parameter/device	MESFET	HEMT	HBT
Maximum frequency	Medium	Highest	Lowest
Phase noise	Medium	Worst	Best
Cost	Lowest	Highest	Medium

112]. This requires a voltage-variable capacitance or inductance in the resonator circuit. Voltage-variable inductors can be formed from transistors using a gyrator topology [51, 52], but these circuits are difficult to stabilize. By far the easiest way is to use one of the junctions of the active devices to form a variable capacitor or varactor. In an FET circuit, the varactor is formed from the gate Schottky junction, and the source and drain contacts are connected together to minimize the access resistance to the junction. Varying the reverse-bias voltage on the Schottky junction changes the width of the depletion region and so varies the capacitance of the junction. It is common to use a combination of a fixed capacitance to set the nominal oscillation frequency and a varactor to tune the frequency by 10% to 20% around this nominal value. Note that the VCO designer normally has to design the chip to have a much broader frequency range than the specification to account for process and temperature variations and the frequency-pulling effects of circuits connected to it. This usually requires a compromise with the other characteristics of the oscillator, such as the phase noise, the flatness of the power versus frequency, and the linearity of the tuning response.

An example VCO circuit using a source-drain FET as a varactor is shown in Figure 5.113. The resonator consists of a series combination of a spiral inductor, a source-drain connected FET, and a fixed MIM capacitor. The gate contact of the source-drain connected FET is held at 0V by the

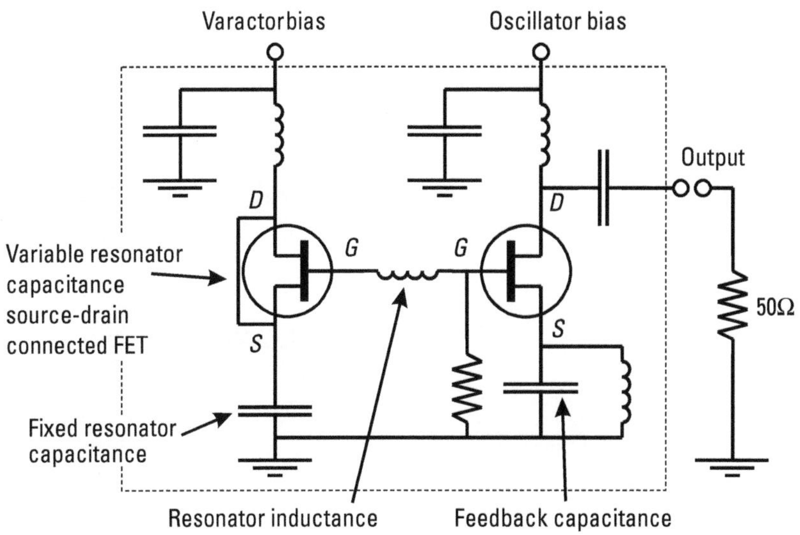

Figure 5.113 Circuit diagram of an MMIC VCO.

resistor connected to ground, and the junction is reverse-biased by a positive bias voltage through a large decoupling spiral inductor. The negative resistance section is formed from the common-source FET with an MIM feedback capacitor to ground. The drain bias current is fed through a large decoupling spiral inductor, and the source current bypasses the feedback capacitor via another spiral inductor connect to ground. The gate of the common source is held at 0V by the resistor connected to ground, so the FET is operating at its drain saturation current (I_{DSS}). The RF output from the oscillator is via a dc-blocking MIM capacitor so that external impedance connected to the oscillator does not upset the dc bias of the circuit. The layout of this VCO MMIC is shown in Figure 5.114.

5.5 Mixers

Mixers are required in heterodyne transmit and receive systems where low-frequency signals are up-converted to higher RF frequencies for transmission, then down-converted back to lower frequencies when received. This unique characteristic of frequency conversion is the primary function of the mixer

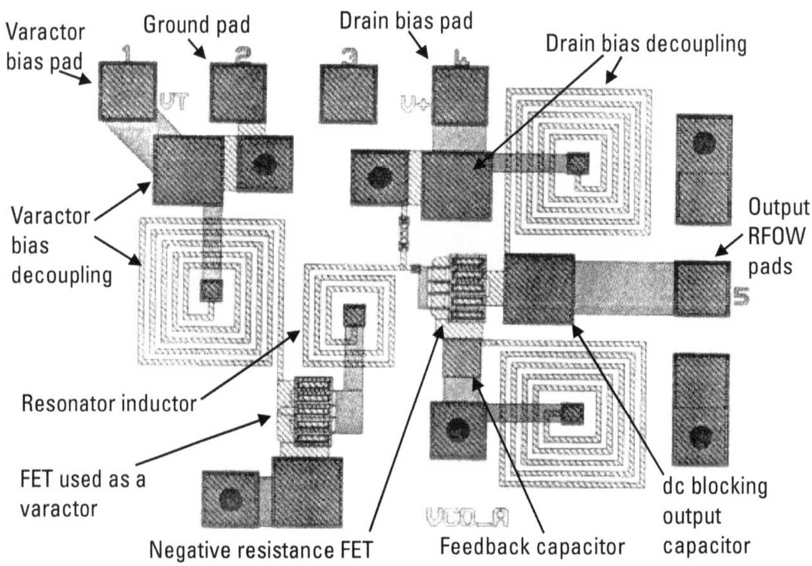

Figure 5.114 Layout of a MESFET VCO. (*Source:* Bookham Inc., 2006. All Rights Reserved.)

and is shown schematically in Figure 5.115. Mixers are nonlinear devices that use their nonlinear characteristics to convert one frequency to another while maintaining any applied amplitude, frequency, or phase modulation. The precise nature of the nonlinearity is not necessarily important as all nonlinearities will create the mixing function, but specific nonlinear characteristics are more efficient at converting frequencies in certain applications. Further information on mixers can be found in [113].

The basic principles of frequency conversion can be understood by reconsidering the analysis of nonlinear elements discussed in Section 5.3.3.1. It was shown for the case of the nonlinear resistor in Figure 5.73, whose resistance (R) is an arbitrary (nonlinear) function of the applied voltage (V), that whatever shapes the curve takes, the current (I) can be expressed as a polynomial expansion of the applied voltage as given in (5.32).

$$I = 1/R_1 V + 1/R_2 V^2 + 1/R_3 V^3 + 1/R_4 V^4 + 1/R_5 V^5 + \ldots \tag{5.32}$$

If the applied signal voltage now contains two frequencies, the LO and the RF, and each is a sinusoidal signal with amplitude A and angular frequency ω, the expression for the applied voltage is given in (5.33):

$$V = A_{LO}\cos(\omega_{LO}t) + A_{RF}\cos(\omega_{RF}t) \tag{5.33}$$

Equation 5.33 may be substituted into (5.32) to give a new expression for the current. This expression for the current in a nonlinear resistor is now very complex and contains many terms, where the sum of $\cos(\omega_{LO}t)$ and $\cos(\omega_{RF}t)$ are raised to the power of 2, 3, 4, and so forth. This complex

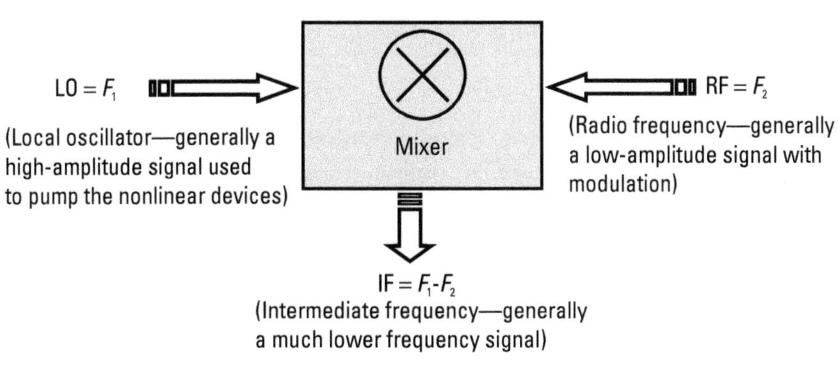

Figure 5.115 Function of a mixer.

expression can be reduced using De Moivre's theorem and standard mathematical expressions to show that the current contains components at all the harmonics of the LO and RF (ω_{LO}, $2\omega_{LO}$, $3\omega_{LO}$, $4\omega_{LO}$, … , ω_{RF}, $2\omega_{RF}$, $3\omega_{RF}$, $4\omega_{RF}$, …), plus the sum and difference products ($n \times \omega_{LO} \pm m \times \omega_{RF}$, where n and m are integers).

The art of mixer design is to maximize the efficiency of converting to the desired sum or difference frequency while filtering and minimizing the amount of unwanted mixing frequencies' appearing at the mixer terminals.

5.5.1 MMIC Mixers

The nonlinear device used in MMIC mixers is either a transistor or a diode, with the diode being constructed from two terminals of a transistor. Mixers are also categorized into two general types, passive or active. As the name suggests, active mixers are made using the active devices, the transistors, and they are biased (operating in their transconductance, or "hot," mode) so that the devices have signal gain. As a consequence, the mixer can be designed to exhibit conversion gain rather than conversion loss, but the drawback is that biasing this type of mixer is quite complex. The other notable characteristic of active mixers is that they are not reciprocal, so the same design cannot function as both an up-converter and a down-converter. Passive mixers are made from either diodes or unbiased transistors (operating in their transresistance, or "cold," mode), with the diode mixers generally being easier to connect to the dc bias supplies.

There are many different topologies for MMIC mixers, and they can be broken down into the following categories, which normally correspond to the number of nonlinear devices used in the design: single-ended, balanced, double-balanced, and image-reject mixers.

5.5.1.1 Single-Ended Mixers

A single-ended mixer consists of a single, nonlinear element, such as a diode, and the LO and RF signals are first combined and fed into the mixer through a high-frequency matching network, as in Figure 5.116. The IF is extracted from the nonlinear device via a low-frequency matching network. The RF/LO matching network must be designed to provide the correct impedance to the IF frequency, and the IF matching network must also provide the correct impedance to the RF and LO signals.

This is the simplest type of mixer, and the nonlinear elements, such as diodes, can be mounted in series or shunt arrangements. The diodes may be

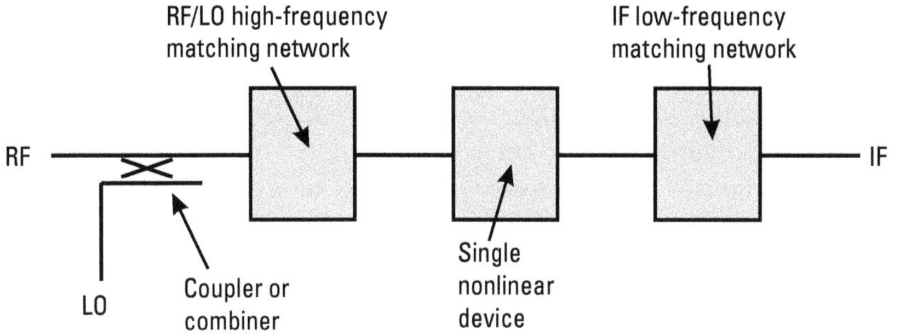

Figure 5.116 Single-ended MMIC mixer schematic.

unbiased, or if low LO power levels are required, they can be biased. The main disadvantage of single-ended mixers is that all the harmonics of the LO and RF and all the spurious signals given by $n \times \omega_{LO} \pm m \times \omega_{RF}$ appear at the mixer output, which may upset the system the mixer is to work in.

Single-Gate FET MMIC Mixers

Designers can utilize the three-terminal active devices as the nonlinear elements within MMIC mixers (i.e., the FETs and bipolar transistors), and these can be used either in their passive or active modes. In the passive mode (also known as the transresistive, conductance, switched, resistive, or cold mode), no drain bias is used, and the resulting mixers are reciprocal (i.e., they can be used for up-conversion as well as down-conversion). Generally, the RF signal is applied to the source contact of the FET, the LO signal is applied to the gate contact, and the IF is extracted from the drain contact, as shown in Figure 5.117.

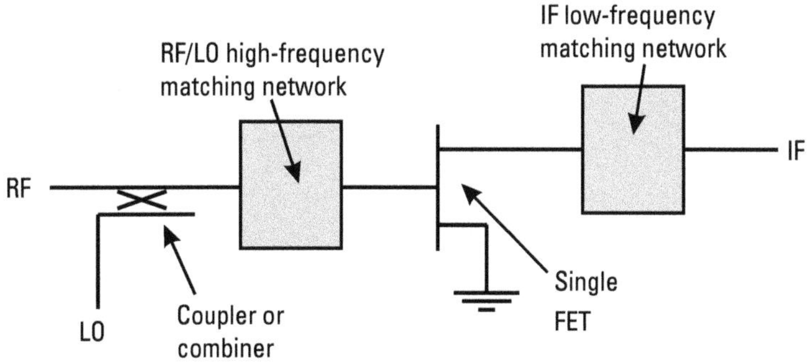

Figure 5.117 Single-ended FET MMIC mixer schematic.

In the active mode (also known as the transconductance, pumped, or hot mode), positive drain bias is applied, and the mixer can be designed to have conversion gain rather than conversion loss. One disadvantage of the active mode is that the mixer is not reciprocal; that is, different mixer designs are required for up- and down-conversion. In the active mode, the RF and LO signals are generally applied to the gate of a common-source device, and the resulting IF signal is extracted from the drain contact. Examples of single-ended MMIC mixers are in the literature [114], and a single-gate active FET mixer MMIC layout using a Lange coupler to combine the on-chip LO and amplified RF signals is shown in Figure 5.118.

Dual-Gate FET MMIC Mixers

The advantage of using a dual-gate FET in the common-source mode is that it has three terminals, one for each mixer signal, so the need for an RF/LO combiner/coupler is eliminated. There is also good inherent RF/LO isolation from the use of two separate gates. Usually, the RF is applied to the first gate, and the LO is applied to the second gate, as in Figure 5.119. The RF and LO matching circuits are designed to act as short circuits to IF frequency, and the IF matching circuit is designed to act as a short circuit to RF and LO signals.

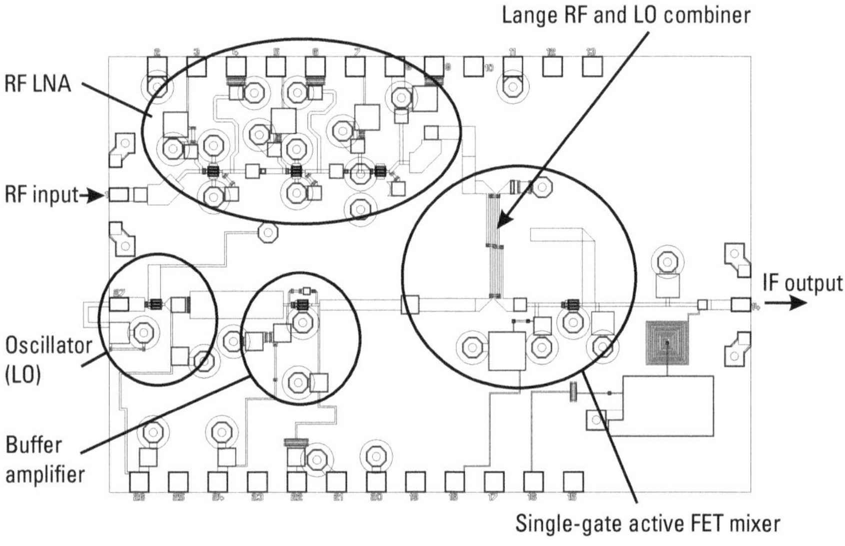

Figure 5.118 Single-gate active FET mixer MMIC layout using a Lange coupler to combine the on-chip LO and amplified RF signals. (*Source:* Bookham Inc., 2006. All Rights Reserved.)

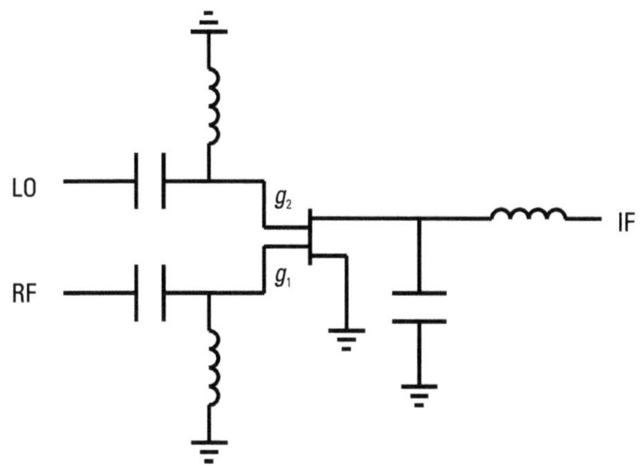

Figure 5.119 Dual-gate FET mixer.

5.5.1.2 Balanced Mixers

A balanced mixer, shown in Figure 5.120, uses two nonlinear devices, which are fed the LO and RF signals through a power splitter or coupler that has the two outputs 90° or 180° out of phase with each other. On the output side, the IF signals from both diodes are combined in phase, so some of the spurious mixing products are suppressed. Again, the diodes can be mounted in a series or shunt arrangement and may be biased or unbiased. This topology also has the effect of canceling the dc mixing products at the IF output,

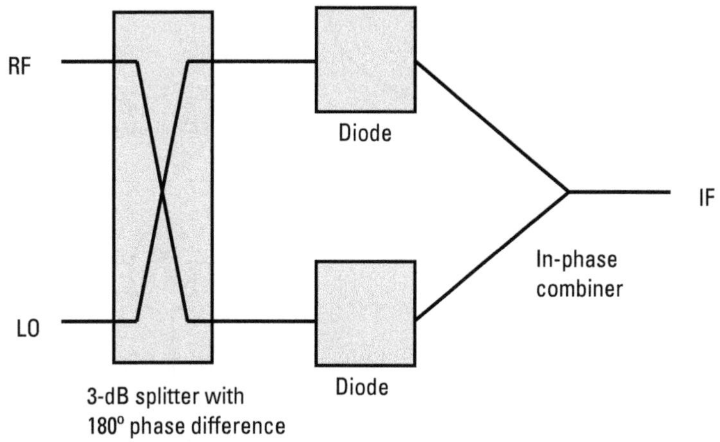

Figure 5.120 Balanced MMIC diode mixer.

and when a 180° coupler is used, the even harmonics of the LO are also suppressed. MMIC balanced mixers have been designed with octave bandwidths at millimeter-wave frequencies [115], and an example balanced diode mixer layout is shown in Figure 5.121, which incorporates a rat-race coupler to combine the on-chip LO with the RF (amplified by an on-chip LNA).

Subharmonically Pumped Mixers

A subharmonically pumped mixer is a type of balanced mixer that uses two antiparallel diodes for the nonlinear elements, as shown in Figure 5.122. The mixer is pumped at one-half (or one-quarter) the usual LO frequency (i.e., a subharmonic of the normal LO frequency) and the IF is the difference frequency between the RF and twice the LO frequency. The diodes are mounted antiparallel, which means they cannot be dc biased, but this arrangement does suppress mixing with the odd-order harmonics of the LO frequency.

The MMIC mixer in Figure 5.122 makes use of quarter-wavelength (at the LO frequency) sections of microstrip transmission line to separate the RF and LO signals. The $\lambda_{LO}/4$ section on the LO side is grounded, so it appears as an open circuit at the diodes and allows the LO frequency to pump the diodes. As the RF is nearly twice the LO frequency, the short appears as a

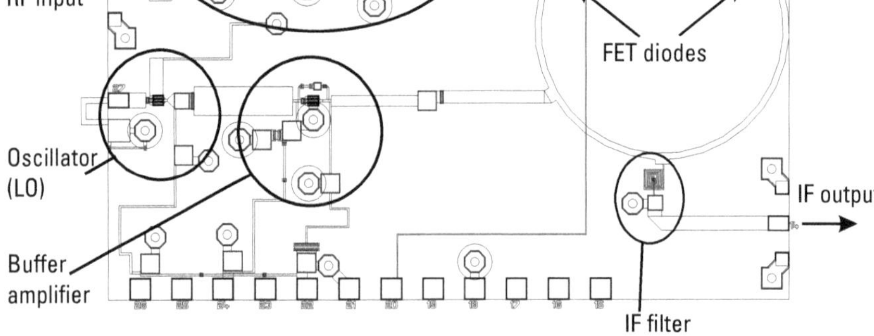

Figure 5.121 Balanced diode mixer using rat-race coupler. (*Source:* Bookham Inc., 2006. All Rights Reserved.)

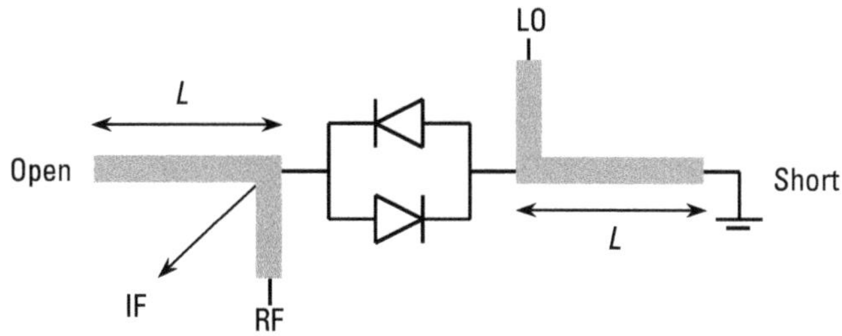

Figure 5.122 Subharmonically pumped MMIC diode mixer where *L* is one-quarter of the wavelength at the LO frequency.

short circuit at the RF frequency and prevents the RF escaping out of the LO port. Conversely, the open circuit section on the RF side is transformed to an open at the RF frequency but a short at the LO frequency, allowing the RF signal to reach the diodes but preventing the LO signal from escaping through the RF port. The IF signal, being much lower than both the LO and RF frequencies, sees simply the ground on the other side of the diodes and can be extracted via a low-pass filter. An example subharmonic mixer MMIC layout is shown in Figure 5.123.

This type of mixer is often used in high-frequency applications, such as space radiometers [116–118], where it is expensive or difficult to generate a fundamental LO, but is also used for MMIC mixer design [119].

5.5.1.3 Double-Balanced Mixers

Double-balanced mixers [120] use four nonlinear devices and are, in effect, two balanced mixers connected in parallel and 180° out of phase with each other, as shown in Figure 5.124. The advantage of this arrangement is that the even harmonics of both the RF and LO frequencies are suppressed, and there is complete isolation between RF, LO, and IF ports. The disadvantage is that the design is becoming much more complicated with the need for baluns on the LO and RF ports.

Quad-FET Mixers

An example of a double-balanced passive FET mixer is the quad-resistive-FET ring mixer shown in Figure 5.125. The major advantage of this type of mixer is its high linearity, which means is produces low levels of intermodulation products.

An example quad-resistive-FET ring mixer is shown in Figure 5.126. This mixer utilizes passive baluns for the LO and RF inputs, which are

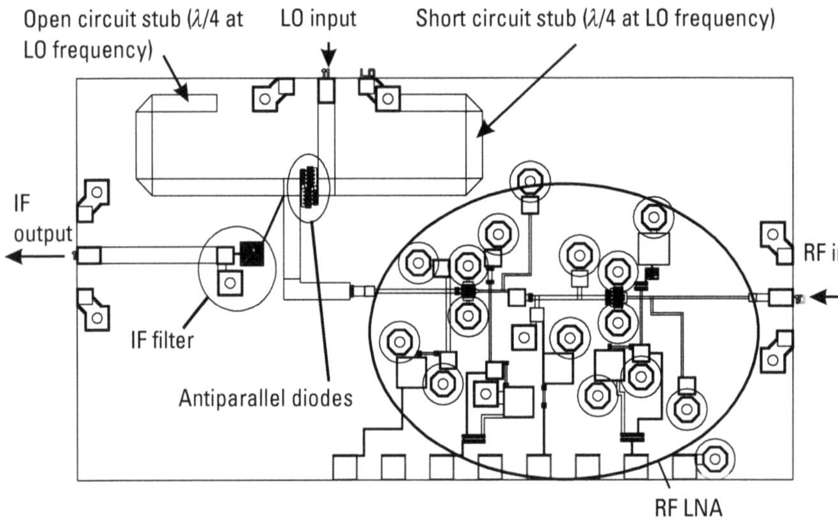

Figure 5.123 Subharmonic mixer MMIC layout with on-chip RF LNA. (*Source:* Bookham Inc., 2006. All Rights Reserved.)

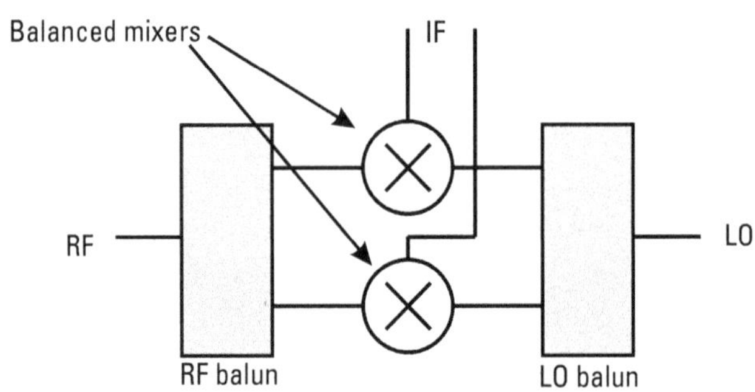

Figure 5.124 Double-balanced MMIC diode mixers.

constructed from lumped-element Wilkinson splitters followed by low-pass filters on one port and high-pass filters on the other port to achieve the 180° phase difference.

Gilbert-Cell Mixers

Gilbert-cell mixers are a good example of active double-balanced MMIC mixers [121]. The classic Gilbert multiplier topology, shown in Figure 5.127, comes originally from the silicon IC industry [122] and is ideally

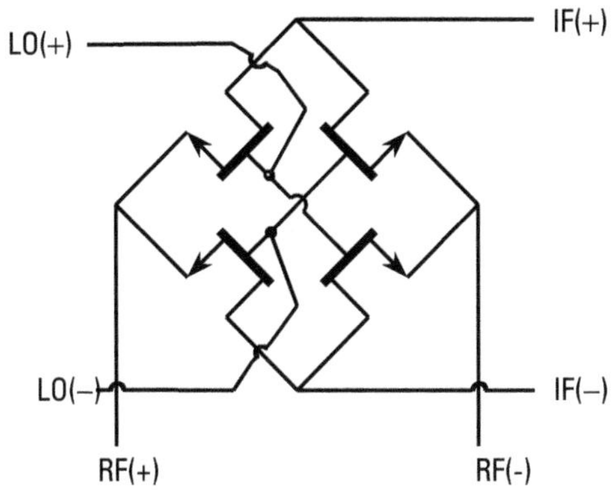

Figure 5.125 Quad-resistive-FET ring mixer topology.

suited not only to GaAs HBT but also to FET technology. This type of mixer also exhibits excellent linearity and is used where low intermodulation products are required. The Gilbert cell is a good example of a circuit that is virtually impossible to reproduce at high frequencies from discrete transistors mounted on a dielectric circuit board, but it is perfect for MMIC fabrication. An example millimeter-wave Gilbert-cell HEMT mixer is shown in Figure 5.128.

5.5.1.4 Image-Reject Mixers

In a simple mixer circuit (single-ended, for example), signals are produced at the IF port that come from $F_{LO} + F_{IF}$, which is known as the upper sideband (USB), and from $F_{LO} - F_{IF}$, which is known as the lower sideband (LSB). Only one of these signals is the desired RF signal; the other is termed the image frequency. The image-reject mixer is designed to prevent signals at the image frequency from reaching the IF output port. Ordinarily, noise from both sidebands and the desired RF signal appear at the IF port, but an image-reject mixer rejects the noise from the image frequency and so improves the noise figure by up to 3 dB. The topology of an image-reject mixer, shown in Figure 5.129, consists of two double-balanced mixers (eight nonlinear devices) with the RF fed by a 90° hybrid splitter and the LO fed with an in-phase power splitter. The IF is extracted via another 90° hybrid combiner that outputs each sideband from a separate port. MMIC image-rejection mixers have been designed to work up to millimeter-wave frequencies [124, 125]. Figure 5.130 shows a 60-GHz passive HEMT image-reject mixer [126] that incorporates a spiral transmission-line transformer, which forms a compact and wide-bandwidth balun to split the LO

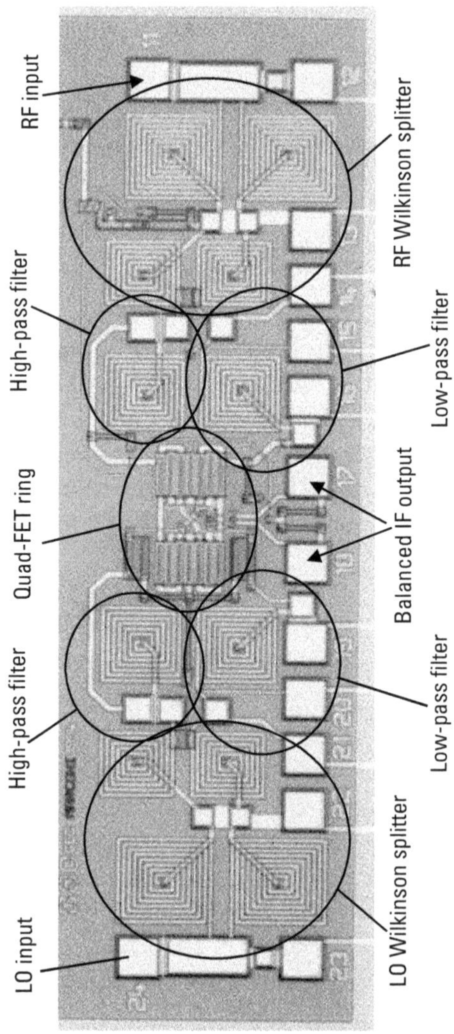

Figure 5.126 A quad-resistive-FET ring mixer using passive baluns. (*Source:* Bookham Inc., 2006. All Rights Reserved.)

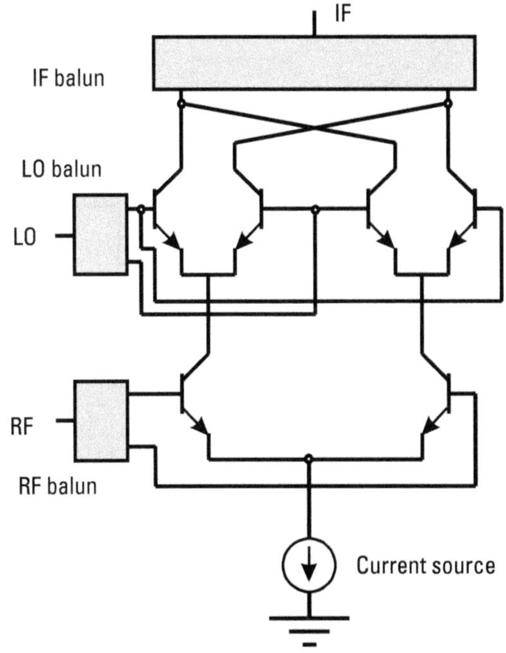

Figure 5.127 Gilbert-cell active double-balanced HBT mixer.

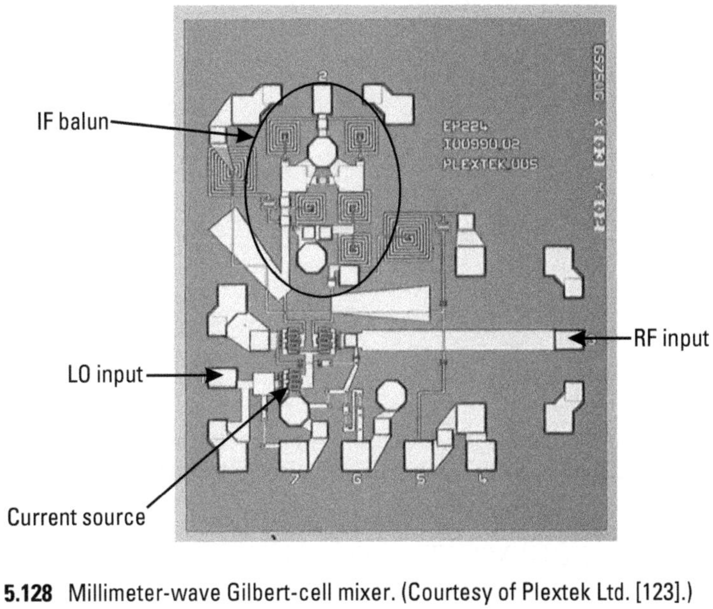

Figure 5.128 Millimeter-wave Gilbert-cell mixer. (Courtesy of Plextek Ltd. [123].)

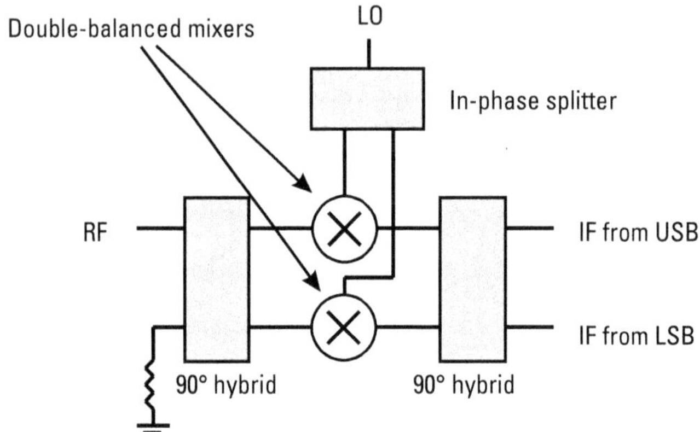

Figure 5.129 Image-reject mixer.

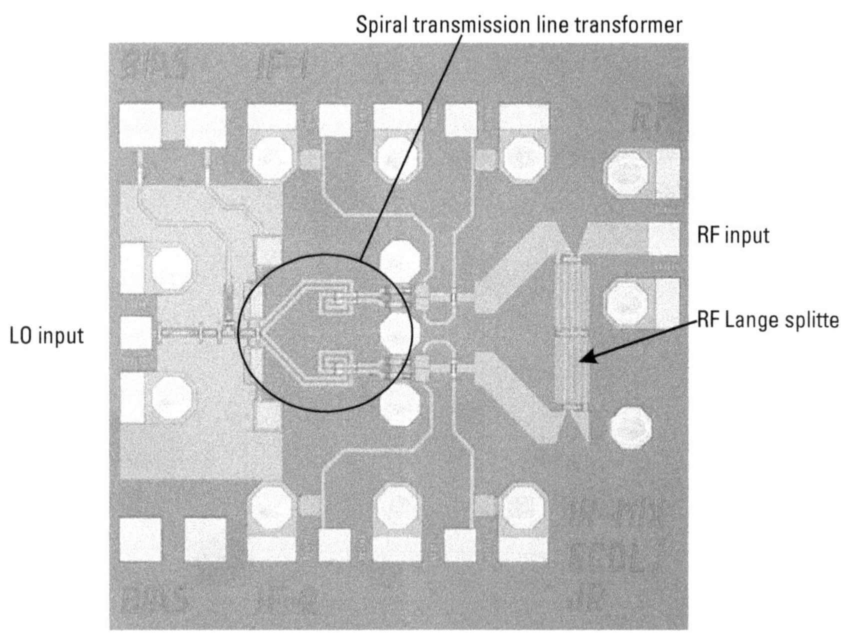

Figure 5.130 A 60-GHz passive HEMT image-reject mixer. (Courtesy of Electronic Circuit Design Laboratory, Helsinki University of Technology, Sweden [127].)

between two passive HEMTs. This mixer exhibits a conversion loss from 13 to 16 dB and an image-rejection ratio of 19 dB.

5.5.1.5 Active Baluns

Passive baluns were discussed in an earlier section, but they can also be made from active devices because of the 180° phase shift between signals on the ports of the transistor. Figure 5.131 shows that active baluns can be constructed from inverters, common-gate/common-source pairs, and differential amplifiers.

5.6 Switches

Circuits that switch RF and microwave signals are very useful, especially for applications such as phased-array radar and instrumentation. The best MMIC technology for switches is the FET because of the inherent isolation between the gate contact and the source and drain contacts, and because the gate draws virtually zero current in both control states (on and off) [128]. The advantages of MMIC FET switches over PIN diode switches [129] are that they have negligible dc power consumption and require much simpler bias and driver circuits.

When used as a switch, the FET is in a passive mode (i.e., there is no dc bias voltage on the drain contact), and it functions as a voltage-controlled resistor. The basic topology is shown in Figure 5.132, where the resistance between the source and drain contacts is controlled by the voltage applied to the gate contact through a large value resistor (typically 4K ohms). This large resistor completely decouples the dc control circuitry from the RF signals and consumes negligible dc power because of the near-zero gate current. When zero bias is applied to the gate, the FET is in its low-impedance state,

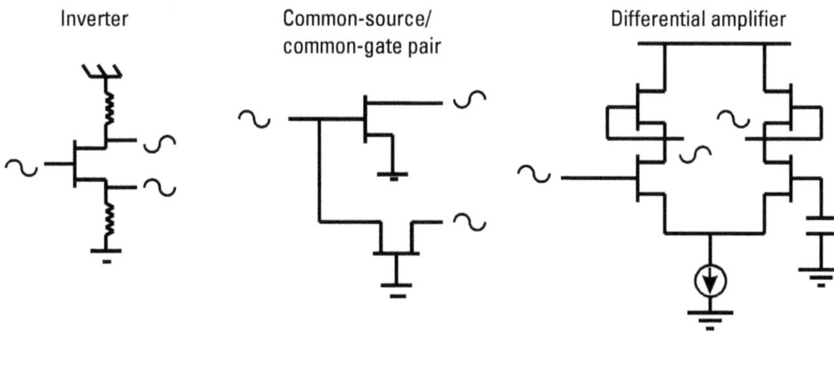

Figure 5.131 Three different active-balun topologies.

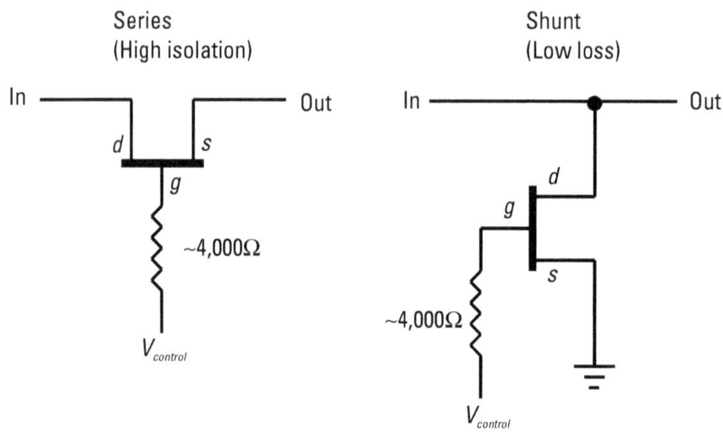

Figure 5.132 Series and shunt configuration FETs operating as single-pole single-throw switches.

and when a negative voltage (greater than the pinch-off voltage, ~ −5volts) is applied to the gate, the FET is in its high-impedance state.

The simplest switch configuration is the single-pole single-throw (SPST) switch, which, as shown in Figure 5.132, can be formed from FETs in series or shunt arrangements [130]. The series SPST switch arrangement is "on" when the FET is in the low-impedance state ($V_{control}$ = 0V) and "off" when the FET is in the high-impedance state ($V_{control}$ = −5V); it exhibits high isolation in the "off" state. The shunt SPST arrangement is "off" when the FET is in the low-impedance state ($V_{control}$ = 0V) and "on" when the FET is in the high-impedance state ($V_{control}$ = −5V); it exhibits low insertion loss in the "on" state. As ideal switches require low "on" loss and high "off" isolation, it is common to use a pair of FETs for each SPST switch, one series FET followed by a shunt FET, as in Figure 5.133. This requires complementary control voltages for both states: SPST "on" requires $V_{control}$ = 0V series, −5V shunt; SPST "off" requires $V_{control}$ = −5V series, 0V shunt.

Placing two SPST switches in parallel can produce a single-pole, double-throw (SPDT) switch [131], as shown by the MESFET SPDT in Figure 5.134. The SPDT switch can be designed so that, in its "off" state, the RF signal is either reflected from the switch or terminated in a matched resistor. Further paralleling of SPDT switches can be used to make single-pole four-throw (SP4T), single-pole six-throw (SP6T), single-pole eight-throw (SP8T) switches [132], and these can be cascaded to make various arrangements, such

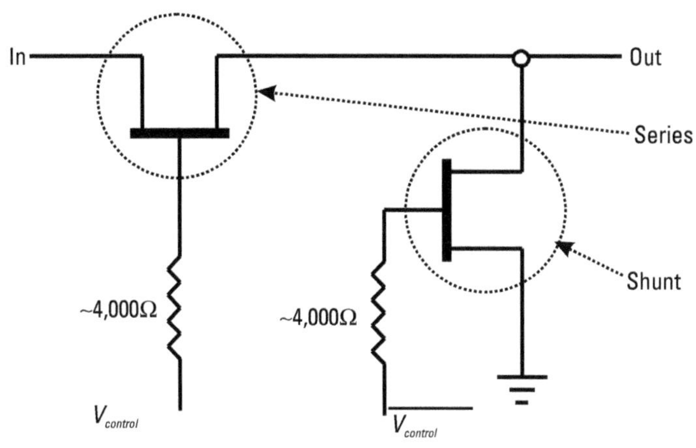

Figure 5.133 SPST series-shunt FET pair.

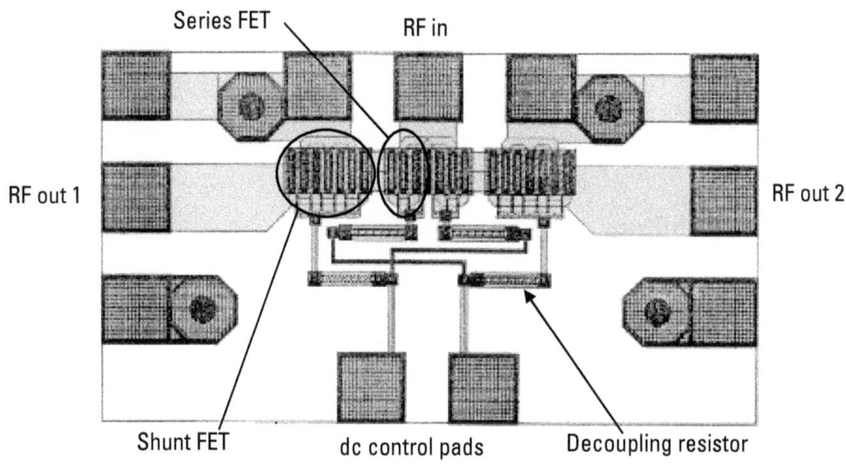

Figure 5.134 Single-pole double-throw MESFET switch layout. (*Source:* Bookham Inc., 2006. All Rights Reserved.)

as single-pole sixteen-throw (SP16T) switches. An example SPDT MMIC switch is shown in Figure 5.135.

If highly linear switches are required for applications, such as Global System for Mobile Communications (GSM) handsets or WLAN, then the nonlinear performance of pHEMT switches can be improved by including source-drain bypass resistors and using a reference voltage on the drains/sources of the switch devices [133].

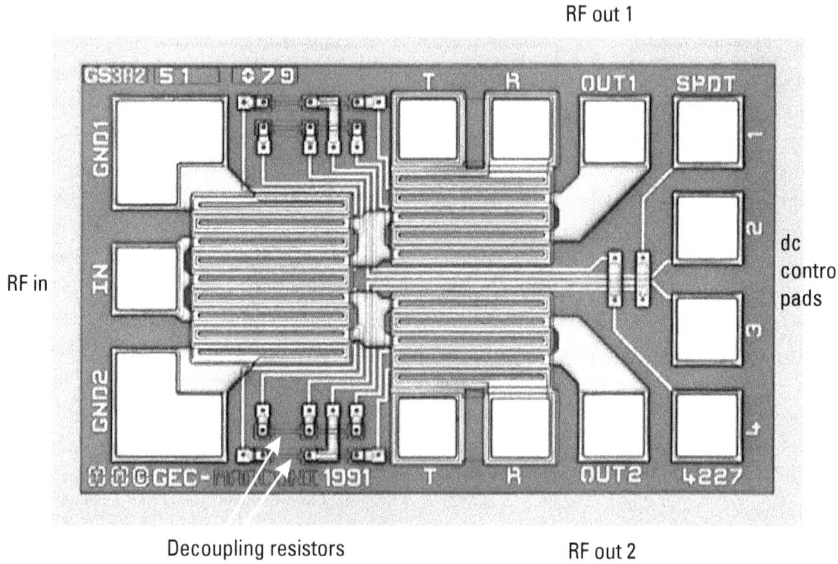

Figure 5.135 A single-pole double-throw switch MMIC. (*Source:* Bookham Inc., 2006. All Rights Reserved.)

5.7 Phase Shifters

Phase shifters are used to delay or shift the phase of an RF signal by a fixed or variable amount. The phase shift can be continuously variable, as in an analog phase shifter, or it can be switched in discrete steps, as in a digital phase shifter. Analog phase-shifter designs include the reflective and the loaded-line styles, and digital phase-shifter designs include the switched-delay-line and switched-filter styles.

5.7.1 Reflective Phase Shifters

The reflective style of phase shifter [134–137] is constructed from a 90° hybrid coupler, such as a Lange coupler, connected to two varactor diodes that are dc-biased through decoupling inductors, as in Figures 5.136 and 5.137. The Lange coupler splits the input power equally between the two diodes and couples the power reflected from the diodes to the output port. The phase of the reflected power from the varactor diodes is a function of their capacitance, which is varied by the control voltage.

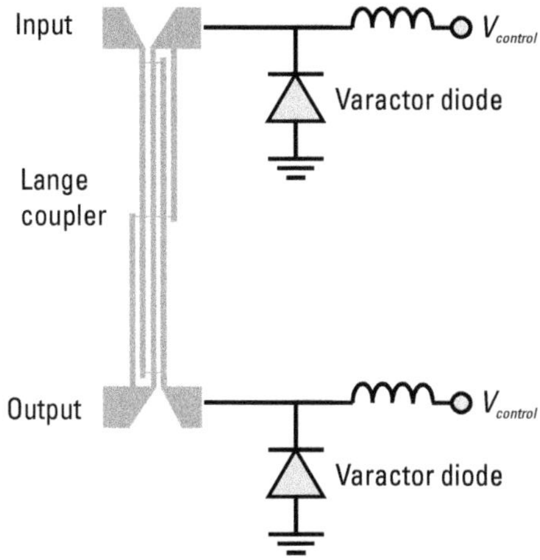

Figure 5.136 Reflective analog phase shifter.

5.7.2 Loaded-Line Phase Shifters

The loaded-line phase shifter [138–141] is constructed from a quarter-wavelength length of transmission line with a reactive load at each end, such as a varactor diode, as shown in Figure 5.138. The phase change along the transmission line varies according to the voltage on the varactors, which determines their capacitance. The varactor diodes at either end of the transmission line can be replaced with a switch connected to different lumped reactances in order to make a digitally controlled phase shifter.

5.7.3 Switched-Delay-Line Phase Shifters

The switched-delay-line phase shifter [142–144] is really the simplest style of digital phase shifter and consists of two different lengths of transmission line between two SPDT switches, as shown in Figure 5.139. The phase difference in degrees is simply the length difference divided by the wavelength and multiplied by 360°. The advantage of this type of phase shifter is that it provides a true-time-delay phase shift that is essentially a frequency-independent time delay of the sort required by phased-array radar applications.

5.7.4 Switched-Filter Phase Shifters

The switched-filter phase shifter [145–148] is a digitally controlled phase shifter that uses two SPDT switches to switch the signal path between a

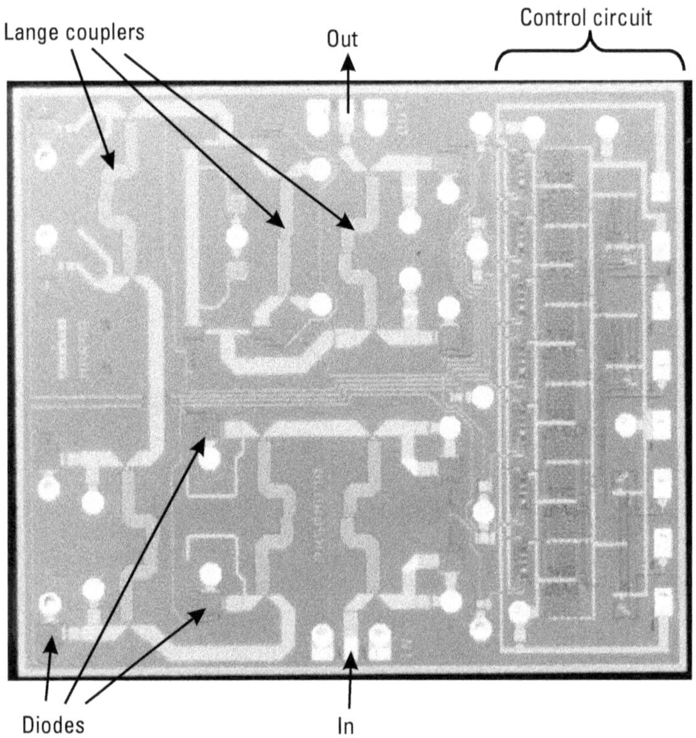

Figure 5.137 Reflective-type phase-shifter MMIC. (Courtesy of Alcatel Alenia Space France.)

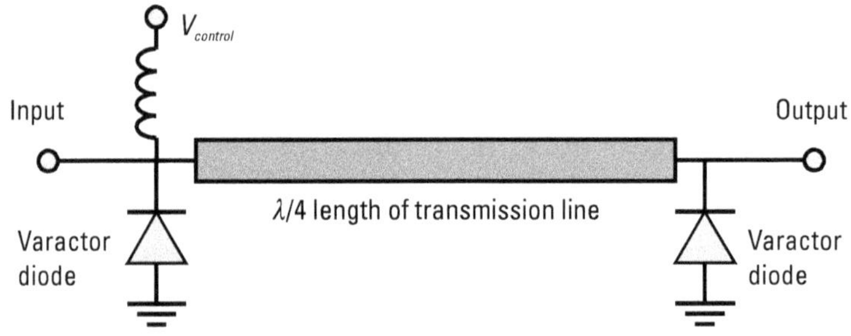

Figure 5.138 Loaded-line analog phase shifter.

high-pass filter and a low-pass filter, as shown in Figure 5.140. Each filter is designed to have equal insertion loss but different phase shifts. If large phase

Figure 5.139 Switched-delay-line phase shifter.

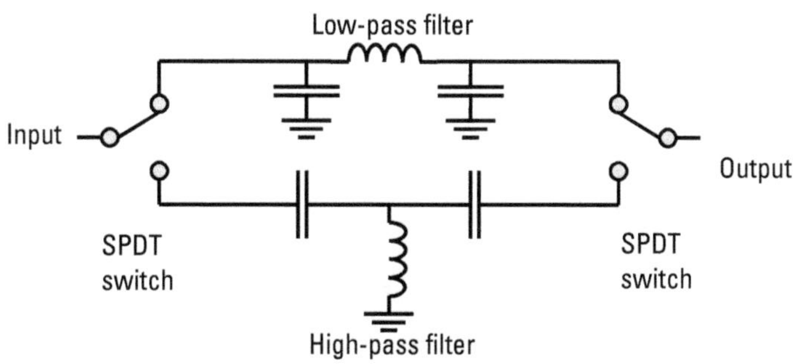

Figure 5.140 Switched-filter phase shifter.

shifts are required, then the filters can be designed to be higher-order filters. An example of a switched-filter 6-bit digital phase-shifter MMIC is shown in Figure 5.141. This MMIC digitally switches the phase in amounts of 180°, 90°, 45°, 22.5°, 11.25°, and 6.125°.

5.8 Switched-Path Attenuators

Switched-path attenuators [149–151] are constructed from two SPDT switches with two paths in between, and each path has a different attenuation, as shown in Figure 5.142. The attenuating path is usually constructed from a resistive-T, Pi, or bridged-T network of resistors [152]. An example 6-bit digital switched-path attenuator MMIC is shown in Figure 5.143, which has a bandwidth from 0.5 to 16 GHz.

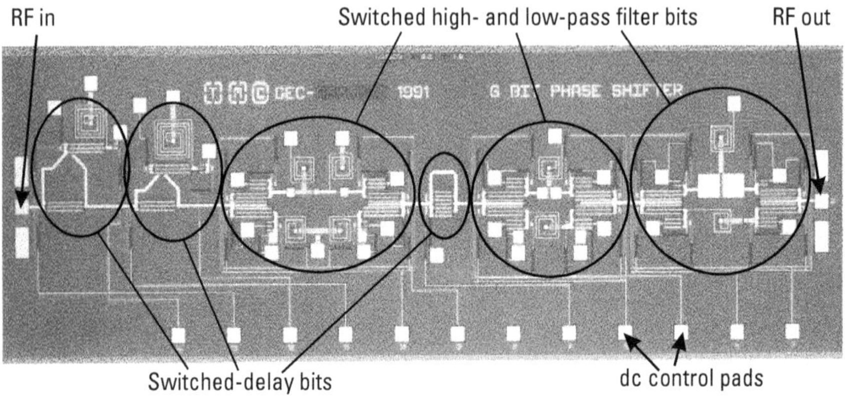

Figure 5.141 A switched-filter 6-bit digital phase-shifter MMIC. (*Source:* Bookham Inc., 2006. All Rights Reserved.)

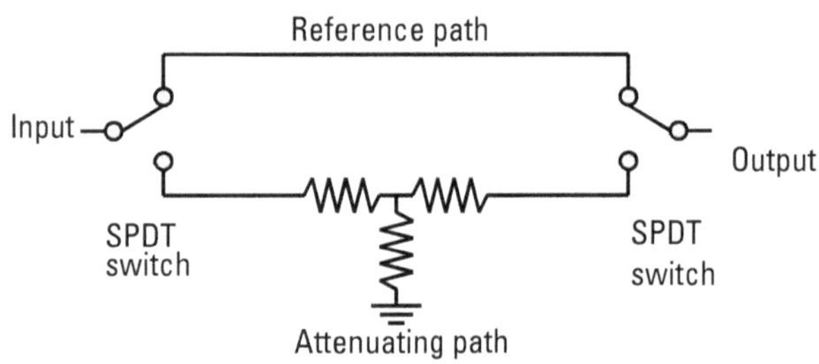

Figure 5.142 Switched-path digital attenuator.

5.9 Circuits with Digital Application

Digital circuits are primarily the domain of silicon CMOS circuits, but with the development of HBT MMIC processes capable of millimeter-wave-frequency performance, more and more circuits with digital applications are being designed on semiconductors other than silicon. The main examples of this are prescalers, log amplifiers, and Darlington amplifiers.

5.9.1 Prescalers

Prescalers are used in frequency-counter circuits to bring the frequency down to a level that can be handled by standard silicon signal-processing circuits.

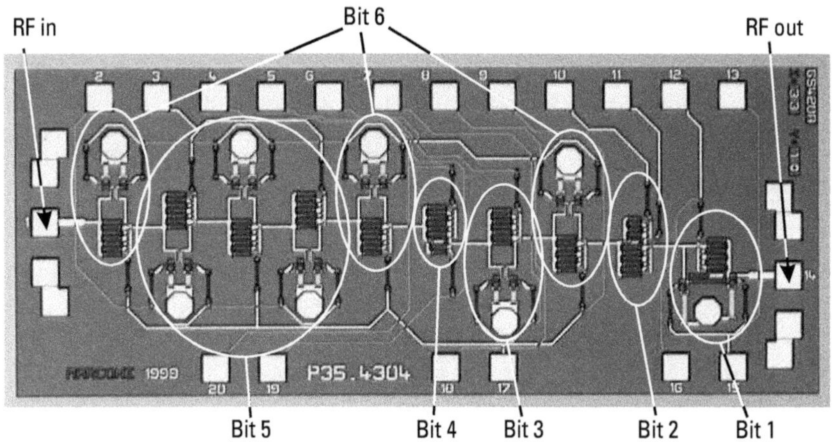

Figure 5.143 A 6-bit digital switched-path attenuator MMIC, 0.5 to 16 GHz. (*Source:* Bookham Inc., 2006. All Rights Reserved.)

For example, a divide-by-two prescaler schematic is shown in Figure 5.144, where the output from a master-slave flip-flop is fed back to its input so that it only changes every second cycle. Virtually the same low-frequency silicon circuit designs for fixed and variable modulus dividers can be reproduced at much higher frequencies using HBT MMIC processes. Examples of prescaler MMICs [153, 154] are in the literature, and a 17-GHz HBT divide-by-two prescaler MMIC is shown in Figure 5.145.

5.9.2 Logarithmic Amplifiers

The output voltage from a logarithmic amplifier is a log function of input voltage [155]. These amplifiers are used in electronic warfare applications to detect very low and very high amplitude signals. A typical MMIC design

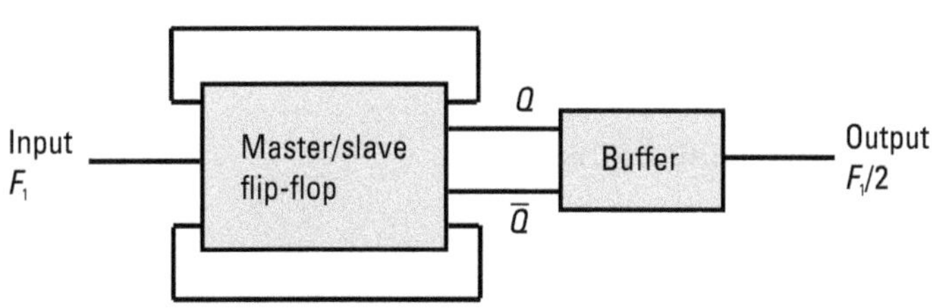

Figure 5.144 Divide-by-two prescaler.

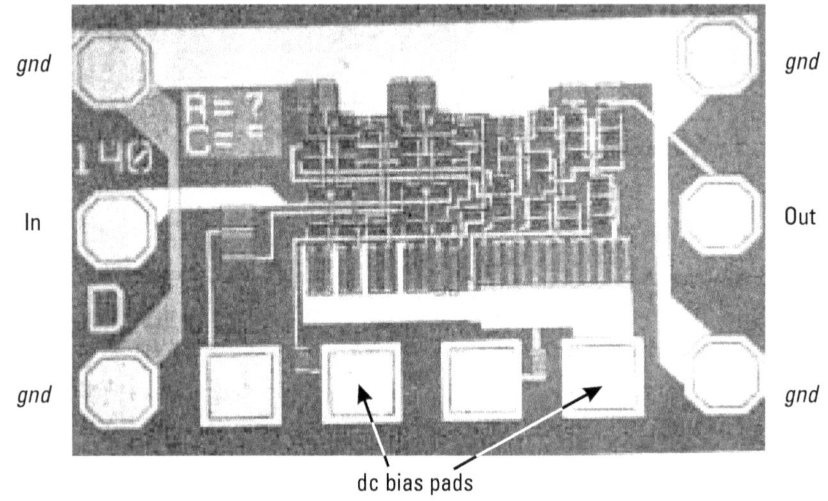

Figure 5.145 An HBT divide-by-two prescaler operating at 17 GHz.

[156] would consist of an HBT operational amplifier with nonlinear feedback from a common-base HBT, as shown in Figure 5.146.

5.9.3 Darlington Pair Amplifiers

In Darlington pairs of bipolar transistors, the amplified current from the first transistor is fed directly into the base contact of the second transistor, multiplying the current gain of two transistors together [157], as shown in Figure 5.147. They are as applicable to HBT circuits [158, 159] as they are to standard bipolar circuits. The advantages of the pair over the single transistor are much higher gain and increased input impedance, leading to larger bandwidths. These are dc-coupled circuits and commonly used in transimpedance amplifiers for optoelectronic communications. An example GaAs HBT Darlington amplifier MMIC is shown in Figure 5.148. This MMIC exhibits 31-dB gain from dc to 3 GHz.

5.10 Millimeter-Wave Circuits

Millimeter-wave circuits are so-called because they operate at frequencies where the free-space wavelength is less than one centimeter and best expressed in millimeters. This means they have frequencies of 30 to 300 GHz, and because the wavelength is reduced by a higher dielectric constant

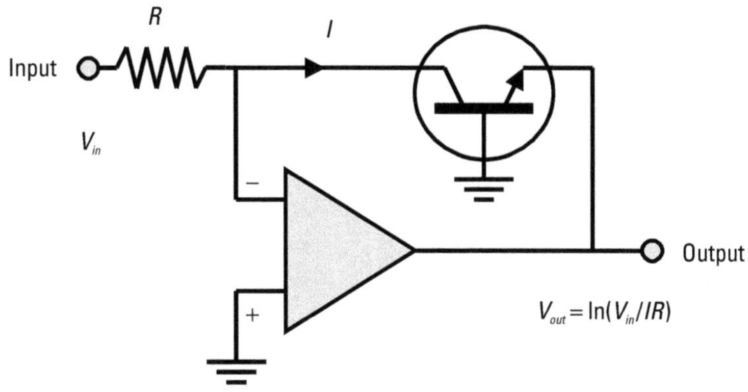

Figure 5.146 Logarithmic amplifier.

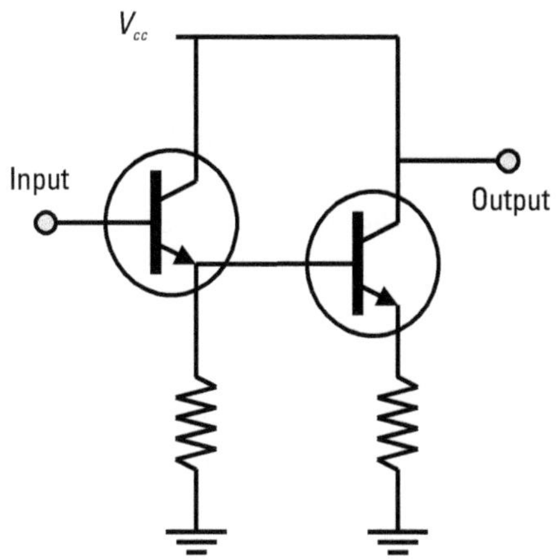

Figure 5.147 Darlington pair.

than free space, the wavelength on a GaAs microstrip is as short as ~3 mm. This has a number of consequences for MMIC design because of the higher frequency; for example, the choice of components changes, and simulation techniques become an issue. These areas are discussed in the following sections.

5.10.1 High-Frequency Effects

A number of MMIC design issues can be neglected at RF and microwave frequencies, but they start to become more important at frequencies higher than

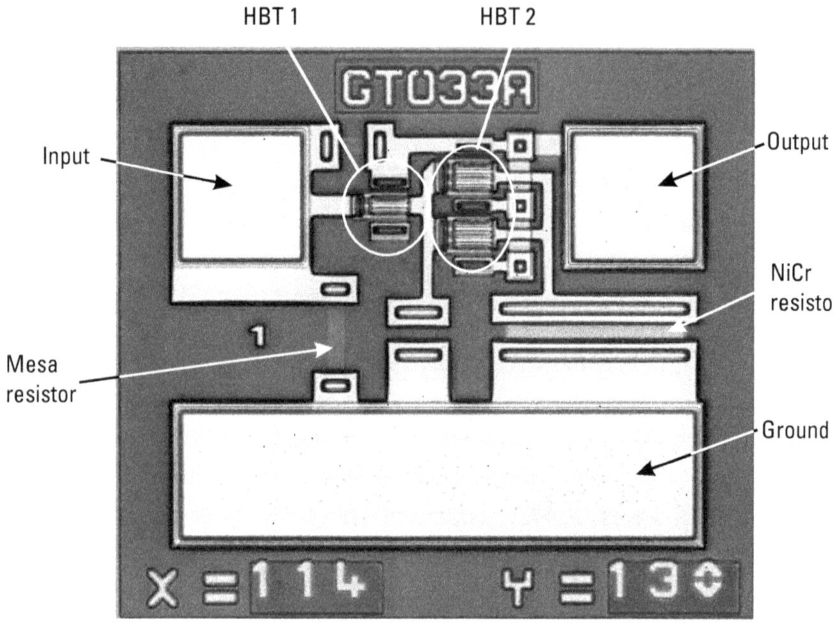

Figure 5.148 HBT Darlington amplifier MMIC. (*Source:* Bookham Inc., 2006. All Rights Reserved.)

30 GHz. This is not only because the frequency is higher and the wavelength is shorter but because the active devices become limited, and parasitics associated with the simulation models and chip-to-system interfaces become more important. In fact, the use of 3D EM analysis tools becomes essential for first-pass design success.

For example, at millimeter-wave frequencies, components like spiral inductors become resonant and no longer look inductive. Figure 5.149 shows that a circuit operating at 40 GHz is only able to use spiral inductors with 2.5 turns or fewer. Also, the thickness of the substrate may allow multiple modes to propagate along a microstrip transmission line, so CPW transmission lines may need to be used instead. Indeed, the transmission lines and other components become dispersive at millimeter-wave frequencies, making broadband designs difficult. This can be seen in Figure 5.150, where the variation of effective dielectric constant is plotted verses frequency for an 80-μm-wide microstrip line on a 100-μm GaAs substrate.

The shorter wavelength at millimeter-wave frequencies (typically < 3mm on GaAs), shown in Figure 5.151, means that the size of matching elements become smaller, closer to the limit of the element equivalent circuit

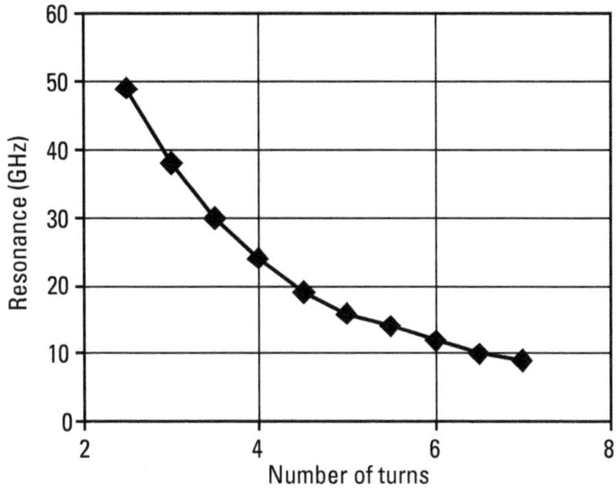

Figure 5.149 Self-resonant frequency of a typical MMIC spiral inductor versus the number of turns of the track.

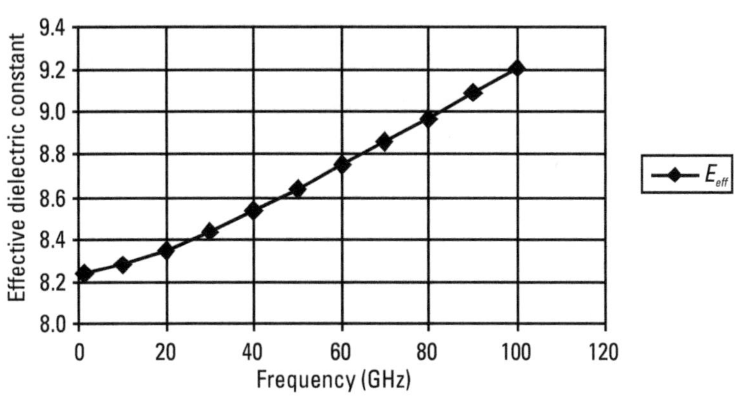

Figure 5.150 Effective dielectric constant of an 80-μm-wide microstrip on a 100-μm-thick GaAs substrate versus frequency.

simulation models. Also, large elements or transmission lines can easily approach fractions of the wavelength, increasing the possibility of component coupling. On the plus side, short wavelengths mean that some alternative components become available to the designer, such as distributed Wilkinson splitters, Lange couplers, and antennas, which would be too large at lower frequencies.

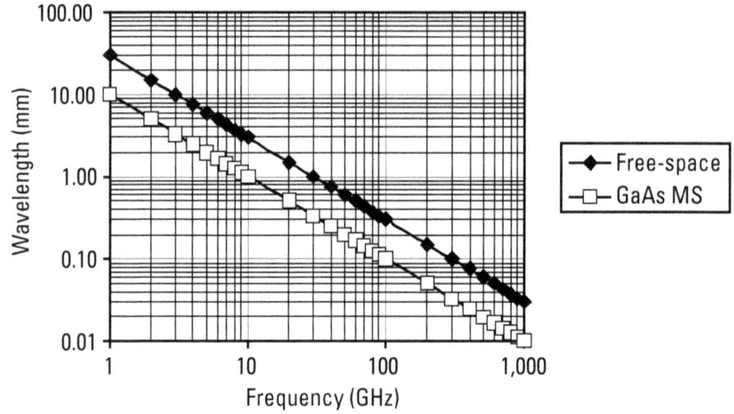

Figure 5.151 Wavelength in free space and on a GaAs microstrip transmission line versus frequency.

Parasitics of the component models and the chip-to-system interface become more important as the frequency increases; they start to cause significant problems at millimeter-wavelengths. For example, the through-substrate via inductance becomes significant (5Ω at 30 GHz from Figure 5.152), no longer looking like a good short circuit; thus, radial stubs are required instead. Because the parasitics associated with lumped elements become a larger proportion of their impedance, using distributed elements such as short or open circuit stubs can give more confidence for matching circuits. In Section 5.3.1.1, it was seen that the chip interface stray capacitance and bond-wire inductance can have a large effect on the

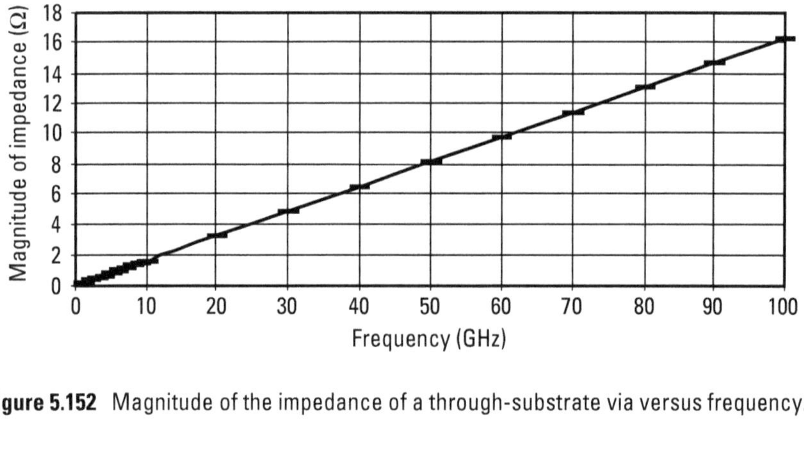

Figure 5.152 Magnitude of the impedance of a through-substrate via versus frequency.

chip performance and make them very sensitive to assembly variation. At millimeter-wave frequencies, this is even more important because, as shown in Figure 5.45, the impedance variation due to the bond-wire length range is about 30Ω at 30 GHz.

At millimeter-wave frequencies, the performance of the active devices also becomes limited. For example, Figure 5.153 shows the G_{max} and stability factor K of a 2 × 75-μm HEMT, and above 30 GHz, the gain is just over 10 dB, dropping down to only 5 dB at 80 GHz. To get this gain at millimeter-wave frequencies, the gain at 1 GHz is over 25 dB, and the stability factor is well below 1 at about 0.1, which means it will be difficult to stabilize this device at RF frequencies, often requiring lossy matching techniques. The active device reproducibility is also worse at these frequencies, leading to greater chip performance variation and lower yields. The difficulty in extracting accurate models also gets worse with frequency, increasing uncertainty and modeling errors.

5.10.2 Component Choice

It was noted earlier that the short wavelength allows alternative components to be used in the MMIC design process, particularly those that require dimensions of one-quarter of the wavelength. These include open and short circuit stubs, antennas, radial stubs, coupled-line couplers and filters, branch-line couplers, rat-race couplers, Lange couplers, Marchand baluns, distributed Wilkinson combiners, distributed transmission-line combiners, traveling-wave combiners, and tapered transmission-line splitters. The other

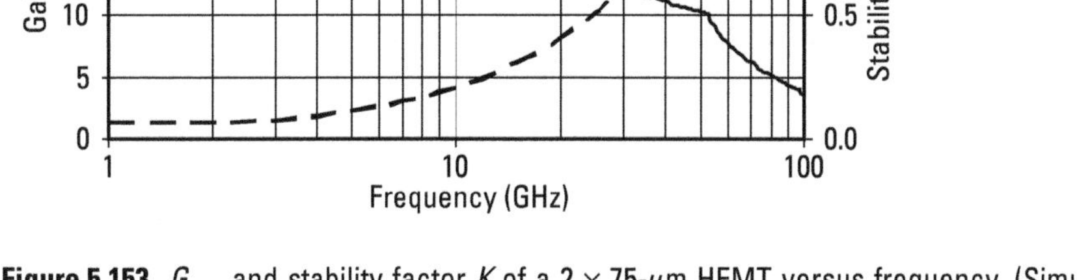

Figure 5.153 G_{max} and stability factor K of a 2 × 75-μm HEMT versus frequency. (Simulated with Agilent ADS and the PH25 Design Kit, courtesy of UMS.)

choice that has to be made early on in the design process is which type of transmission line to use, microstrip or CPW. There are advantages and disadvantages with both transmissions lines, each being preferred in different applications. The next section looks at some of the differences between their characteristics and how this affects their applicability to millimeter-wave MMIC circuit design.

5.10.2.1 Microstrip versus CPW Transmission Lines

The choice between using microstrip and CPW for the transmission lines in an MMIC design is not always clear or easy to make. Figure 5.154 shows a schematic diagram of the two transmission-line types. There are differences in not only the transmission-line properties but also in the characterization and modeling, the processing costs, the ease of connecting to ground, and the density of components on the chip surface.

Transmission Line and Components

As a pure transmission-line, CPW has certain advantages over microstrip. First, CPW exhibits a quasi-TEM mode of propagation to much higher frequencies than microstrip, which, for a 100-μm-thick substrate, is no longer single-moded above about 60 GHz. An example 100-GHz amplifier that utilizes CPW to ensure a single mode is shown in Figure 5.155. The

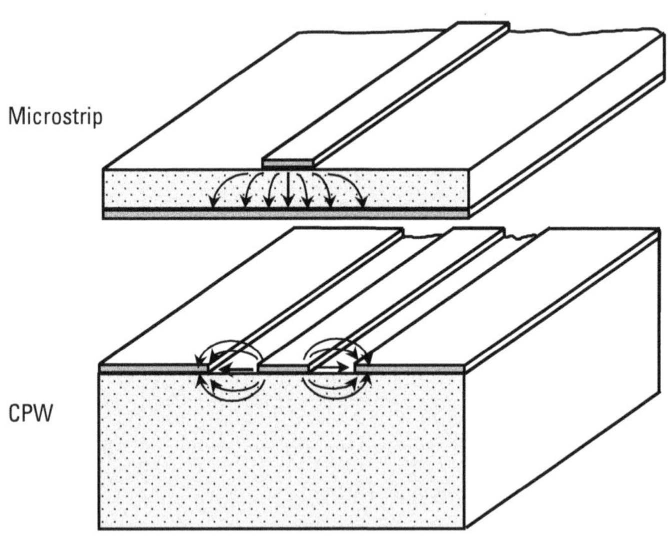

Figure 5.154 Microstrip and CPW transmission lines.

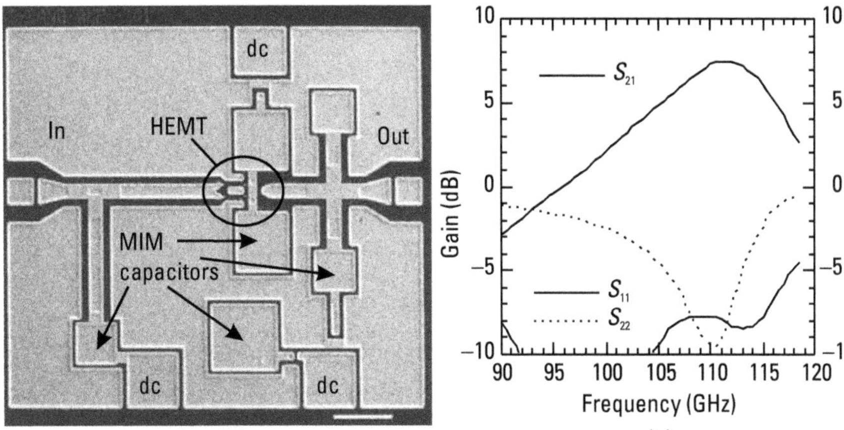

(a) (b)

Figure 5.155 (a) Photo and (b) frequency response of a CPW dual-gate HEMT 100-GHz amplifier. (Courtesy of Katholieke Universiteit, Leuven/IMEC [80].)

characteristic impedance of microstrip is also limited to values between 40Ω and 130Ω, whereas CPW has a wider range of characteristic impedances and the gap-to-track-width ratio, which determines the impedance, can be scaled together to make tapers where the impedance is kept constant. The other advantage of CPW is that it has lower dispersion than microstrip, which eases broadband design, and the odd- and even-mode propagation velocities of coupled lines are closer, so broadband coupler design is easier. Baluns are also easier to make using CPW and unbalanced slotline, whereas microstrip baluns are difficult to design in a compact manner.

Microstrip has some small advantages in that the loss of microstrip is slightly lower than CPW, and the effective dielectric constant of the microstrip mode is higher than that of CPW, which gives slightly smaller distributed components due to a shorter wavelength.

The biggest disadvantage of CPW is that odd modes can be generated at discontinuities, and although the bridges inhibit their propagation, closely spaced discontinuities can couple by other modes except the desired quasi-TEM even CPW mode. The implication is that the transmission-line elements, discontinuities, and lumped elements need powerful modeling tools and many man-hours to be characterized sufficiently. This single fact puts off most foundries from offering CPW MMIC processes. Microstrip elements are much better behaved and can be reliably and quickly modeled from measured *s*-parameters.

Processing

The number of processing steps, hence the cost of CPW, is less than that for microstrip. This is because with microstrip the signal is on the top surface, while the ground is on the back surface, so the wafer must be thinned to get the correct distance between the signal and ground, the back surface must be metalized, and through-substrate vias must be created to connect components to the ground plane. CPW has the signal and ground on the same top surface, so the wafer does not need to be thinned or metalized, and through-substrate vias are not needed.

The extra processing that CPW does require is the production of bridges between the ground planes on both sides of the signal line to eliminate the propagation of unwanted odd modes at discontinuities. These are usually air bridges, but they can also be dielectric bridges, as shown in Figure 5.156.

Ground Connection

CPW also has advantages because the coplanar ground plane is so close and easy to connect to. Figure 5.157 shows that the CPW shunt capacitor is

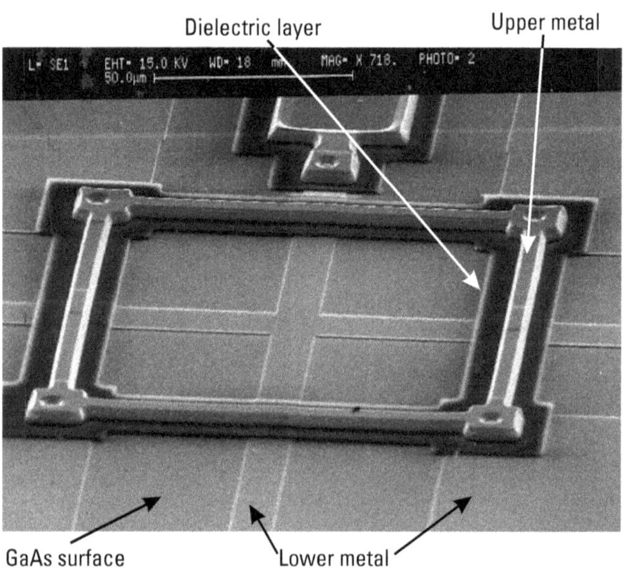

Figure 5.156 Bridges of metal, supported by a dielectric layer, between the ground planes of CPW at a cross-junction discontinuity to eliminate unwanted odd modes. (*Source:* Bookham Inc., 2006. All Rights Reserved.)

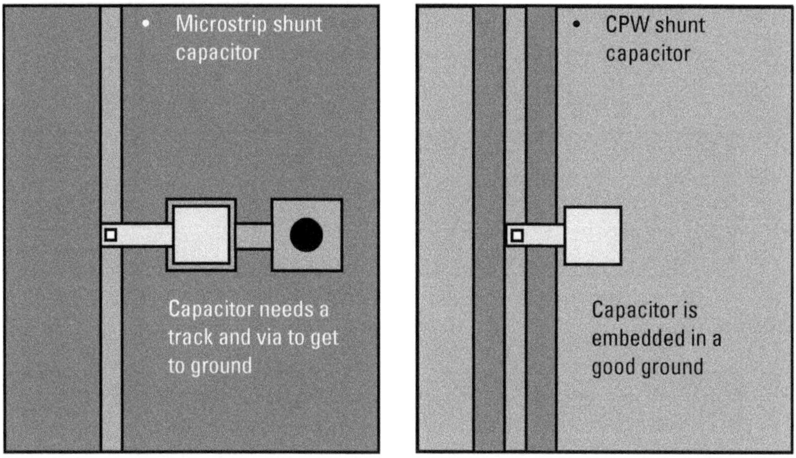

Figure 5.157 Microstrip and CPW shunt capacitor configurations.

actually embedded in the ground plane with only a short track to the signal line, so it looks like a very good low-impedance ground up to very high frequencies. On the other hand, the microstrip shunt capacitor requires a through-substrate via that does not have very low impedance at millimeter-wave frequencies. Similarly, the ground connections of the active devices are very short, so the device maintains good gain at high frequencies, while the through-substrate vias of the microstrip active devices introduce inductive source feedback that reduces their high-frequency gain. The close ground connection offers similar advantages when the chip is being tested RFOW. A typical RFOW probe, as in Figure 5.158, is already in a CPW configuration, and calibration can simply place a measurement reference plane at the probe tips. With microstrip, through-substrate vias are required to contact to the ground plane, so it is best to calibrate with test structures that include the RFOW pads and vias and place the measurement reference plane at the start of the microstrip transmission line.

Another advantage of the close ground connection is that the layout of active devices in the cascode arrangement is very compact, especially using dual-gate FETs, and grounding has very little inductance (Figure 5.159). With microstrip, a cascode layout is possible, but the layout is larger and has higher inductance to ground. The cascode dual-gate FET can be laid out in a T configuration, as in Figure 5.160, or in a Pi configuration, as in Figure 5.161. An example CPW distributed amplifier MMIC using dual-T-gate FETs and operating from 5 to 55 GHz is shown in Figure 5.162.

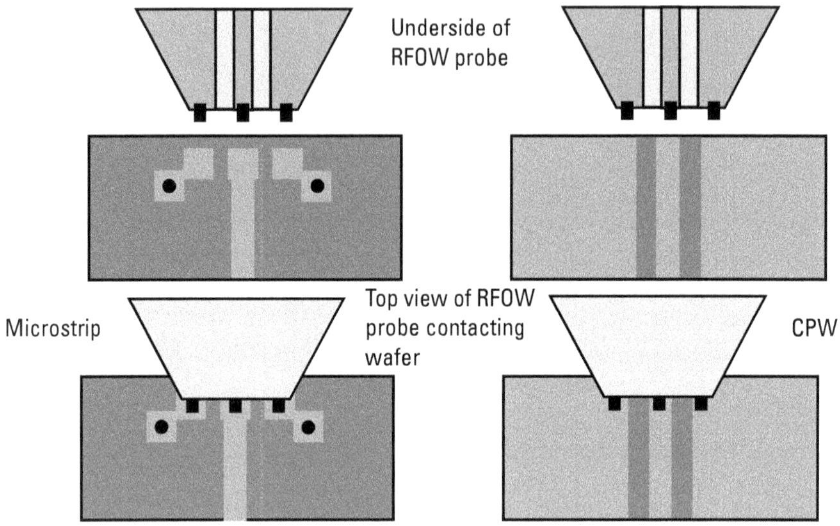

Figure 5.158 Top and underside view of RFOW probes contacting microstrip and CPW transmission lines.

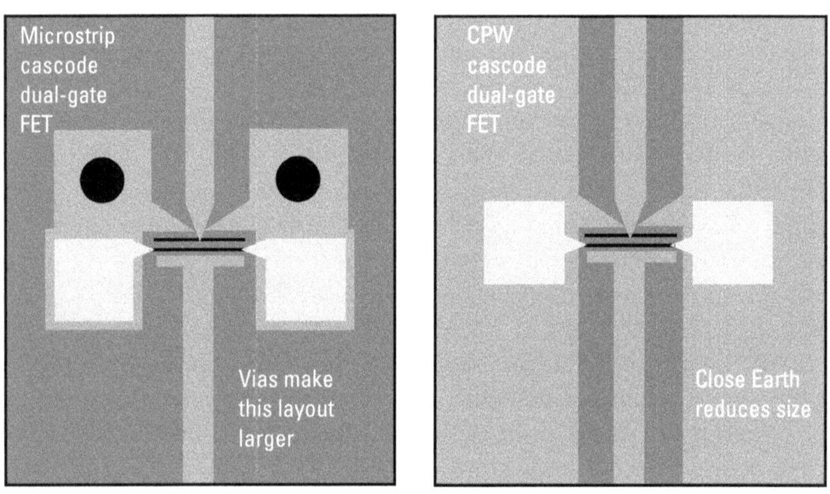

Figure 5.159 Dual-gate cascode FETs in microstrip and CPW layout.

Surface Density

Both types of transmission line have certain advantages when it comes to the packing density of components on the chip surface for miniaturization and

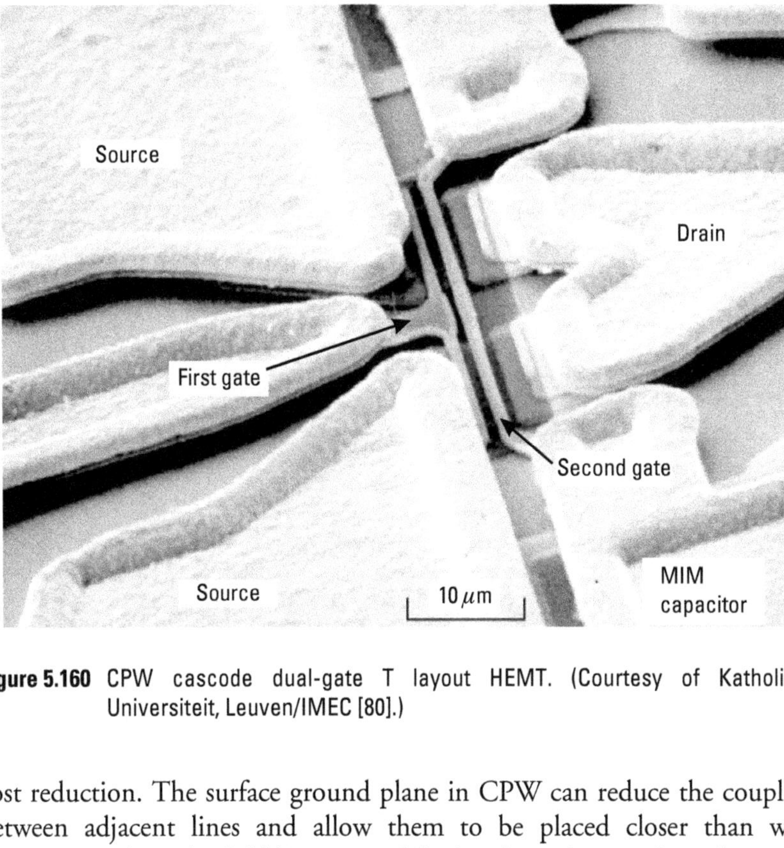

Figure 5.160 CPW cascode dual-gate T layout HEMT. (Courtesy of Katholieke Universiteit, Leuven/IMEC [80].)

cost reduction. The surface ground plane in CPW can reduce the coupling between adjacent lines and allow them to be placed closer than with microstrip, where the field lines extend further from the signal conductor, as shown in Figure 5.163. However, in dense and complex CPW layouts, such as the example in Figure 5.164, there is sometimes so little surface ground plane that its performance as a good ground is impaired. Microstrip grounding, because it is a continuous sheet of metal on the back surface, is unaffected by the circuit density or complexity. Indeed, CPW power amplifier design suffers from the same difficulties because the high dc current these amplifiers utilize must be routed through the noncontinuous top ground plane back to the power supply without fusing the air bridges. An example CPW power amplifier is shown in Figure 5.165, where a ground island is formed near the large power devices, and the dc ground return path from this island is via the CPW air-bridge connections. CPW power amplifiers also have difficulty dissipating the heat they generate because the thermal conductivity through their typically 625-μm-thick substrate is poor, and flip-chip mounting must be considered. Flip-chip mounting also has its own

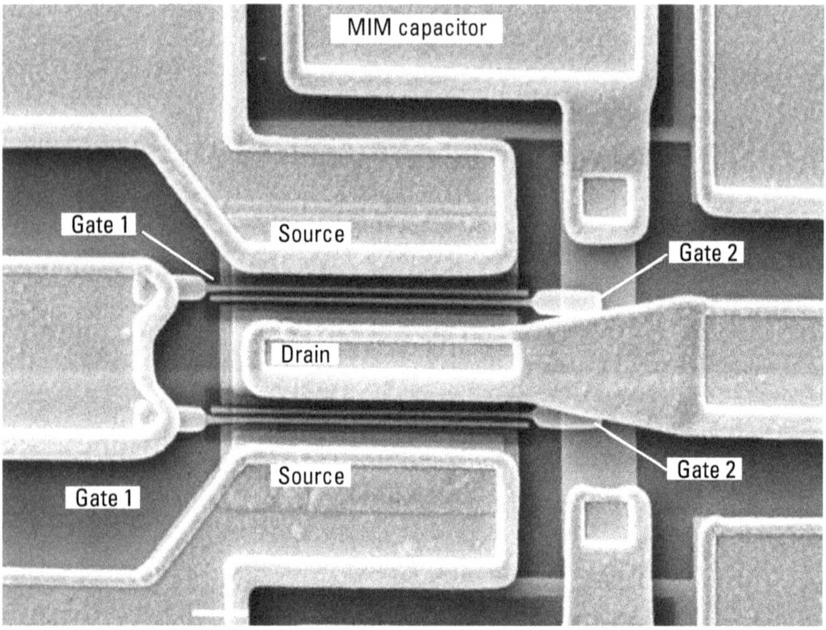

Figure 5.161 CPW cascode dual-gate Pi layout HEMT. (Courtesy of Katholieke Universiteit, Leuven/IMEC [80].)

difficulties: solder connections must be provided on the active device to connect to a thermal heat-sink, and the chips are face down, so the solder joints cannot be visually inspected. Microstrip has the advantage for power amplifiers because the substrate thickness can be as low as 50 μm, giving good thermal dissipation.

5.10.3 Simulation Issues

At millimeter-wave frequencies, there is always the need for 2.5D or 3D electromagnetic (EM) simulation of part or all of the MMIC layout to ensure an accurate design. This is partly because the wavelength is small, therefore there is a greater chance of radiating RF energy, and also because the effect of parasitics is larger, and the connecting together of close discontinuities using *s*-parameter models may not take into account all of the modes of interaction between them.

Current flow in shared components, such as through-substrate vias, is also uncertain at millimeter-wave frequencies. Consider the through-substrate via shown in Figure 5.166, which is between the large FETs in the

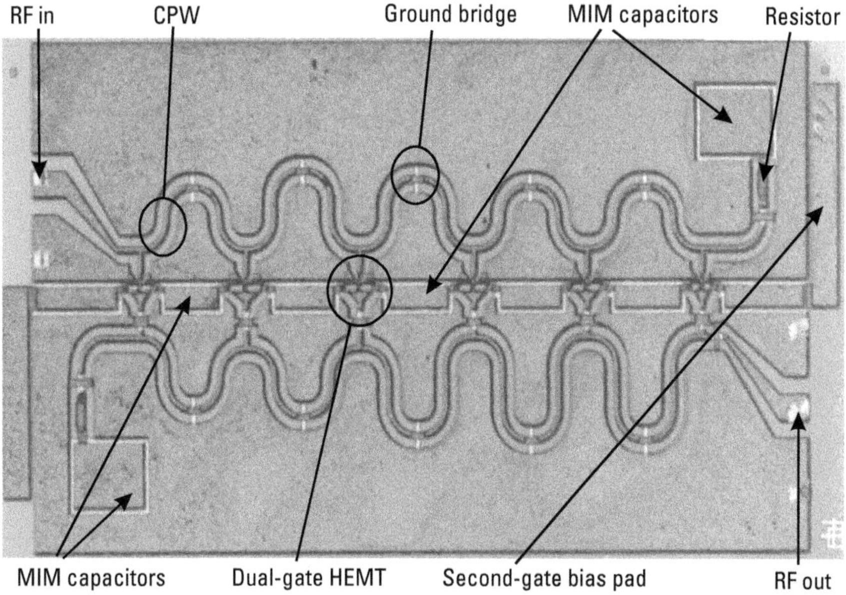

Figure 5.162 A 5–55-GHz CPW distributed dual-gate amplifier. (Courtesy of Katholieke Universiteit, Leuven/IMEC [80].)

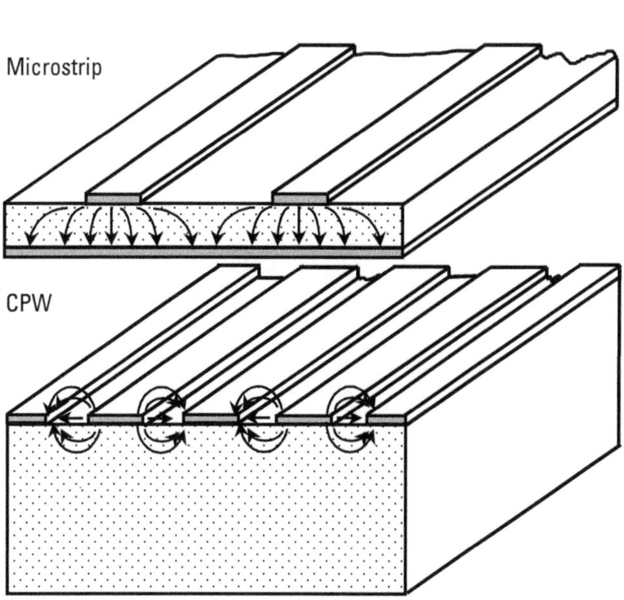

Figure 5.163 Fields between adjacent microstrip and CPW transmission lines.

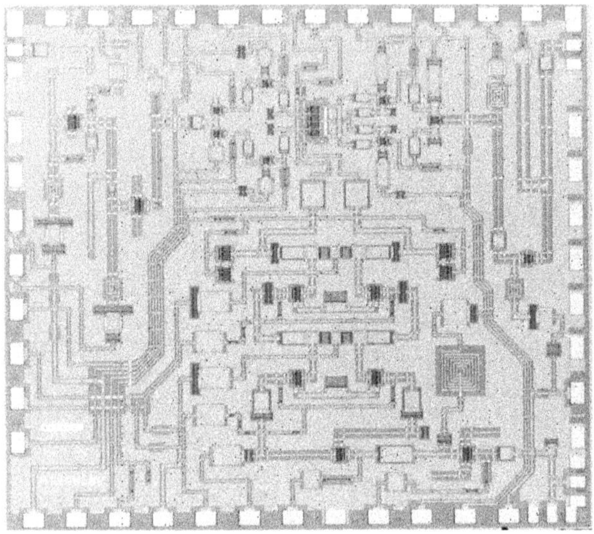

Figure 5.164 Densely packed CPW mixer MMIC [160]. (Courtesy of Alcatel Alenia Space France.)

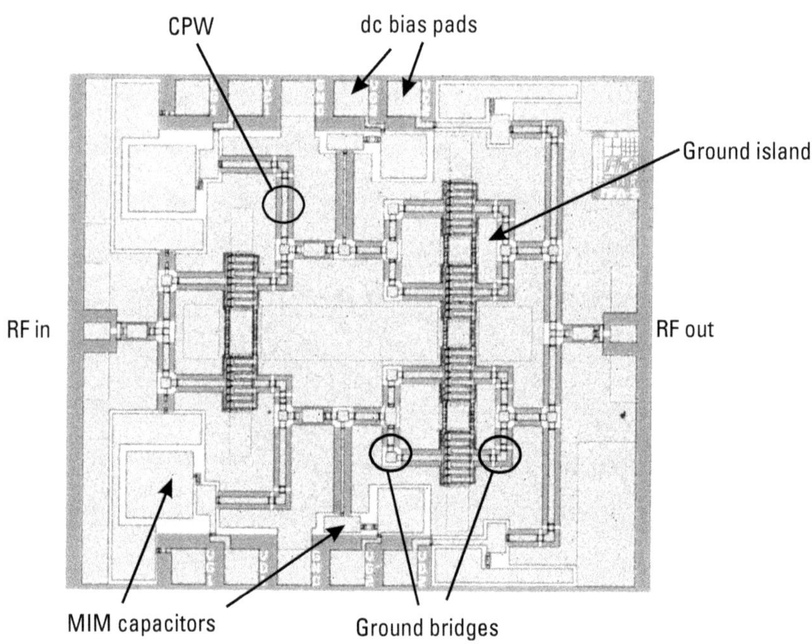

Figure 5.165 CPW 35–42-GHz GaAs pHEMT power amplifier.

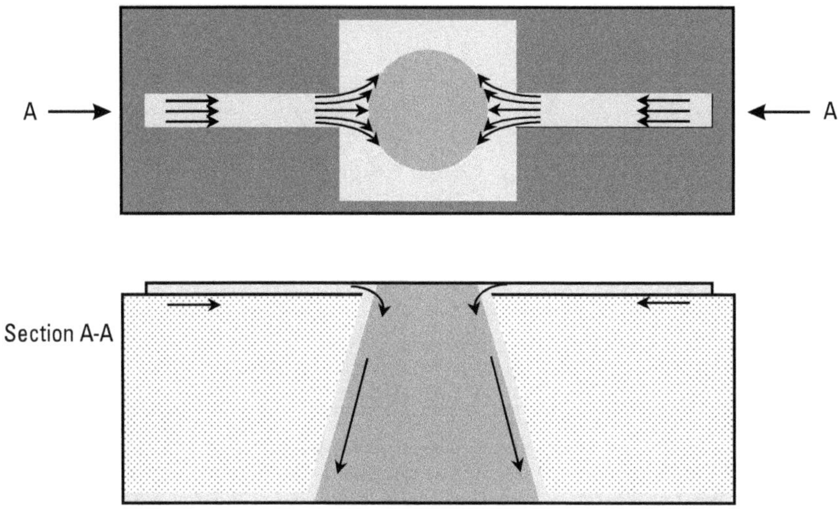

Figure 5.166 Millimeter-wave-frequency current flow near a through-substrate via.

output stage of a power amplifier, such as the one shown in Figure 5.167. One may assume that when the via was characterized, the current flowed evenly in the conically shaped metal and that when shared by two FETs, each one would see increased inductance from the via. In fact, the current tends to flow down the side of the via facing the signal, so two FETs can share the same via without looking more inductive [161]. This effect needs to be checked with 3D EM simulation at the required design frequency.

At millimeter-wave frequencies, the wavelength is much shorter, so the possibility of odd modes between the far-spaced output FETs (in Figure 5.167) is more likely. If this is the case, then bus-bar-type power combiners must be used to help prevent odd modes; otherwise, complex odd-mode analysis of the amplifier is required to ensure that it will not oscillate when operating.

5.11 Yield Improvement

To understand how to improve the yield of an MMIC, it is necessary to define what is meant by the MMIC yield because different people mean different things when referring to MMIC yield. Then, a number of techniques can be used during the design stage to give the MMIC design the best chance of having a high yield when processed.

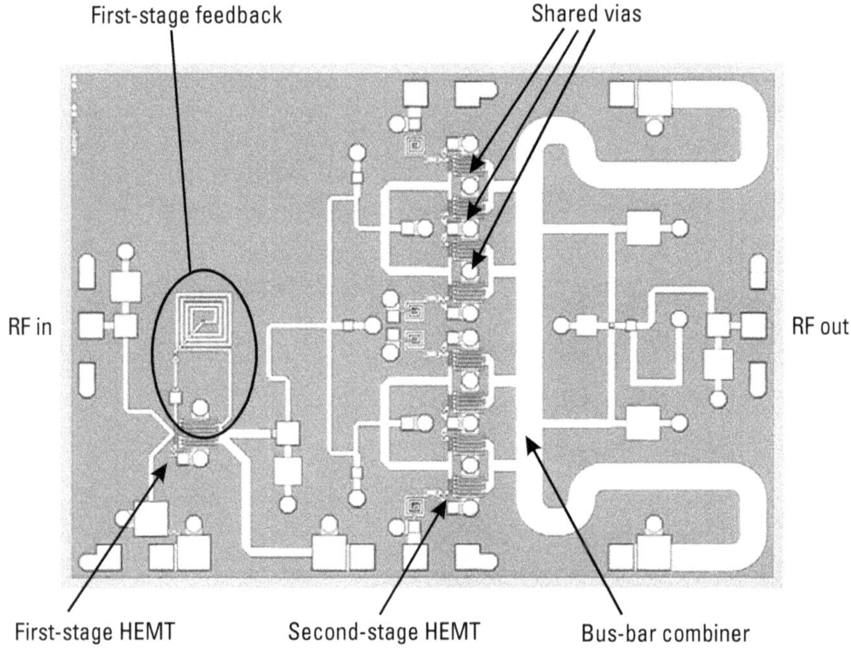

Figure 5.167 Power amplifier MMIC with shared through-substrate vias in the output stage. (*Source:* Bookham Inc., 2006. All Rights Reserved.)

5.11.1 What Is the MMIC Yield?

MMIC yield may mean one thing to a processing engineer and something quite different to a test engineer, so to look at the different yields that may be referred to, we need to reexamine the chip-production process and look at the many reasons we may not get a chip that will function to specification in the intended application. The chip-rejection reasons are generally broken down into the following areas: processing yield, dc and RF test yield, saw and dicing yield, visual yield, and packaging yield.

5.11.1.1 Processing Yield

Wafers may get damaged during processing, either by machine malfunctions, raw processing material that is not to specification, or human error on the part of the equipment operators. Processing yield tends to refer to the number of wafers that complete the process route compared to the number at the start of the process.

5.11.1.2 dc and RF Test Yield

Process control devices are often tested at dc after the front-face processing is complete to check that the devices conform to specification. Also, after the back-face processing is complete, some circuits may be tested again for dc functionality. Depending on the type of circuit, the chips will probably then be tested RF-on-wafer (RFOW) against the performance requirement. Many factors may cause the devices to fail these tests, including defects in the original substrate material, design errors, mask-writing errors, process variations, and material quality. The dc and RF test yields normally refer to the number of devices that pass the dc and RF performance criteria.

5.11.1.3 Saw and Dicing Yield

Wafers are either sawn up with a diamond saw or scribed with a diamond and cleaved along the scribe lines using physical pressure to separate the individual MMIC chips. The wafers or chips may be damaged during this process, lowering the expected number of working dies. Saw or dicing yield usually refers to the average percentage of undamaged chips that complete this process.

5.11.1.4 Visual Yield

Wafers are often sawn or diced while attached to an adhesive film to keep the die in place during this process. After sawing or dicing, the film is expanded to increase the space between the dies, and the chips are picked off either manually using tweezers or by an automated machine. The chips are then examined visually using a microscope and sorted into different classes according to predetermined guidelines. Visual defects can include cracks in the substrate or surface layers, substrate defects, material that is not fully removed or deposited during processing, defects due to dirt particles, surface scratches, or extraneous material. The visual yield usually comprises the number of chips sorted into the top class of the specification.

5.11.1.5 Packaging Yield

It is common for the bare chips to be packaged for protection in plastic or ceramic. Assembly of the chips into the packages can include bonding the chip to the package, wire-bonding or "flip-chip" connection of the chip ports to the package lead frame, sealing and marking the package, and cropping the package leads. Each of these activities can fail, reducing the eventual number of working packaged circuits. The packaging yield refers to the percentage of circuits that pass successfully through the packaging process.

5.11.1.6 Overall MMIC Yield

The overall yield is the product of all these individual yields, and if care is not exercised at each stage, then the overall yield can be quite low. Yield information is commercially sensitive, so foundries will not disclose this information without first ensuring that a nondisclosure agreement is in place. The yield of any MMIC does improve with wafer size as less area is lost to drop-in process control monitor (PCM) areas and edge sites. Generally speaking, for a high-volume product, yield does increase as the process is exercised continuously and operators become more experienced with the particular product.

5.11.2 Yield-Improvement Techniques

A number of actions can be taken during the design stage to improve the chances of achieving a high-yielding MMIC product. These include a number of design techniques specifically aimed at making the circuit performance robust against process variations and some methods that indicate how changing the process could improve the yield. Sensitivity analysis and tolerance analysis are important tools for the MMIC designer, and an example of the Taguchi tolerance-analysis technique is detailed at the end of this section.

5.11.2.1 Circuit Design for Process Tolerance

Process variations are a fact of life for the MMIC designer, so every circuit must be designed to be as insensitive as possible to these variations. Some of the design techniques to do this are fairly commonsensical, such as designing to a wider frequency range than the specification, keeping components well spaced, and avoiding using the models near their limits, but using feedback in amplifier stages and avoiding MIM capacitors for matching networks can also help a lot. These next sections go into a bit more detail about these design techniques and also describe one particular strategy to ensure that power amplifier designs pass the specification with the first design.

Frequency Guard Bands

The performance of measured MMICs can differ from the simulation in terms of their gain level and match values, but experience has shown that the largest and most common differences are in the frequency of the response, with the desired gain and match response moving down or up in frequency by a small amount. Chip rejects due to this shift can be minimized by designing the MMIC for at least 20% wider bandwidth than the required specification, as in Figure 5.168, in other words, with guard bands on either side of the specified frequency band.

Good Element Spacing

Most active and passive elements are characterized between fairly long lengths of transmission line (typically 200 μm), which allow any evanescent modes to decay and not propagate to the RFOW probes [Figure 5.169(a)]. In this way, they can be accurately represented by the measured *s*-parameters or a model fitted to the *s*-parameters. If components are connected together with insufficient lengths of transmission line between them [Figure 5.169(b)], evanescent modes can propagate between them such that their behavior is no longer represented by the *s*-parameters or the model.

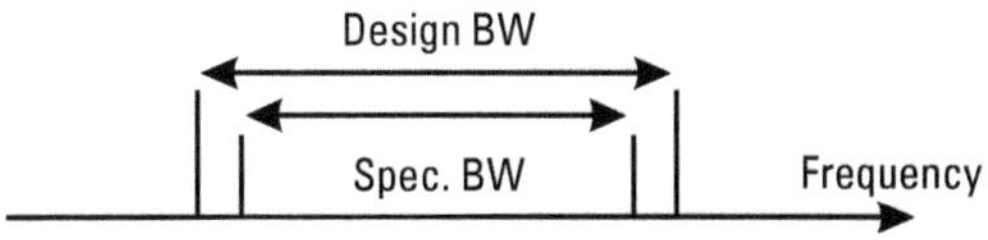

Figure 5.168 Design bandwidth compared to the specification bandwidth.

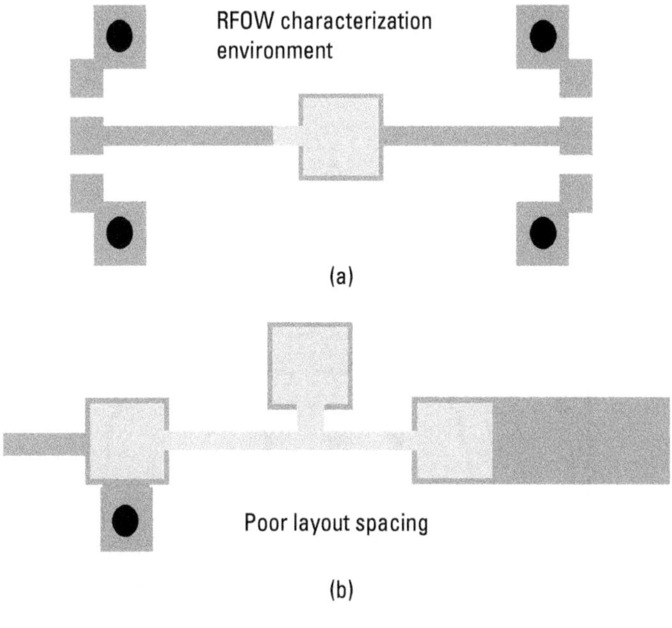

Figure 5.169 (a) RFOW characterization environment and (b) poor layout spacing.

The electric and magnetic fields extend from microstrip transmission lines and other components further than their physical layout, as in Figure 5.170, so if unconnected components are too close, they will be coupled by the fringing fields. Where microstrip lines are concerned, coupling is minimal if the separation is more than five strip widths or more than one substrate height from the side of the conductor strip.

Good spacing between elements of the MMIC design, whether they are connected or adjacent, unconnected elements, will produce a design with closer agreement between the simulated and measured performances.

Avoiding Model Limits

Active and passive device models are often scaleable for ease of use by the designer. This means that the model may have been fitted to a wide range of element sizes and the model parameters fitted by polynomial series to be scaleable against the device size. When elements become very small or very large, the parasitic parts of the model can start to vary exponentially against the feature size. This can lead to larger errors at the size-range limits, which can accentuate the effect of process variation. It is good practice to use scaleable elements in the middle of their range and avoid using them at the limits. Measured *s*-parameters of discrete sizes of each element type may be available from the foundry, so if the design requires a component near the model limit, the specific measurement of that size element should be used in the simulation.

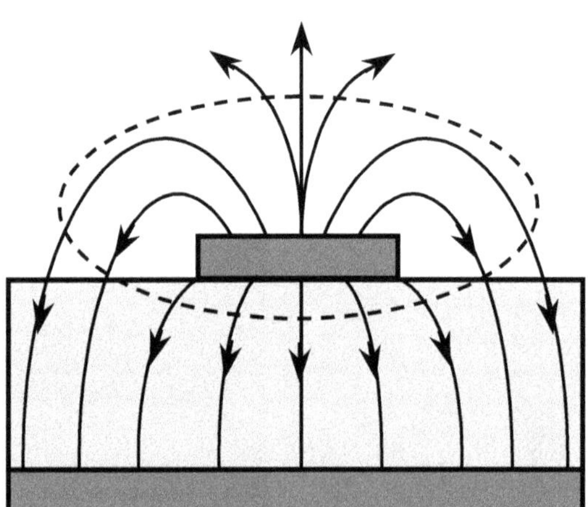

Figure 5.170 Fringing electric and magnetic fields around a microstrip transmission line.

Amplifiers with Feedback

An amplifier with feedback is less sensitive to the gain of the active device than one without. This is because the gain with feedback $A_F = A/(1 + \beta A)$ of the amplifier in Figure 5.171, when βA is a lot less than unity, is given by $A_F = 1/\beta$; hence, the gain can be made dependent on the passive feedback network and independent of the device gain A, which will vary with processing. An example of a two-stage amplifier MMIC with feedback on each stage is shown in Figure 5.172.

Matching with Only Transmission Lines

Transmission lines on MMICs are very insensitive to process variations. The possible process variations that could affect their RF characteristics include their physical dimensions, the metal thickness, and the substrate thickness. The dimensional variation of metal features in an MMIC process is less than a micron, which is very small compared to the size of transmission lines, and the metal thickness is well controlled and has only a second-order effect on the transmission-line loss and impedance. Variation of the thickness of the substrate has the greatest effect, but this only affects tracks with widths approaching the substrate thickness. As the substrate thickness in MMIC processes is controlled within a few percent, this again is only a second-order effect.

This means that an MMIC that only uses transmission lines for its matching elements, like the example in Figure 5.173, will be relatively insensitive to process variations.

Capacitors for dc Blocking Only

The capacitance of metal-insulator-metal (MIM) capacitors is a strong function of the insulator/dielectric thickness, and this can vary from wafer to wafer by a few percentage points. If capacitors are used for matching

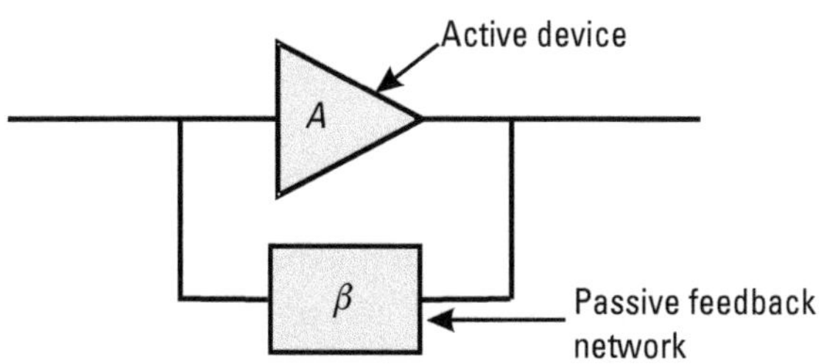

Figure 5.171 Amplifier stage with feedback.

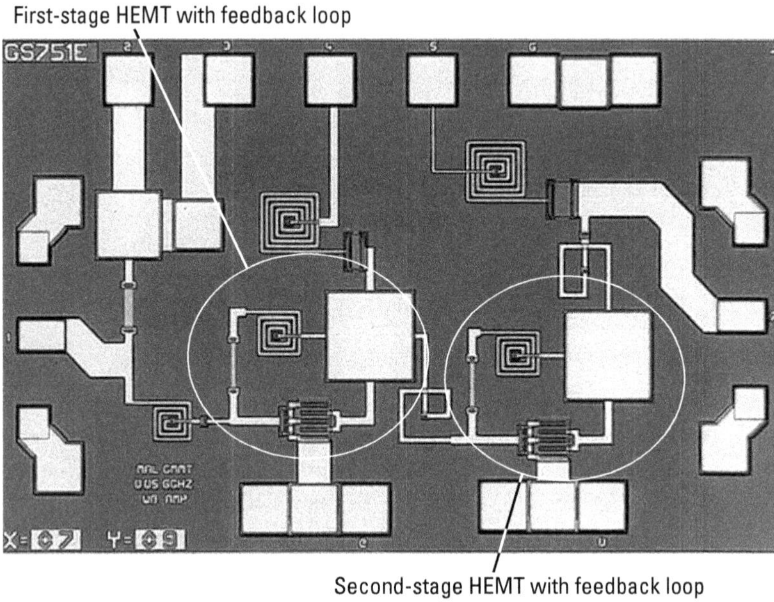

Figure 5.172 A dc –7-GHz MMIC wideband amplifier. (Courtesy of TM R&D, Malaysia).

elements, then the impedance of that element can vary by the same amount, leading to performance variation. Variations in the chip performance can be minimized by only utilizing MIM capacitors for low-impedance dc blocking where their effect on the RF performance is negligible or for other areas of the circuit that have been found to be relatively insensitive to variation. This is also shown in Figure 5.173.

Resistor Design

If a resistor is needed with a very low value or an extreme aspect ratio such that the tolerance variation would be large, it is sometimes better to replace this resistor with two or more larger-valued resistors in parallel, which reduces the tolerance variation.

First-Pass Design Strategies

A first-pass design strategy aims to meet the required product specification with the first MMIC design iteration as opposed to the common occurrence where two or three design, processing, and test cycles are needed before the specification is met. One particular first-pass design strategy for power amplifiers has been reported in the literature [162] and is described here.

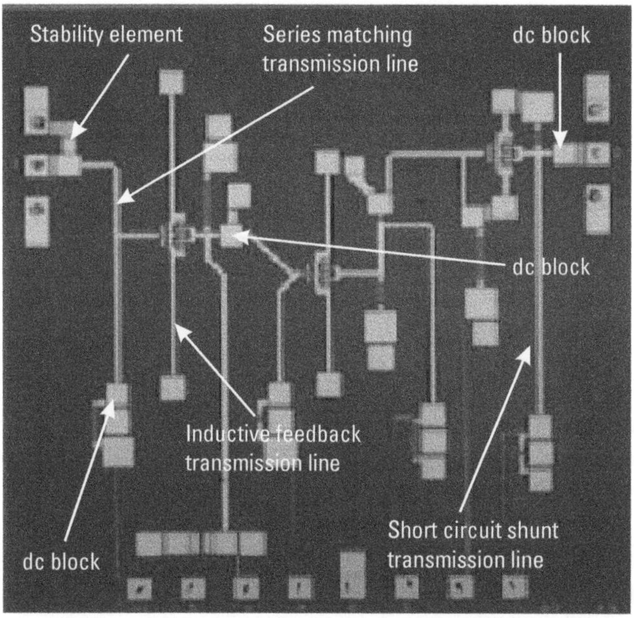

Figure 5.173 Example 14-GHz LNA using transmission lines for matching and MIM capacitors for dc blocks. (*Source:* Bookham Inc., 2006. All Rights Reserved.)

The first task is to select the optimum transistor size and then to perform parameter extraction and equivalent circuit modeling on just the selected device size, instead of developing scalable models. The next step is experimental verification of the large-signal behavior of the model under optimum load conditions by comparing the model to measurements of the actual devices on a load-pull test bench. Lastly, lumped simulation models for passive components are derived from electromagnetic simulation of their structures; these simulate at high speed, allowing quick statistical analysis of the whole chip design.

5.11.2.2 Process Design for Product Yield

MMIC foundry processes are normally fixed, and there is no chance for the MMIC designer to alter the process parameters. However, if a very-high-volume product is predicted to be the major design going through a particular process, then economic pressures may allow the process to be adjusted slightly if it means an improvement in chip yield. Software has been developed that links the physical structure of the active device produced by the process to the performance of the chip product. In this way, the device structure or the processing technique may be modified to improve the yield of a high-volume product. Also, techniques such as Taguchi tolerance analysis can indicate which of the active device parameters are causing the most failures and show which area of the process needs a tighter tolerance. Statistical process control (SPC) can be used to maintain a constant process, reduce process variation ranges, and tune the process to different nominal values. Each of these techniques can be used to change a process and improve the chip yield it produces.

5.11.2.3 Sensitivity Analysis

Sensitivity analysis is used to determine the circuit performance changes against *unquantifiable variations* within the circuit, such as element model characterization errors and close-component interaction. The analysis process is to vary all the matching elements individually by a fixed percentage, such as 5% or 10%, and to simulate the resulting performance variation. Varying the insensitive matching elements will not have a large effect on the simulated performance, but varying the sensitive elements will show a large effect. When the most sensitive elements are identified, they can be laid out in a "safest" way, or the matching topology can be altered to a less sensitive arrangement. A "safer" layout entails well-spaced components (greater than the substrate height), connected in the same way as they were characterized, and in the middle of their size-scaling range.

5.11.2.4 Tolerance Analysis

Tolerance analysis is used to determine the circuit performance changes caused by *quantifiable variations*, such as the process variations determined and controlled by SPC. Monte Carlo or Taguchi [163, 164] statistical techniques can be used in standard simulators to predict the product yield against these known variations. Monte Carlo is good for getting an accurate value of the yield, but the Taguchi technique is quicker, gives an approximate value for the yield, and gives more information about the causes of the nonyielding chips. An example Taguchi tolerance analysis of an MMIC design [165] is described in the next section.

Taguchi Tolerance-Analysis Example

In this section, the Taguchi design of experiment (DOE) tolerance-analysis technique is demonstrated on a two-stage power amplifier MMIC to show the yield of the chip for known process variations. This example also shows which measured performance parameter is most commonly failing to meet the required specification and which of the process variations are causing this failure. Finally, the same Taguchi analysis can calculate how, by tightening certain process variations, the MMIC yield can be improved.

Taguchi design-of-experiments (DOE) have long been used for problem solving, process tuning, and tighter process control in many fabrication areas [166, 167] and can be adapted both for tolerance analysis [166], as in this example, and for sensitivity analysis [168]. The Taguchi principle is to perform a small number of balanced MMIC simulations in which the process parameters are varied to high and low values within their process spreads. The combinations of the high and low values of each parameter for each simulation are precisely defined by standard orthogonal arrays. The orthogonal nature of the array allows the high and low performance values for each process parameter to be extracted and averaged separately from each other. The result is a list of the partial MMIC performance spreads due to each process parameter. The overall performance spread is the root-sum-of squares of each partial spread, so only the large values are significant. In this way, the key parameters causing the largest spread in the performance are clearly identified. Knowing the overall performance spread and mean value, one can also calculate the MMIC yield against a specification limit. Halving a process parameter spread will also halve the partial performance spread, so improvements in MMIC yield can be directly calculated from improved process control.

This technique is demonstrated on a 14–14.5-GHz two-stage power amplifier MMIC design. The process parameters analyzed are the two MIM capacitor tolerances (nitride and polyimide) and the five main intrinsic FET parameters, C_{gs}, C_{gd}, C_{ds}, R_{ds}, and G_m. Table 5.3 shows the seven parameters, their associated process tolerance limits, and the high and low values used in the Taguchi Analysis.

Table 5.4 shows the L8 orthogonal array, which selects the variable values to be used in each of the eight circuit trial simulations.

Table 5.5 shows the results of the eight trials in terms of S_{21}, S_{11}, and S_{22} at the frequency-band edges of 14 and 14.5 GHz.

The overall average and standard deviation can be used to predict a pass-rate yield against a performance specification limit using the standard statistical formula [z = (LIMIT – MEAN)/STD DEV], and using a lookup table, the z value is converted to a pass rate. In this example, the specification limit is set at greater than 10 dB for S_{21} and less than −10 dB for S_{11} and S_{22}. The corresponding pass rates are calculated in Table 5.6.

Table 5.3
Taguchi Variables and the High and Low Analysis Values

Variable	Nominal	Range (%)	Estimated STD DEV (%)	Taguchi High Value	Taguchi Low Value
NIT	1.0	±13	6.5	1.065	0.935
POLY	1.0	±13	6.5	1.065	0.935
C_{gd}	0.065 pF	±14	7.0	0.0696	0.0605
C_{gs}	0.93 pF	±14	7.0	0.995	0.865
C_{ds}	0.182 pF	±13	6.5	0.194	0.170
R_{ds}	10.2×10^{-3}	±14	7.0	10.9×10^{-3}	9.49×10^{-3}
G_m	92.3×10^{-3}	±14	7.0	98.8×10^{-3}	35.8×10^{-3}

Table 5.4
L8 Orthogonal Array of Parameter Values for Each of the Eight Trial Simulations

Trial Number	NIT	POLY	C_{gd}	C_{gs}	C_{ds}	R_{ds}	G_m
1	0.935	0.935	0.0605	0.865	0.170	9.49×10^{-3}	85.8×10^{-3}
2	0.935	0.935	0.0605	0.995	0.194	10.9×10^{-3}	98.8×10^{-3}
3	0.935	1.065	0.0696	0.865	0.170	10.9×10^{-3}	98.8×10^{-3}
4	0.935	1.065	0.0696	0.995	0.194	9.49×10^{-3}	85.8×10^{-3}
5	1.065	0.935	0.0696	0.865	0.194	$9.49 \times 10–3$	98.8×10^{-3}
6	1.065	0.935	0.0696	0.995	0.170	10.9×10^{-3}	85.8×10^{-3}
7	1.065	1.065	0.0605	0.865	0.194	10.9×10^{-3}	85.8×10^{-3}
8	1.065	1.065	0.0605	0.995	0.170	9.49×10^{-3}	98.8×10^{-3}

The overall pass rate is the product of the individual pass rates. Thus, the percentage of passing the S_{21}, S_{11}, and S_{22} specification at 14.0 GHz is 86.9%, and at 14.5 GHz, it is 83.3%; the overall pass rate is 72%.

A Monte Carlo analysis was also performed using identical data and circuit files and produced a pass-rate figure of 80% with 2,000 trials. In a Monte Carlo analysis, the G_m value could be set at the low end of its range, and the simulation may fail on low gain at both ends of the frequency band because they are correlated, but this counts only as one fail. The simple multiplicative combination of the Taguchi pass rates assumes no correlation

Table 5.5
Trial Simulation Measures and Their Values for Each Trial Simulation

Trial Number	14 GHz			14.5 GHz		
	S_{21}	S_{11}	S_{22}	S_{21}	S_{11}	S_{22}
1	11.91	−14.98	−19.26	11.73	−16.07	−15.06
2	11.54	−12.57	−26.09	11.52	−15.36	−15.76
3	12.29	−12.82	−31.71	12.25	−11.67	−17.91
4	10.34	−14.18	−20.73	10.14	−14.97	−13.48
5	13.79	−14.88	−17.61	13.59	−13.46	−12.45
6	9.50	−11.47	−29.70	9.25	−13.23	−19.50
7	11.15	−10.86	−30.59	10.98	−10.68	−16.04
8	12.45	−13.49	−23.07	12.15	−15.81	−14.85
Center value	11.90	−15.41	−28.60	11.60	−14.61	−16.09
Overall average	11.62	−13.15	−24.85	11.45	−13.91	−15.63
Overall STD DEV	1.33	1.52	5.46	1.34	1.99	2.27

Table 5.6
Pass Rates for the Six Performance Measures

14 GHz			14.5 GHz		
S_{21}	S_{11}	S_{22}	S_{21}	S_{11}	S_{22}
$Z=\frac{10-11.62}{1.33}$	$Z=\frac{10-13.15}{1.52}$	$Z=\frac{10-24.85}{5.46}$	$Z=\frac{10-11.45}{1.34}$	$Z=\frac{10-13.91}{1.99}$	$Z=\frac{10-15.63}{2.27}$
= 1.22	= 2.07	= 2.72	= 1.08	= 1.96	= 2.48
= 88.95%	= 98.1%	= 99.7%	= 85.99%	= 97.5%	= 99.3%

between specification requirements and cannot take into account the fact that a large number of the devices failing on S_{21} at 14 GHz also failed at 14.5 GHz, giving two fails in the Taguchi analysis. However, if we assume absolute correlation between passing S_{21} at 14 and 14.5 GHz (i.e., the 10% failing at 14 GHz are within the 14% failing at 14.5 GHz), then the predicted pass rate becomes 81%. This is now in good agreement with the Monte Carlo analysis and requires only eight simulations. Nevertheless, given that simulations are relatively quick and cheap, Monte Carlo analysis is more accurate for predicting the overall pass rate and requires no specialist knowledge of the circuit performance.

The major advantage of the Taguchi analysis is that it shows clearly the contribution from each of the input parameter tolerance limits to the overall spread of each output RF characteristic (i.e., it reveals the major contributors to the final performance spread, hence, to the yield). For example, to analyze the contributions to the spreads in S_{21}, S_{11}, and S_{22} at 14.5 GHz, we proceed as follows: Take the average value of S_{21} for all the high and then all the low values of the input variables. The standard deviation contribution of that factor is half the difference between the high and low value: for example, at 14.5 GHz, the S_{21} standard deviation for nitride tolerance $s = (11.49 - 11.41)/2 = 0.04$ dB. Thus, the standard deviation of the nitride tolerance causes a standard deviation in S_{21} at 14.5 GHz of 0.04 dB. We can do the same for all parameters to derive the second column in Table 5.7. Similarly, we can complete Table 5.7 for S_{11} and S_{22} to show the factors causing their spread.

The overall standard deviation of S_{21} at 14.5 GHz is the root sum of the squares of the individual standard deviations. Because of this, the smaller values become negligible, and the large values indicate the major contributors. In the example, we can see that the major contributors to the S_{21} spread are G_m, C_{gs}, and R_{ds}, and this agrees with an understanding of the FET model and its circuit interactions. This also shows the major contributions to S_{11} are R_{ds} and C_{gs}, followed by NIT, POLY, and C_{gd}, and the major contributions to S_{22} are R_{ds} and C_{ds}. These results again agree with our understanding of the FET model and instill confidence in this technique.

If we were now in the position of being able to tighten the input parameter tolerances, we can identify the major contributions and calculate how to improve our circuit yield. For example, most failures are due to S_{21} at 14 GHz. The spread of S_{21} at 14 GHz is due mainly to G_m, C_{gs}, and R_{ds}. If we can halve the present spread of G_m, C_{gs}, and R_{ds} (to ±7%), we can halve their contribution to the performance spread. The performance spread of S_{21} now becomes the root sum of these squares, as shown in Table 5.8.

We now apply the new performance spread to the formula to predict the new yield against S_{21}. This is calculated [$z = (10 - 11.45)/0.66 = 2.20$] to give a 98.6% pass rate.

Multiplying this by the other pass rates (except S_{21} at 14 GHz), we obtain a new overall chip pass rate of 93.4%.

To summarize this example, the Monte Carlo analysis predicted an overall MMIC yield of 80% against the required specification, while the

Table 5.7
Partial Performance Measure Spreads at 14.5 GHz for Each Variable

Parameter	S_{21} (dB)	S_{11} (dB)	S_{22} (dB)
NIT	0.04	0.61	0.08
POLY	0.07	0.63	0.06
C_{gd}	0.15	0.58	021
C_{gs}	0.69	0.94	0.27
C_{ds}	0.11	029	1.20
R_{ds}	0.45	1.17	1.67
G_m	0.93	0.17	0.39
	1.26	1.86	2.12 overall STD DEV

Table 5.8
Improved Partial Measure Performance Spread Due to Halving the Variation Range of Three Parameters

Parameter	Standard Deviation on S_{21}/dB
Nitride	0.04
Poly	0.07
C_{gd}	0.15
C_{gs}	0.35 (0.69/2)
C_{ds}	0.11
R_{gd}	0.23 (0.45/2)
G_m	0.47 (0.93/2)
	0.66 STD DEV

Taguchi analysis predicted a 72% to 83% pass yield, depending on the interpretation of the result. More significantly, the Taguchi analysis showed that the majority of the specification failures were due to the S_{21} gain variation and identified the key parameters affecting the gain spread as G_m, C_{gs}, and R_{ds}. Further calculations showed that halving the process spread only on these parameters would improve the overall pass yield to 94%.

5.12 Questions

5.1 Which MMIC transistor technology has an operating frequency down to dc?

5.2 Which MMIC transistor technology is most suited to low-noise millimeter-wave applications?

5.3 Which MMIC transistor technology can operate over 100 GHz?

5.4 What is a good figure for the gain per stage in a multistage amplifier MMIC?

5.5 What range of input match is reasonable to expect for a 10-GHz MMIC?

5.6 Where are the dc bias pads usually placed, and what characteristics of their placement are generally a foundry standard?

5.7 In a chain of amplifiers, which amplifier has the most impact on the overall noise figure, and what characteristic of this amplifier can reduce the contributions from the other amplifiers in the chain?

5.8 What technique is employed to ensure good input match and low noise figure simultaneously?

5.9 Which is the other important amplifier parameter, as well as noise figure, gain, and matches versus frequency?

5.10 How are oscillations prevented without degrading the noise figure?

5.11 In the distributed amplifier, where is the wave traveling from and to?

5.12 If a distributed amplifier is to be constructed with transistors with an input capacitance of 0.10 pF, what element is needed to connect their inputs together, and what is its value to match to 50Ω characteristic system impedance?

5.13 What three advantages come with using cascode FETS?

5.14 Which circuit technique trades less current for more voltage?

5.15 Which circuit technique reduces the number of different supply voltages required by FET MMICs?

5.16 A 4 × 100 FET can be biased at 50% I_{dss} with $V_{gs} = -0.5$V. If I_{dss} is 400 mA per millimeter of gate, what self-bias resistor value is required?

5.17 What is the main drawback for both the self- and stack-biasing techniques?

5.18 What is the definition of a linear device?

5.19 What assumption is made for large-signal operation?

5.20 Power amplifier design is concerned with handling the effects of mainly strong or weak nonlinearities?

5.21 What are the four main effects of increasing the input power level into an FET with a nonlinear transconductance?

5.22 What are the three steps in the power amplifier design methodology described in this chapter?

5.23 Which MMIC transistor technology is best for an efficient 2.5-GHz power amplifier, and how is the process capability rated?

5.24 As FET device size increases, which parameters trade off against each other, and what determines the ideal device size?

5.25 In a multistage power amplifier, which stage has most influence over the overall efficiency?

5.26 Which stages should operate linearly, and how far should the output power from them be backed off to ensure this?

5.27 Why do we need power splitters and not just one large device?

5.28 Which are the most common power-splitting or power-combining techniques?

5.29 Complete the power budget in Figure 5.174. The requirement specification is that P_{out} = +34 dBm at P_{1dB}. FET A has 11-dB

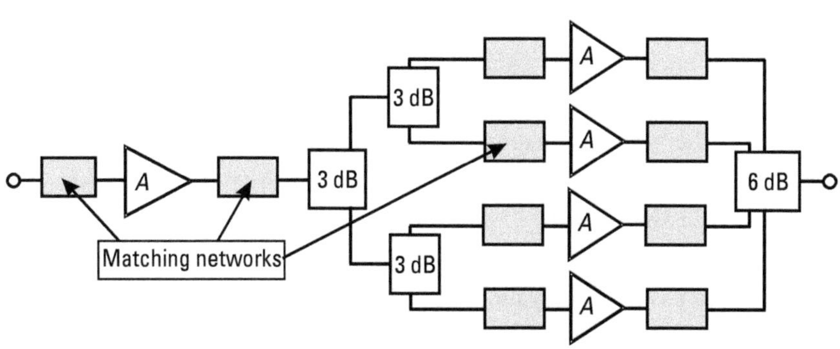

Figure 5.174 Power amplifier architecture for Question 5.29.

small-signal gain and P_{1dB} of 29 dBm. (a) What input power is needed? (b) Will this architecture meet the specification?

5.30 Are the matching circuits in a power amplifier designed for the frequency response using small- or large-signal simulation?

5.31 As well as the RF response, what other aspects of the MMIC design must the matching circuits cater to?

5.32 What parameter is traded off for better efficiency in bias modes other than Class A?

5.33 For large-signal optimization, which technique is based on the dc *I*/*V* characteristics of the FETs?

5.34 Which large-signal optimization technique is based on real measurements?

5.35 Which large-signal optimization techniques predict the final output power?

5.36 Are soft or hard compression characteristics more desirable for a power amplifier?

5.37 Which MMIC transistor technology allows design of oscillators with the lowest phase noise?

5.38 Which MMIC transistor technology allows design of oscillators with the lowest cost?

5.39 For a general oscillator, what are the three sections that the circuit can be broken down into?

5.40 For high-Q or low-phase-noise applications, which part of the oscillator could be off chip?

5.41 Which component can produce negative resistance?

5.42 What is the main function of mixers?

5.43 What do LO, RF, and IF stand for?

5.44 In addition to the harmonics of the LO and RF frequencies, what other frequency components are present in a general mixer?

5.45 Which devices are used in an active mixer, and which are used in a passive mixer?

5.46 Of active or passive mixers, which can be reciprocal, which can have conversion gain, and which are easier to bias?

5.47 What are the advantages of using a balanced mixer?

5.48 How many nonlinear devices are used in a doubly balanced mixer?

5.49 When might you use an antiparallel diode pair?

5.50 What is the image frequency, and how does its suppression affect the mixer noise figure?

5.51 What are the two modes of operation for FET mixers?

5.52 Which of the two FET mixer modes gives the best linearity, and which gives conversion gain?

5.53 Which of the following is a balanced mixer, a dual-gate FET mixer, or a Gilbert-cell mixer?

5.54 What is the function of the balun?

5.55 What bias voltages are applied to an FET acting as a switch?

5.56 What are the advantages of FET switches over PIN diode switches?

5.57 What are four main issues that start to become more important for millimeter-wave MMICs?

5.58 What limits the number of turns of track in a spiral inductor that a designer can use in an MMIC design?

5.59 What is the effective dielectric constant for an 80-μm-wide line on a 100-μm substrate at 30 GHz?

5.60 What is the wavelength in 50Ω GaAs microstrip at 40 GHz?

5.61 What is the impedance of a via through a 100-μm substrate at 60 GHz, and what can be used instead of a through-substrate via for a ground at 60 GHz?

5.62 What are the two common transmission-line types used at millimeter-wave frequencies?

5.63 Which millimeter-wave transmission-line type has the lowest cost?

5.64 Which millimeter-wave transmission type is better for power amplifiers, and what are the two main reasons why?

5.65 Which millimeter-wave transmission-line type has less shunt parasitics, and what is the advantage of this?

5.66 Which millimeter-wave transmission-line type is easily characterized by *s*-parameter blocks and does not require 3D simulation?

5.67 Which millimeter-wave transmission-line type is limited in the range of transmission-line characteristic impedance?

5.68 What fraction of a wavelength along a lossless transmission-line transforms an open circuit to a short circuit?

5.69 What are the two main reasons for needing 3D simulations at millimeter-wave frequencies?

References

[1] Attributed (unconfirmed) to Professor Moritz von Jacobi of St. Petersburg (1801–1874), at http://www.du.edu/~jcalvert/tech/jacobi.htm.

[2] Harter, A., "LNA Matching Techniques for Optimizing Noise Figures," *RF Design,* February 2003, pp. 20–30.

[3] Gonzalez, G., *Microwave Transistor Amplifiers,* Upper Saddle River, NJ: Prentice Hall, 1984.

[4] White, J. F., "Approach Increases Amplifier Gain," *Microwaves & RF,* ED Online ID #8835, September 2004.

[5] Polkinghorn, F. A., "Philip Smith, Electrical Engineer," an oral history conducted in 1973, *IEEE History Center,* Rutgers University, New Brunswick, NJ.

[6] Bowick, C., *RF Circuit Design,* Newnes Publications, Burlington, MA: Elsevier Science, pp. 70–71.

[7] Matthaei, G. L., L. Young, and E. M. T. Jones, *Microwave Filters, Impedance-Matching Networks and Coupling Structures,* New York: McGraw-Hill, 1964.

[8] Pitzalis, O., and R. A. Gilson, "Tables of Impedance Matching Networks Which Approximate Prescribed Attenuation Versus Frequency Slopes," *IEEE Trans. Microwave Theory Tech.,* Vol. 19, No. 4, April 1971.

[9] Vinding, J. P., "Radial Line Stubs as Elements in Strip Line Circuits," *NEREM Record,* 1967, pp. 108–109.

[10] Giannini, F., R. Sorrentino, and J. Vrba, "Planar Circuit Analysis of Microstrip Radial Stub," *IEEE Trans. Microwave Theory Tech.,* Vol. 32, No. 12, December 1984, pp. 1652–1655.

[11] Edwards, T. C. *Foundations for Microstrip Circuit Design,* Chichester, United Kingdom: Wiley, 1991.

[12] Gupta, K. C., R. Garg, and I. J. Bahl, *Microstrip Lines and Slotlines,* Dedham, MA: Artech House, 1979.

[13] Gardiol, F. E., *Microstrip Circuits,* New York: Wiley, 1994.

[14] Lau, D. K. Y., et al., "Simplified Design Technique for High Performance Microstrip Multi-Section Couplers," *IEEE Trans. Microwave Theory Tech.,* Vol. 46, No. 12, December 1998, pp. 2507–2513.

[15] Taylor, J. L., and D. D. Arigel, "Wiggly Phase Shifters and Directional Couplers for Radio-Frequency Hybrid-Microcircuit Applications," *IEEE Trans. Parts, Hybrids, and Packaging,* Vol. 12, No. 4, December 1976.

[16] Tserng, H. Q., and S. R. Nelson, "5–18 GHz, 3dB Hybrid Coupler," *Electronics Lett.,* Vol. 17, No. 7, April 1981, pp. 258–259.

[17] Cristal, E. G., and L. Young, "Theory and Tables of Optimum Symmetrical TEM-Modes Coupled-Transmission Line Directional Coupler," *IEEE Trans. Microwave Theory Tech.,* Vol. 13, No. 5, September 1965.

[18] Pozar, D. M., *Microwave Engineering,* 2nd ed., New York: Wiley, 1998, pp. 379–383.

[19] Muraguchi, M., T. Yukitake, and Y. Naito, "Optimum Design of 3-dB Branch-Line Couplers Using Microstrip Lines," *IEEE Trans. Microwave Theory Tech.,* Vol. 31, August 1983, pp. 674–678.

[20] Howe, H., *Stripline Circuit Design,* Norwood, MA: Artech House, 1974.

[21] Chang, K., and L.-H. Hsieh, *Microwave Ring Circuits and Related Structures,* 2nd ed., New York: Wiley, 2004.

[22] Hirota, T., A. Minakawa, and M. Muraguchi, "Reduced-Size Branch-Line and Rat-Race Hybrids for Uniplanar MMICs," *IEEE Trans. Microwave Theory Tech.,* Vol. 38, March 1990, pp. 270–275.

[23] Lange, J., "Interdigital Stripline Quadrature Hybrid," *IEEE Trans. Microwave Theory Tech.,* Vol. 17, December 1969, pp. 1150–1151.

[24] Tsai, M. C., "A Wideband Class B Amplifier Using Tapered Interdigital Power Combiners," *Microwave and Optical Tech. Lett.,* Vol. 4, No. 11, October 1991, pp. 455–459.

[25] Wilkinson, J., "An *N*-Way Hybrid Power Divider," *IRE Trans. Microwave Theory Tech.,* Vol. 8, 1960, pp. 116–118.

[26] Kother, D., et al., "MMIC Wilkinson Couplers for Frequencies up to 110 GHz," *Proc. IEEE MTT Symposium,* Vol. 2, Orlando, FL, May 16–20, 1995, pp. 663–666.

[27] Binari, S. C., et al., "Millimetre-Wave Monolithic Passive Circuit Components," *RCA Review,* Vol. 45, December 1984, pp. 579–586.

[28] Arai, S., et al. "Ka-Band 2-Watt Power GaAs MMIC," *Proc. IEEE MTT Symposium,* Vol. 3, Long Beach, CA, June 13–15, 1989, pp. 1105–1108.

[29] Bert, A. G., and D. Kaminsky, "The Travelling Wave Divider/Combiner," *IEEE Trans. Microwave Theory Tech.,* Vol. 28, No. 12, December 1980, pp. 1468–1473.

[30] Willems, D., I. Bahl, and B. Kruger, "A Quasi-Microstrip Traveling-Wave Power Divider/Combiner for Use in High-Density Packages," *IEEE Microwave Guided Wave Lett.,* Vol. 3, May 1993, pp. 148–149.

[31] Bahl, I. J., "Design of Power MMICs and Power Combining Techniques," *IEEE Third International Workshop on Integrated Nonlinear Microwave and Millimeterwave Circuits,* Duisburg, Germany, October 1994, pp. 71–91.

[32] Russel, K. J., "Microwave Power Combining Techniques," *IEEE Trans. Microwave Theory Tech.,* Vol. 27, May 1979, pp. 472–478.

[33] Yau, W., and J. M. Schellenberg, "An *N*-Way Broadband Planar Power Combiner/Divider," *Microwave Journal,* Vol. 29, No. 11, November 1986, pp. 147–151.

[34] Nagai, N., E. Matkawa, and K. Ono, "New *N*-Way Hybrid Power Dividers," *IEEE Trans. Microwave Theory Tech.,* Vol. 25, No. 12, December 1977, pp. 1008–1012.

[35] Salegh, A. A. M., "Computation of the Frequency Response of a Class of Symmetric *N*-Way Power Dividers," *The Bell System Technical Journal,* Vol. 59, No. 8, October 1980, pp. 1493–1512.

[36] Morgan, M., and S. Weinreb, "A Monolithic HEMT Diode Balanced Mixer for 100–140 GHz," *Proc. IEEE MTT Symposium,* Vol. 1, Phoenix, AZ, May 20–25, 2001, pp. 99–102.

[37] Bischof, W., et al., "An Extremely Wide Band Balanced Amplitude Controller in Coplanar Waveguide Technology," *Proc. 25th European Microwave Conference,* October 1995, pp. 921–925.

[38] Schiffman, B. M., "A New Class of Broad-Band Microwave 90-Degree Phase Shifters," *IRE Trans. Microwave Theory Tech.,* April 1958, pp. 232–237.

[39] Marchand, N., "Transmission Lines Conversion Transformers," *Electronics,* Vol. 17, No. 12, December 1944, pp. 142–145.

[40] Valletta, E., L. C. N. de Vreede, and J. N. Burghartz, "Design of Planar Marchand Balun for MMIC Applications," *Proc. SAFE 2002,* Veldhoven, the Netherlands, November 2002.

[41] Chen, T. H., et al., "Broadband Monolithic Passive Baluns and Monolithic Double-Balanced Mixer," *IEEE Trans. Microwave Theory Tech.,* Vol. 39, December 1991, pp. 1980–1986.

[42] Pavio, A. M., and A. Kikel, "A Monolithic or Hybrid Broadband Compensated Balun," *Proc. IEEE MTT Symposium,* 1990, pp. 483–486.

[43] Devlin, L. M., et al., "A Monolithic, 2 to 18 GHz Upconverter," *Proc. IEEE MTT Symposium,* Seattle, Washington, June 2002.

[44] Cohn, S. B. "Parallel-Coupled Transmission-Line-Resonator Filters," *IRE Trans. Microwave Theory Tech.,* April 1958, pp. 223–231.

[45] Kinayman, N., and I. Aksun, *Modern Microwave Circuits,* Norwood, MA: Artech House, 2005.

[46] Williamson, S., and A. Dearn, "Low Cost Microstrip Filters and Mixers at 43 GHz," *IEEE Colloquium on MM-Wave Circuits and Technology for Commercial Applications,* London, United Kingdom, March 1999.

[47] Presser, A., "Varactor Tunable High Q Microwave Filter," *RCA Review,* Vol. 42, 1973, p. 961.

[48] Chang, C. Y., and T. Itoh, "A Varactor Tuned, Active Microwave Band Pass Filter," *Proc. IEEE MTT Symposium,* Dallas, TX, May 8–10, 1990, p. 499.

[49] Pauker, V., "GaAs Monolithic Active Gyrator," *Proc. IEEE GaAs IC Symposium,* Grenelefe, FL, 1986, p. 82.

[50] Suwaki, H., and I. Ohira, "A Very Small MMIC Variable Filter Based on a New Active Filter Design Concept," *Proc. IEEE GaAs IC Symposium,* New Orleans, LA, October 7–10, 1990, p. 93.

[51] Arnold, R. G., and S. P. Marsh, "A Microwave Active Bandstop Filter with Tuneable Centre Frequency," *Proc. IEEE MTT Symposium,* Atlanta, GA, June 14–18, 1993.

[52] Marsh, S. P., and R. G. Arnold, "MMIC Gyrator Bandstop Filter with Ultra-Wideband Tuning," *Proc. IEEE GaAs IC Symposium,* Philadelphia, PA, October 16–19, 1994.

[53] Fusco, V. F., et al., "Silicon and Soft-Board Millimetre Wave Antennas for Broad-Band Mobile Wireless," *Proc. IEEE 2000,* pp. 67–70.

[54] Öjefors, E., et al., "24 GHz ISM-Band Antennas on Surface Micromachined Substrates for Integration with a Commercial SiGe Process," *Proc. ECWT,* Munich, Germany, October 9–10, 2003.

[55] Pozar, D. M., and D. H. Schaubert, *Microstrip Antennas: The Analysis and Design of Microstrip Antennas and Arrays,* New York: IEEE Press, 1995.

[56] Hammad, H. F., Y. M. M. Antar, and A. P. Freundorfer, "Wide Band Antenna Suitable for MMIC Applications," *Proc. PIERS,* Nantes, France, July 13–17, 1998, p. 70.

[57] Jackson, D. R., J. T. Williams, and A. K. Bhattacharyya, "Microstrip Patch Designs That Do Not Excite Surface Waves," *IEEE Trans. Antennas Propagat.,* Vol. 41, August 1993, pp. 1026–1037.

[58] Weast, B. C., *Handbook of Chemistry and Physics,* 44th ed., Cleveland, OH: Chemical Rubber Publishing Co., 1962.

[59] Budka, T. P., "Wide Bandwidth Millimeter-Wave Bond-Wire Interconnects," *IEEE Trans. Microwave Theory Tech.,* Vol. 49, April 2001, pp. 715–718.

[60] Kärkkäinen, M., et al., "Integrated Amplifier Circuits for 60 GHz Broadband Telecommunication," *Analog Integrated Circuits and Signal Processing,* Springer Science + Business Media, Inc., Vol. 42, No. 1, January 2005, pp. 37–46.

[61] Kärkkäinen, M., et al., "Integrated Amplifier Circuits for 60 GHz Broadband Telecommunication," *Analog Integrated Circuits and Signal Processing,* Springer Science + Business Media Inc., Vol. 42, No. 1, January 2005, pp. 37–46.

[62] Robinson, G. H., "Resonance Frequency Calculations for Microstrip Cavities," *IEEE Trans. Microwave Theory Tech.,* Vol. 19, No. 7, July 1971, pp. 665–666.

[63] Marcuvitz, N., *Waveguide Handbook,* London: Peregrinus, 1993.

[64] Johnson, J. B., "Thermal Agitation of Electricity in Conductors," *Phys. Rev.,* Vol. 32, 1928, p. 97.

[65] Nyquist, H., "Thermal Agitation of Electric Charge in Conductors," *Phys. Rev.,* Vol. 32, 1928, p. 110.

[66] Friis, H. T., "Noise Figures of Radio Receivers," *Proc. IRE,* Vol. 32, July 1944, pp. 419–422.

[67] Friis, H. T., "Discussion on 'Noise Figures of Radio Receivers,'" *Proc. IRE,* Vol. 33, February 1945, pp. 125–127.

[68] Lehmann, R. E., and D. D. Heston, "X-Band Monolithic Series Feedback LNA," *IEEE Trans. Microwave Theory Tech.,* Vol. 33, No. 12, December 1985, pp. 1560–1566.

[69] Murphy, M., "Applying the Series Feedback Technique to LNA Design," *Microwave Journal,* Vol. 32, No. 11, November 1989.

[70] Henkes, D., "LNA Design Uses Series Feedback to Achieve Simultaneous Low Input VSWR and Low Noise," *Applied Microwave & Wireless,* Vol. 10, October 1998, pp. 26–32.

[71] White, J. F., *High Frequency Techniques, An Introduction to RF and Microwave Engineering,* Hoboken, NJ: John Wiley & Sons, 2004.

[72] Kucera, J. J., and U. Lott, "Low-Power Silicon BJT LNA for 1.9 GHz," *IEEE Microwave and Guided Wave Lett.,* Vol. 8, No. 3, March 1998, pp. 136–137.

[73] Dearn, A. W., et al., "A Universal GaAs HBT PA with Active Bias Circuitry, Covering 4.9–6 GHz," *Proc. 12th GAAS Symposium,* Amsterdam, Netherlands, October 11–12, 2004, pp. 155–158.

[74] Percival, W. S., "Thermionic Valve Circuits," British Patent No. 460,562, 1937.

[75] Ginzton, E. L., et al., "Distributed Amplification," *Proc. IRE,* Vol. 36, August 1948, pp. 956–969.

[76] Masuda, S., et al., "An Over 110-GHz InP HEMT Flip-Chip Distributed Baseband Amplifier with Inverted Microstrip Line Structure for Optical Transmission Systems," *Proc. IEEE GaAs IC Symposium,* Monterey, CA, October 20–22, 2002, pp. 99–102.

[77] Beyer, J. B., et al., "MESFET Distributed Amplifier Design Guidelines," *IEEE Trans. Microwave Theory Tech.,* Vol. 32, No. 3, March 1984, pp. 268–275.

[78] Ayasli, Y., et al., "A Monolithic GaAs 1–13-GHz Travelling-Wave Amplifier," *IEEE Trans Microwave Theory Tech.,* Vol. 30, No. 7, July 1982, pp. 976–981.

[79] Devlin, L. M., et al., "A 2–18 GHz ESM Receiver Front-End," *Proc. European Microwave Conf.,* Milan, Italy, September 2002.

[80] Baeyens, Y., "Monolithic Microwave Integrated Circuits Using GaAs and InP Based Heterojunction Field-Effect Transistors," Ph.D. thesis, Katholieke Universiteit, Leuven/IMEC, January 1997.

[81] Miller, J. M., "Dependence of the Input Impedance of a Three-Electrode Vacuum Tube upon the Load in the Plate Circuit," *Scientific Papers of the Bureau of Standards,* Vol. 15, No. 351, 1920, pp. 367–385.

[82] Gray, P. E., and C. L. Searle, *Electronic Principles: Physics, Models, and Circuits,* New York: Wiley, 1969, pp. 523–524.

[83] Ramachandran, R., and A. F. Podell, "Segmented Cascode HBT for Microwave-Frequency Power Amplifiers," U.S. Patent No. 5,066,926, filed June 26, 1990, issued November 1991.

[84] Bayraktaroglu, B., and M. Salib, "Unconditionally Thermally Stable Cascode GaAs HBTs for Microwave Applications," *IEEE Microwave and Guided Wave Lett.,* Vol. 7, No. 7, 1997, pp. 187–189.

[85] Maruhashi, K., K. Ohata, and M. Madihian, "A 60-GHz-Band Monolithic HJFET LNA Incorporating a Diode-Regulated Self-Bias Circuit," *IEEE Trans. Microwave Theory Tech.,* Vol. 45, No. 12, December 1997, pp. 2256–2260.

[86] Cripps, S. C., *RF Power Amplifiers for Wireless Communications,* Norwood, MA: Artech House, 1999.

[87] De Moivre, A., born on May 16, 1667, in Vitry (close to Paris), France, died November 27, 1754, in London.

[88] Niclas, K. B., "Planar Power Combining for Medium Power GaAs FET Amplifiers in X/Ku-Bands," *Microwave Journal,* Vol. 22, June 1979, pp. 79–84.

[89] Healy, M., B. J. Minnis, and L. M. Devlin, "Wideband Power Combiners for GaAs MMICs," *Proc. 21st European Microwave Conference,* Stuttgart, Germany, September 1991, pp. 1035–1040.

[90] McCullagh, M., et al., "An Overview of High Frequency Circuit Design on the GMMT GaAs HBT Process," *Proc. IEEE EDMO,* 1995, pp. 148–154.

[91] Marsh, S. P., et al., "5.4 Watt GaAs MESFET MMIC for Phased Array Radar Systems," *Proc. IEEE EDMO,* November 1997, pp. 169–174.

[92] Freitag, R. G. "A Unified Analysis of MMIC Power Amplifier Stability," *Proc. IEEE MTT-Symposium,* Albuquerque, NM, 1992, pp. 297–300.

[93] Lim, J., and S. Eom, "A New 3-Way Power Divider with Various Output Power Ratios," *Proc. IEEE MTT-Symposium,* San Francisco, CA, June 17–21, 1996, pp. 785–788.

[94] Cripps, S. C., "A Theory for the Prediction of GaAs FET Load-Pull Power Contours," *Proc. IEEE MTT-Symposium,* Boston, MA, May 31–June 3, 1983, pp. 221–223.

[95] Marsh, S. P., "Direct Extraction Technique to Derive the Junction Temperature of HBTs under High Self-Heating Bias Conditions," *IEEE Trans. Electron Devices,* Vol. 47, No. 2, February 2000, pp. 288–291.

[96] Lu, K., and C. M. Snowden, "Analysis of Thermal Instability in Multi-Finger Power AlGaAs/GaAs HBT's," *IEEE Trans. Electron Devices,* Vol. 43, November 1996, pp. 1799–1805.

[97] Gao, G.-B., "Emitter Ballasting Design for, and Current Handling Capability of, AlGaAs/GaAs Power Hetero-Junction Bipolar Transistors, *IEEE Trans. Electron Devices,* Vol. 38, 1991, p. 185.

[98] Arnold, R. P., and D. Zoroglu, "A Quantitative Study of Emitter Ballasting," *IEEE Trans. Electron Devices,* Vol. 21, 1974, pp. 385.

[99] Hamilton, S., "Microwave Oscillator Circuits," *Microwave Journal,* Vol. 21, 1978, pp. 63–86.

[100] Kurokawa, K., "Some Basic Characteristics of Broadband Negative Resistance Oscillator Circuits," *Bell Systems Technical Journal,* Vol. 48, July 1969, pp. 1937–1955.

[101] Dearn, A., "Oscillators," *Proc. IEE Training Event—How to Design RF Circuits,* London, U.K., April 2000.

[102] Ho, C. Y., and T. Kajita, "DRO State of the Art," *Applied Microwave Wireless,* Spring 1990, pp. 69–80.

[103] Bernard, O., and R. Croston, "Simulate and Build a Ku-Band DRO," *Microwaves & RF,* Vol. 39, No. 5, May 2000, pp. 103–115.

[104] Bernard, O., "Model and Build a Ku-Band DRO," *Microwaves & RF,* Vol. 39, No. 6, June 2000, pp. 85–89.

[105] Papp, J. C., and Y. Y. Koyano, "An 8–18-GHz YIG-Tuned FET Oscillator," *IEEE Trans. Microwave Theory Tech.,* Vol. 28, No. 7, July 1980, pp. 762–767.

[106] Trew, R. J., "Design Theory for Broad-Band YIG-Tuned FET Oscillators," *IEEE Trans. Microwave Theory Tech.,* Vol. 27, No. 1, January 1979, pp. 8–14.

[107] Wilson, P. G., "Monolithic 38 GHz Dielectric Resonator Oscillator," *Proc. IEEE MTT Symposium,* Boston, MA, July 10–14, 1991, pp. 831–834.

[108] Schott, M., et al., "38 GHz Push-Push GaAs-HBT MMIC Oscillator," *Proc. IEEE MTT Symposium,* 2002, pp. 839–842.

[109] Wang, H., et al., "Low Phase Noise Millimeter-Wave Frequency Sources Using InP-Based HBT MMIC Technology," *IEEE J. Solid-State Circuits,* Vol. 31, No. 10, October 1996, pp. 1419–1425.

[110] Keller, M. G., A. P. Freundorfer, and Y. M. M. Antar, "A Single-Chip Coplanar 0.8-μm GaAs MESFET K/Ka-Band DRO," *IEEE Microwave Guided Wave Lett.,* Vol. 9, No. 12, December 1999, pp. 526–528.

[111] Smith, D. M., J. C. Canyon, and D. L. Tait, "25–42 GHz GaAs Hetero-Junction Bipolar Transistor Low Phase Noise Push-Push VCOs," *IEEE MTT Symposium,* Long Beach, CA, June 13–15, 1989, pp. 725–728.

[112] Ikematsu, H., et al., "A 40-GHz-Band Fully Monolithic VCO with a One-Wave Length Microstrip Resonator for Accurate Oscillation Frequency," *Proc. IEEE MTT Symposium,* Seattle, WA, June 2–7, 2002, pp. 843–846.

[113] Maas, S. A., *Non-linear Microwave Circuits,* New York: IEEE Press, 1997.

[114] High IP3 GaAs MMIC MIXER, part number HMC214MS8, Hittite Microwave Corporation, Chelmesford, MA, USA, at http://www.hittite.com/product_info/product_specs/mixers/hmc214ms8.pdf.

[115] Kamozaki, K., et al., "50–100 GHz Octave Band MMIC Mixers," *Proc. IEEE RFIC Symposium,* Denver, CO, June 8–11, 1997, pp. 95–98.

[116] Cronin, N. J., S. P. Marsh, and P. H. Reece, "MM-Wave Subharmonic Mixer Using Integrated Planar Diodes," *Proc. IEE Colloquium on 'Millimeter Wave Component Design,'* Digest No. 1986/109, November 1986.

[117] Marsh, S. P., N. J. Cronin, and P. H. Reece, "183 GHz Monolithic Subharmonic Mixer," *Proc. 12th Int. Conf. on Infrared and Millimeter Waves,* Orlando, FL, December 1987, pp. 121–122.

[118] Cronin, N. J., S. P. Marsh, and P. H. Reece, "Planar Subharmonic Mixer for 183 GHz," *Proc. 13th Int. Conf. on Infrared and Millimeter Waves,* Waikiki Beach, Honolulu, HI, December 1988.

[119] Hwang, Y.-J., et al., "A 78–114 GHz Monolithic Subharmonically Pumped GaAs-based HEMT Diode Mixer," *IEEE Microwave and Wireless Components Lett.,* Vol. 12, No. 6, June 2002, pp. 209–211.

[120] Freundorfer, A. P., and Y. Jamani, "A Ka-band GaInP/GaAs HBT Double Balanced Downconvert Mixer Using Lumped Elements," *Proc. IEEE APS Int. Symp/URSI Meeting,* Vol. 1, July 1997, pp. 578–581.

[121] Tzeng, B., et al., "A 1–17-GHz InGaP-GaAs HBT MMIC Analog Multiplier and Mixer with Broad-Band Input-Matching Networks," *IEEE Trans. Microwave Theory Tech.,* Vol. 50, No. 11, November 2002, pp. 2564–2568.

[122] Gilbert, B., "A Precise Four-Quadrant Multiplier with Subnanosecond Response," *IEEE J. Solid-State Circuits,* Vol. 3, No. 4, December 1968, pp. 365–373.

[123] Dearn, A. W., and L. M. Devlin, "A MM-Wave Monolithic Gilbert-Cell Mixer," *Proc. IEEE RFIC Symposia,* Boston, MA, June 11–16, 2000.

[124] Alfredsson, M., et al., "Design Considerations for an Image Rejection Mixer in a Broadband Digital Beamforming Phased Array System," *Proc. Asia-Pacific Microwave Conference,* Sydney, Australia, December 3–6, 2000, pp. 241–245.

[125] 13.0–25.0-GHz GaAs MMIC Image Rejection Mixer, Part No. 18KWR0327, Mimix Broadband Inc., Houston, TX, at http://www.mimixbroadband.com/PDFfiles/18WKR0327.pdf.

[126] Varonen, M., et al., ""Resistive HEMT Mixers for 60 GHz Broadband Telecommunication," *IEEE Trans. Microwave Theory Tech.,* Vol. 53, No. 4, April 2005, pp. 1322–1330.

[127] Varonen, M., et al, "Resistive HEMT Mixers for 60 GHz Broadband Telecommunication," *IEEE Trans. Microwave Theory Tech.,* Vol. 53, No. 4, April 2005, pp. 1322–1330.

[128] Ayasli, Y., "Microwave Switching with GaAs FETs," *Microwave Journal,* Vol. 25, No. 11, November 1982, pp. 61–74.

[129] Devlin, L. M., A. W. Dearn, and G. A. Pearson, "Low Loss Monolithic MM-Wave SP4Ts," *Proc. Workshop on Design for Broadband Wireless Access,* Cambridge, U.K., May 2001.

[130] Mun, J. K., et al., "Design and Fabrication of pHEMT MMIC Switches for IEEE 802.11.a/b/g WLAN Applications," *Semicond. Sci. Technol.,* Vol. 20, 2005, pp. 677–684.

[131] Williamson, S., et al., "A Monolithic, Dual Channel, DC to 20 GHz SPDT," *Proc. Wireless Design Conference,* London, United Kingdom, May 2002.

[132] Smuk, J., and M. Shifrin, "Monolithic GaAs Multi-Throw Switches with Integrated Low-Power Decoder/Driver Logic," *Proc. IEEE RFIC Symposium,* Denver, CO, June 8–11, 1997, pp. 47–50.

[133] Dearn, A., and L. Devlin, "Design Trade-offs for High Linearity PHEMT Switches," *Proc. ARMMS Conference,* Abingdon, U.K., November 19–20, 2004.

[134] Hardin, R. N., E. J. Downey, and J. Munushian, "Electronically Variable Phase Shifter Utilizing Variable Capacitance Diodes," *Proc. IRE,* Vol. 48, May 1960, pp. 944–945.

[135] Ellinger, F., R. Voqt, and W. Bachtold, "Compact Reflective-Type Phase-Shifter MMIC for C-Band Using a Lumped-Element Coupler," *IEEE Trans. Microwave Theory Tech.,* Vol. 49, No. 5, May 2001, pp. 913–917.

[136] Bialkowski, M. E., and N. C. Karmakar, "Design of Compact L-Band 180° Phase Shifter," *Microwave Opt. Technol. Lett.,* Vol. 2, July 1999, pp. 144–148.

[137] Paolo, F. D., "A Simple, High Yield 6 to 18 GHz GaAs Monolithic Phase Shifter," *Microwave Journal,* April 1997, pp. 92–104.

[138] Nagra, A. S., et al., "Monolithic GaAs Phase Shifter Circuit with Low Insertion Loss and Continuous 0–360 Phase Shift at 20 GHz," *IEEE Microwave and Guided Wave Lett.,* Vol. 9, No. 1, January 1999, pp. 31–33.

[139] Ellinger, F., H. Jäckel, and W. Bächtold, "Varactor-Loaded Transmission-Line Phase Shifter at C-Band Using Lumped Elements," *IEEE Trans. Microwave Theory Tech.,* Vol. 51, No. 4, April 2003, pp. 1135–1140.

[140] Hayashi, H., T. Nakagawa, and K. Araki, "A Miniaturized MMIC Analog Phase Shifter Using Two Quarter-Wavelength Transmission Lines," *IEEE Trans. Microwave Theory Tech.,* Vol. 50, January 2002, pp. 150–154.

[141] Nam, S., A. W. Payne, and I. D. Robertson, "RF and Microwave Phase Shifter Using Complementary Bias Techniques," *Electron Lett.,* Vol. 37, No. 18, August 2001, pp. 1124–1125.

[142] Pillans, B., et al., "Ka-Band RF MEMS Phase Shifters," *IEEE Microwave and Guided Wave Lett.,* Vol. 9, No. 12, December 1999, pp. 520–522.

[143] Willms, J. G., et al., "A Wideband GaAs 6-Bit True-Time Delay MMIC Employing On-Chip Digital Drivers," *Proc. GAAS Conference,* Paris, France, October 2–6, 2000.

[144] Ouacha, A., and B. Carlegrim, "MMIC Self-Switched Time Shifter for Broadband Applications," *Microwave and Optical Technology Lett.*, Vol. 19, No. 1, September 1998, pp. 15–20.

[145] Schindler, J., et al., "Monolithic 6–18 GHz 3 Bit Phase Shifter," *Proc. IEEE GaAs IC Symposium,* Monterey, CA, November 12–14, 1985, pp. 129–132.

[146] Ayasli, Y., et al., "Wide-Band Monolithic Phase Shifter," *IEEE Trans. Microwave Theory Tech.,* Vol. 32, No. 12, December 1984, pp. 1710–1714.

[147] Gupta, R., et al., "A 0.05- to 14-GHz MMIC 5-Bit Digital Attenuator," *Proc. IEEE GaAs IC Symposium,* Portland, OR, October 1987, pp. 231–234.

[148] Anderson, A. K., and J. S. Joshi, "Wideband Constant Phase Digital Attenuators for Space Applications," *Microwave Engineering Europe,* February 1994, pp. 25–30.

[149] Bayruns, J., P. Wallace, and N. Scheinberg, "A Monolithic DC-1.6 GHz Digital Attenuator," *Proc. IEEE MTT Symposium,* Long Beach, CA, June 13–15, 1989, pp. 1295–1298.

[150] Sjogren, L., et al., "A Low Phase-Error 44-GHz HEMT Attenuator," *IEEE Microwave & Guided Wave Lett.,* Vol. 8, No. 5, May 1998, pp. 194–195.

[151] Gupta, R., et al., "A 0.05- to 14-GHz MMIC 5-Bit Digital Attenuator," *Proc. IEEE GaAs IC Symposium,* Portland, OR, October 1987, pp. 231–234.

[152] Howard, W., *Reference Data for Radio Engineers,* Sam's Engineering, Indianapolis, IN, October 1999.

[153] Peniket, N. A., "GaAs HBT MMIC Technology: A Commercial Reality," *GEC Review,* Vol. 14, No. 2, 1999, pp. 95–102.

[154] Geissberger, A. E., et al., "Fabrication, RF Performance and Yield of a Combined Limiting Amplifier and Dual-Modulus Prescaler GaAs IC Chip," *IEEE Trans. Microwave Theory Tech.,* Vol. 36, December 1988, pp. 1706–1713.

[155] Hughes, R. S., *Logarithmic Amplification,* Dedham, MA: Artech House, 1986.

[156] Gorman. G. M., et al, "A GaAs HBT Monolithic Logarithmic IF (0.5 to 1.5 GHz) Amplifier with 60 dB Dynamic Range and 400 mW power Consumption," *Proc. IEEE MTT Symposium,* Long Beach, CA, June 13–15, 1989, pp. 537–540.

[157] Hodges, D. A., "Darlington's Contributions to Transistor Circuit Design," *IEEE Trans. Microwave Theory Tech.,* Vol. 46, No. 1, January 1999, pp. 102–104.

[158] GaAs InGaP HBT MMIC Darlington Amplifier, Part No. HMC315/315E, Hittite Microwave Corp., Chelmesford, MA, at http://www.hittite.com/product_info/product_specs/amplifiers/hmc315.pdf.

[159] Lin, C., et al., "A Monolithic HBT Broadband Darlington Amplifier Using Modified Triple Darlington Pair," *Proc. GAAS2004 Conference,* Amsterdam, the Netherlands, October 11–12, 2004.

[160] Villemazet, J. F., et al., "Novel Compact Double Balanced Coplanar Active Mixer. Application to a Single Chip MMIC Receiver for Satellite Repeater," *Proc. GAAS 2001 Symposium,* London, United Kingdom, September 2001, pp. 24–25.

[161] Swanson, D. G., and W. J. R. Hoefer, *Microwave Circuit Modeling Using EM Field Simulation,* Norwood, MA: Artech House, 2003.

[162] van den Bogaart, F. L. M., and A. P. de Hek, "First-Pass Design Strategy for High-Power Amplifiers at X-Band," *Proc. IEE Tutorial Colloquium on Design of RFICs and MMICs,* Savoy Place, London, U.K., November 1997.

[163] Taguchi, G., "Introduction to Quality Engineering, Design Quality into Products and Processes," *Asian Productivity Organisation,* 1986.

[164] Taguchi, G., *Systems of Experimental Design,* White Plains, NY: Unipub/Kraus International Publications, 1987.

[165] Marsh, S. P., and S. D. Wadsworth, "Taguchi Tolerance Analysis of MMICs," *Proc. GaAs MANTECH Conference,* San Francisco, CA, June 2–5, 1997, pp. 118–121.

[166] Green, T. J., and R. G. Launsby, "Using Doe to Reduce Costs and Improve the Quality of Microelectronic Manufacturing Processes," *Proc. ISHM '94,* Boston, Massachusetts, November 1994, pp. 60–65.

[167] Bandurek, G. A., M. L. Hughes, and D. Crouch, "The Use of Taguchi Methods in Performance Demonstrators," *Quality and Reliability Engineering International,* Vol. 6, 1990.

[168] Marsh, S. P., A. P. Long, and S. D. Wadsworth, "Statistical Techniques for MMIC Design Sensitivity and Chip Yield Analysis," *Proc. GAAS' 98 Conference,* Amsterdam, the Netherlands, October 5–6, 1998, pp. 551–556.

6

Layout

MMIC design requires that the circuit components be laid out on the planar surface of the semiconductor substrate, and this layout process is the mechanism of going from a schematic circuit design to the physical positioning of the many different layers that make up each component. For example, Figure 6.1 shows a portion of a schematic diagram in a CAD simulator and the associated layout plot displaying a number of the various layers, together with a photo of the resulting chip.

The physical position of all the features on a particular layer, which is stored in the layout CAD tool, is then used to write the photolithographic processing masks. These masks are then used in the wafer fabrication facility to form all of the features of the components physically on the substrate-wafer surface.

This chapter begins by describing the format of layout data files and the type of CAD tools that create them, then goes on to discus the typical process of laying out a chip design. The chapter continues by outlining the circuit arraying procedure, where multiple chip designs are formed into an array of circuits that is then stepped and repeated over the whole wafer surface, and concludes with some comments about mask manufacturing.

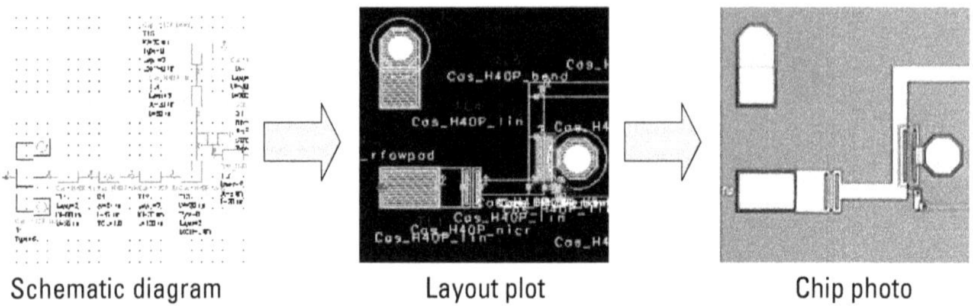

Figure 6.1 Pictorial representation of the MMIC layout process.

6.1 Layout Files

The physical features making up the components of an MMIC are either solid objects, such as a metal area or track, or holes in a continuous layer, such as the holes in the final dielectric passivation layer that allow probe needles to contact the bond-pads. These features can be defined as two-dimensional structures (i.e., in terms of the area they occupy) and associated with a particular layer (i.e., metal, dielectric). For example, Figure 6.2 shows that solid/metal features can be defined as a metal-layer polygon border with the "fill" area defined "inside" and dielectric holes or windows can be defined as dielectric-layer polygon borders with the "fill" area defined "outside." Therefore, layout data files have a format that defines polygons (the border) with associated layers and fill definitions. When the layout data is used to make the chrome on quartz masks, a layout layer that has clear or unfilled background areas (or fields) creates what is known as a positive polarity mask. Conversely,

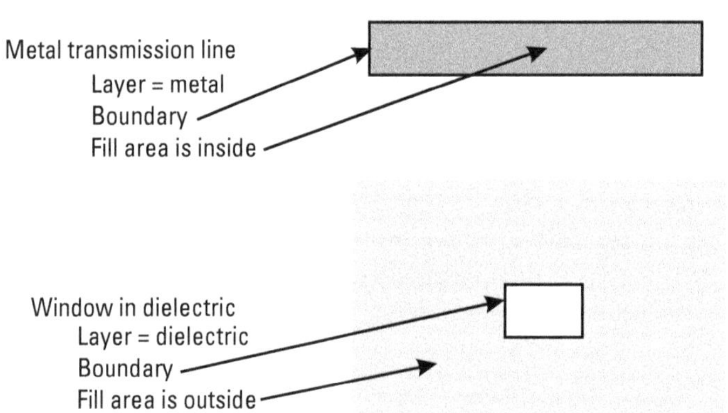

Figure 6.2 Examples of component features defined as polygons.

a layout layer that has opaque or filled background areas (or fields) creates what is known as a negative polarity mask.

The polygon-based format is different from the line-based drawing format and means that certain topologies are invalid (primarily because they cannot physically be made). For example, zero-width lines are not allowed unless they are part of a continuous boundary of a polygon. Also, boundaries of individual polygons may not cross (as in a figure eight) as this creates an ill-defined "fill" function.

Layout data files also support hierarchy, which means a group of individually defined polygons (on any number of layers) can be grouped into one object known as a cell. The cell can then be copied and reproduced at many points in the layout with the link to the original cell retained so that when it is changed, all copies are changed as well. A foundry library is typically constructed with cells for all the commonly used components. Cells can also be grouped with other cells and polygons to form higher-level cells for subcircuits, such as whole gain stages, and the cell that contains all the components for a chip design is known as the top-level cell for that circuit. The process of IC design with linked schematic and layout designs is known as schematic-driven layout (SDL) and the amount of cell hierarchy or grouping depends on the complexity of the circuit (which is typically much higher with silicon ICs) [1].

The industry-standard format for these types of two-dimensional IC layout files, Graphical Data System II (GDSII) [2–3],[1] was first developed by the Calma company for their Graphical Data System computer-layout tool and is now owned by Cadence Design Systems [4]. Some CAD layout tools are using a newer and more compact [5] format known as the Open Artwork System Interchange Standard (OASIS), which is a trademark of the Semiconductor Equipment and Materials Institute, known as SEMI.[2]

As mentioned in Chapter 4, many of the RF and microwave circuit simulation tools, such as Agilent ADS[3] and AWR Microwave Office,[4] can produce layout plots that are directly linked or synchronized to the schematic circuit diagram page, which helps remove some of the errors during the layout process. These linked systems can import foundry libraries, but they do have some limitations in that each component can only be connected at

1. http://www.xs4all.nl/~kholwerd/interface/bnf/gdsformat.html.
2. http://www.semi.org.
3. Agilent Technologies, USA, http://eesof.tm.agilent.com.
4. Applied Wave Research, USA, http://www.appwave.com/products/mwoffice.

predetermined points on its edges. Other tools, such as WaveMaker from Barnard Microsystems,[5] can import foundry libraries so that the component cells are correctly drawn but are much more flexible in terms of how they can be connected to each other. Linked systems are ideal for inexperienced MMIC designers because they will not allow incorrect connection of components; however, experienced designers often require more flexibility to connect components in more innovative ways and rely on their own judgment that the MMIC will work as simulated.

6.2 Circuit Layout Process

Each foundry has its own way of producing the active and passive components on the wafer surface in a reliable manner; this includes the detailed specification of the maximum and minimum separations between, and overlays of, different layers within each component. This is shown in Figures 6.3 and 6.4, which show the foundry-defined layer separations for a silicon nitride MIM capacitor and a HEMT, respectively. In both examples, the components are produced from five separately overlaid layers, each manufactured by a specific process step, but this will vary from foundry to foundry. The foundry will not want these changed, and the designer does not want to

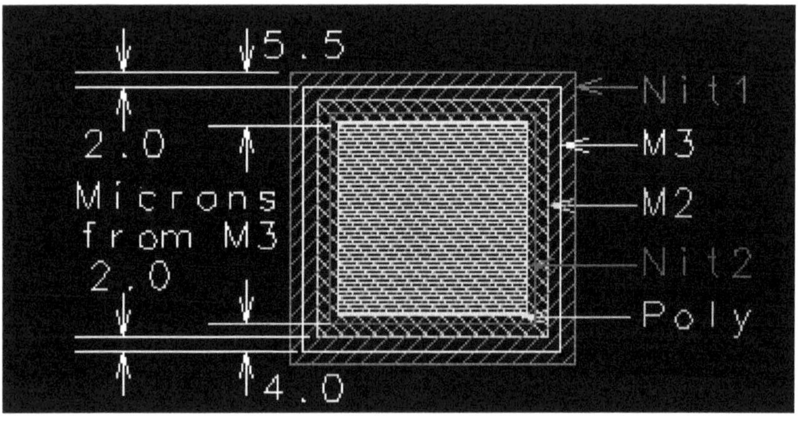

Figure 6.3 Foundry-defined layer position and spacing for a silicon nitride MIM capacitor. (*Source:* Bookham Inc., 2006. All Rights Reserved.)

5. http://www.barnardmicrosystems.com.

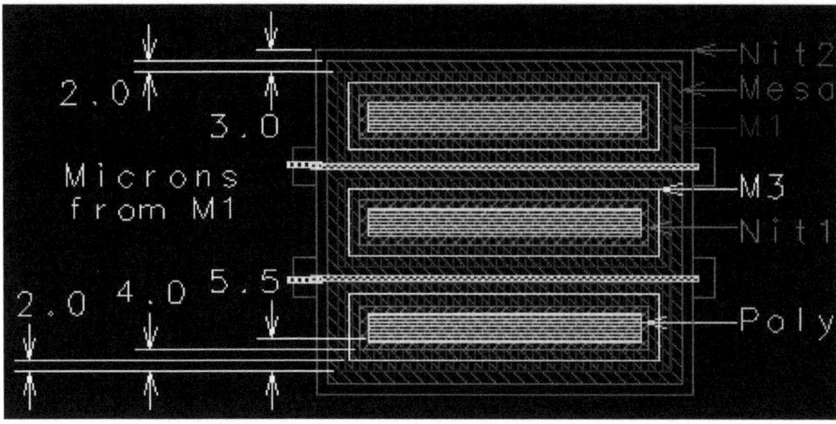

Figure 6.4 Foundry-defined layer position and spacing for a HEMT. (*Source:* Bookham Inc., 2006. All Rights Reserved.)

have to draw each component, so the usual layout process is to use "standard cells" from an imported foundry library of components, leaving the designer the much simpler task of joining those together electrically using transmission lines. The foundry library will contain the layouts of all the commonly used components, such as the transistors, capacitors, resistors, and so forth, and they may be scalable in size if linked to scalable electrical models on the schematic page. The connecting transmission lines are sometimes provided as foundry components that are scalable in length and width, but they can also be drawn manually as polygons or paths on the appropriate metal layer.

Moving from the schematic circuit to the planar circuit layout often brings to light problems that are not obvious on the schematic page, such as overlapping components and other physical spacing issues. For this reason the design and layout is normally an iterative process (Figure 6.5) that requires many cycles of optimizing the RF response on the schematic page, laying out the components and increasing spacing where necessary, then reoptimizing the RF response. This iterative design process continues when the dc and RFOW pads are added to the circuit, and the designer must consider how to compact the layout into the minimum area to bring down the manufacturing cost of the chip. This often requires putting bends in the transmission lines and reoptimizing the RF performance to adjust to the different effective electrical length they represent. It is also common to place a bounding rectangle around the extremities of the layout on a layer that is not used to make a mask as an aid to the layout process. This helps the designer

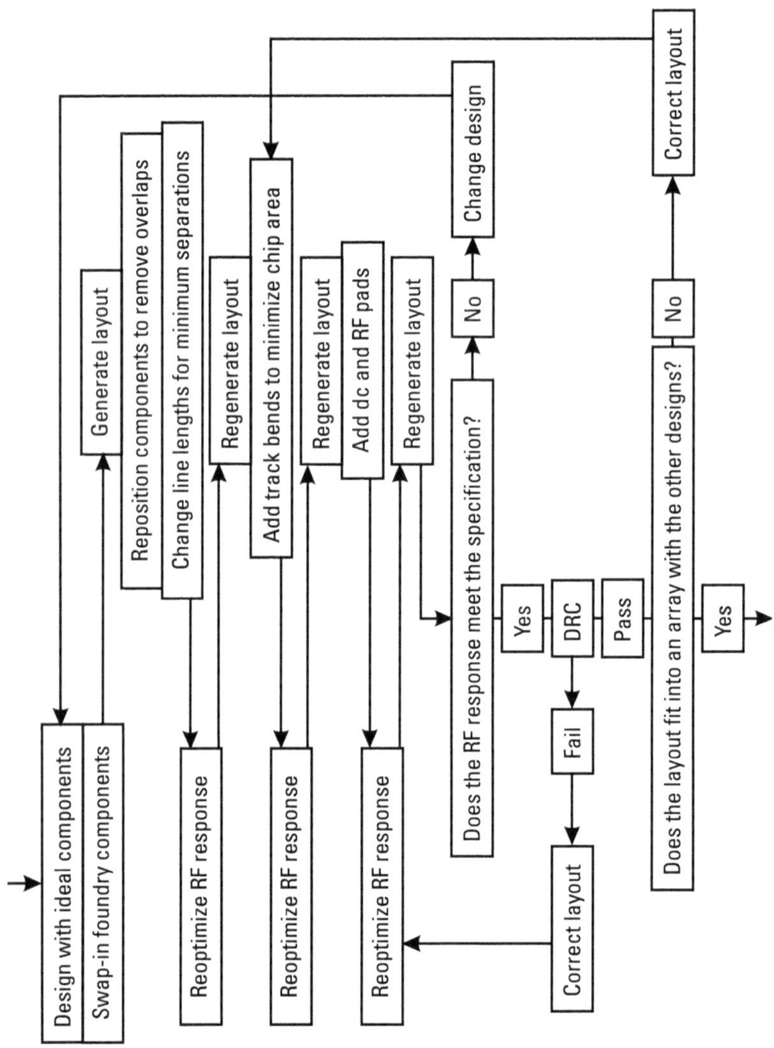

Figure 6.5 Flow diagram representing the iterative nature of the schematic-driven layout process.

notice if any space is being wasted and acts as a size reference during the chip arraying process.

6.3 Layout Checking

Layout checking is one of the most important tasks during MMIC design because it is the last point before significantly large amounts of money and time are invested in the design. If mistakes are spotted at this stage, there will likely only be a small delay before they are corrected in the layout data and the design can progress. If an error gets past this stage, the masks are written, and one or more batches of wafers may be processed before the error is discovered, wasting considerable time and money. Therefore, it is in both the foundry's and the designer's interest to perform rigorous checking of the final layout. This section highlights the importance of checking the layout and gives some examples of the layout design rules that an MMIC designer may encounter, then goes on to discuss some of the checking methods, such as design rule checking (DRC), electric rule checking (ERC), layout versus schematic (LVS) checking, and reverse engineering.

6.3.1 Layout Design Rules

Layout design rules are primarily determined by the wafer-processing methods (which is why understanding the processing technology in Chapter 7 is a vital tool for the MMIC designer) because they determine the size and tolerance of the features on the chip. Examples of layout design rules are given in the following sections.

6.3.1.1 Minimum Feature Sizes and Spacing

The design rules take into account the processing capability. For example, there may be a rule that the minimum gap between metal tracks is 6 μm because if you were to try to make the gap less than 6 μm, the two tracks could merge into one, creating a short circuit. Similarly, placing through-substrate via holes too close together weakens the wafer and could cause it to crack.

The designer should be aware that the foundry usually specifies the minimum features and spacing that the process will tolerate, but these minimum sizes and spacing should be avoided wherever possible in order to enhance circuit yield, reproducibility, and reliability.

6.3.1.2 Metallization Constraints

The interconnect metallization is often defined by a lift-off process. This technique does not allow closed features [Figure 6.6(a)] to be defined because of the risk of incomplete removal of the isolated central material. Annular features or isolated slots must therefore be broken on at least two sides [Figure 6.6(b)] to attach the central region to the external metal for reliable lift-off. For similar reasons, long narrow gaps in either interconnect metal [Figure 6.6(c)] should be avoided and will have a maximum acceptable aspect ratio constraint.

6.3.1.3 FET Orientation on Mesa

FETs and diodes are often constructed with the gate feed metal running up a positive slope onto the mesa of the active semiconductor. In the case of a GaAs wafer, the crystallographic properties means that the mesa edges perpendicular to this have a negative slope and do not allow a continuous metal contact up onto the mesa. For this reason, all components with gate contacts (FETs and diodes) must be laid out with the gates aligned parallel to each other, as shown in Figure 6.7. This rule allows the processing engineers to ensure that all of the gates are aligned with the correct mesa etch (gate axis) of the wafer.

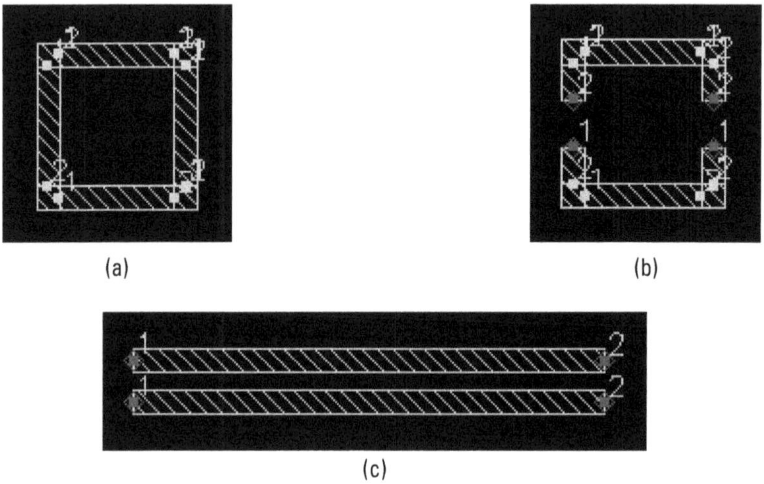

(a) (b) (c)

Figure 6.6 Metallization constraints for a lift-off process: (a) closed structure not allowed, (b) open structure allowed, and (c) maximum aspect ratio of 200:1 (typical).

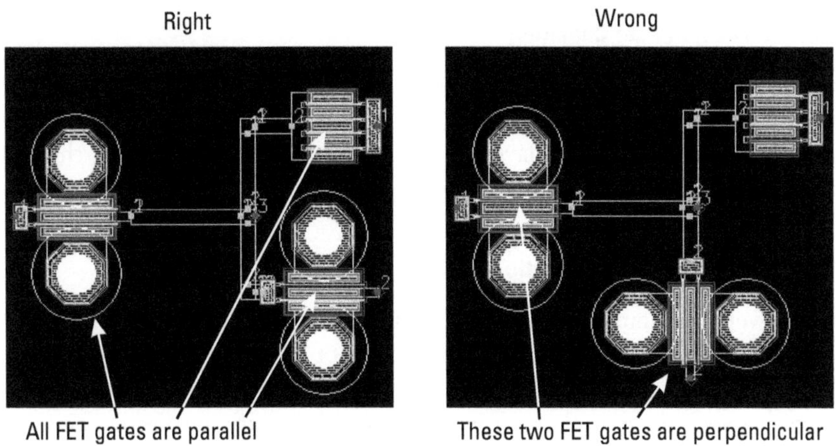

Figure 6.7 FET gate orientation rules for a GaAs mesa process.

6.3.2 Design Rule Checking

Design rule checking (DRC) [6] is performed by a program that scans the layout data file and measures the distances from every polygon border to every other polygon border, then compares these distances to the specified design rules. It can also perform Boolean functions to interpret design rules that are not simply measurement related. It then flags where on the layout the rule has been broken, and either this can be corrected or an exception can be conceded by the quality-assurance department.

The DRC cannot pick up errors that are electrical in nature if they do not break a processing rule. Consider the example of a designer intending to connect two components but unintentionally leaving a small gap between them, as shown in Figure 6.8. The DRC will flag this if the gap is smaller than the allowable spacing (a typical value is 4 μm). If the gap is, say, 6 μm then no processing rule is broken, and the computer must assume that the spacing is intentional. Electrically, the gap is a dc open circuit, which is obviously catastrophic in a bias path. This is another reason why great care must be taken at the MMIC layout stage. One method that can highlight this error is electrical rule checking.

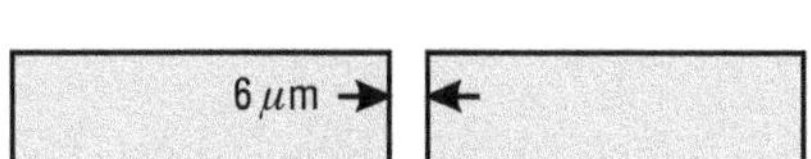

Figure 6.8 A 6-μm gap in metal track that would pass the DRC but fail electrically.

6.3.3 Electrical Rule Checking

Electric rule checking (ERC) examines the layout for inconsistencies in the electrical connectivity of the circuit [7]. A typical ERC program will check either the schematic or the layout data file for unconnected inputs, shorted outputs, and sensible ground and power connections.

6.3.4 Layout Versus Schematic

Layout versus schematic (LVS) [8] checking is a program or part of a CAD tool that compares the sizes and connectivity of the components on the layout page to the components on the schematic page. If any parameters are different between the two, then error messages will identify the discrepancy.

6.3.5 Reverse Engineering

Reverse engineering is where an independent MMIC designer takes the final layout data and recreates a schematic design in the CAD system with no input from the original designer. The simulated RF responses generated by the independent engineer can then be compared to the original designer's simulated response and also to the specification. This helps to show up any assumptions that the original designer may have made and allows any compromises to be discussed.

6.3.6 Visual Checking

The MMIC layouts can be inspected visually by plotting them onto large sheets of paper after they have been magnified up to 1,000 times their physical size. For example, a 1×1 mm chip will be plotted with dimensions of 1m $\times$ 1m. This means that gaps or errors of the order of 1 μm or 2 μm are now of the order of 1 or 2 mm and can be spotted by careful visual inspection.

6.4 Chip Arraying

Arraying is the process of taking multiple MMIC design layouts and constructing an orthogonal array or reticule of the designs. This is shown in Figure 6.9, where MMIC design layout "C" is assembled together with eight other MMIC design layouts ("A" to "I") in an array of nine layouts. If contact lithography is being used, masks for each layer are written by repeating the array over an area large enough to cover the whole semiconductor substrate wafer. If stepper

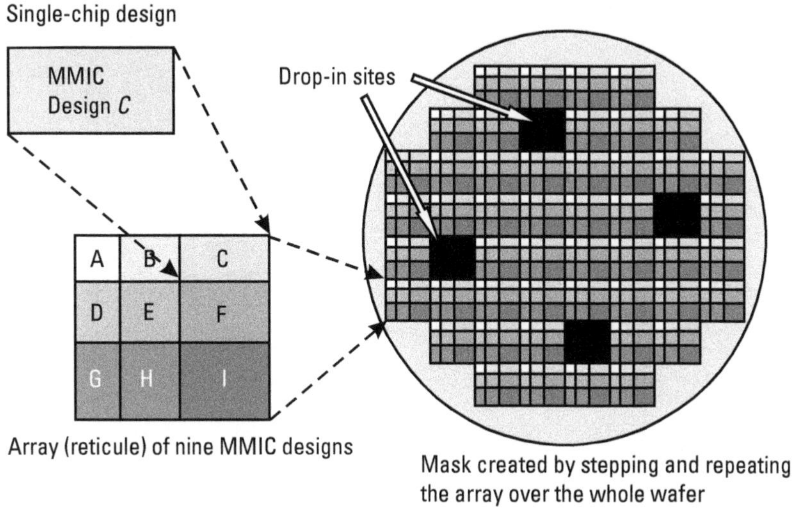

Figure 6.9 The MMIC design arraying process

lithography is being used, the array data is used to write masks for a single array, and the stepper will step and repeat exposures of the array over the whole of the wafer. An example array of sixteen different MMIC design layouts is shown in Figure 6.10.

It is common practice to place process control monitor (PCM) structures either in place of one MMIC layout in each array or in place of several arrays across the wafer. The areas where they are placed are known as drop-in sites, shown in Figure 6.9, and may contain other test and calibration structures other than the PCMs.

An example drop-in array is shown in Figure 6.11. This is exactly the same size as the main circuit array and contains the PCM cell layout, microstrip calibration structures, and other test structures, such as subsections of the MMIC designs in the main circuit array.

6.4.1 Chip Identifiers

When the individual MMIC design layouts have been checked and corrected, they require the addition of identifying marks before they are arrayed. These will usually consist of references to the mask-set name, the individual chip design, and its *X*/*Y* coordinates on the mask, as shown in Figure 6.12. This procedure is normally performed by the foundry and starts with its creating a chip-legend border, which is typically 30 μm to 50 μm larger than

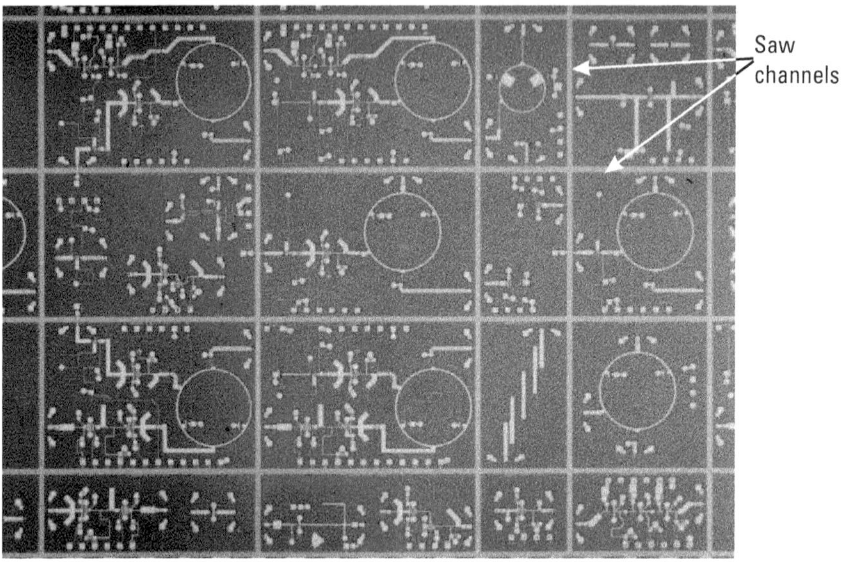

Figure 6.10 Example array of sixteen MMIC layouts. (*Source:* Bookham Inc., 2006. All Rights Reserved.)

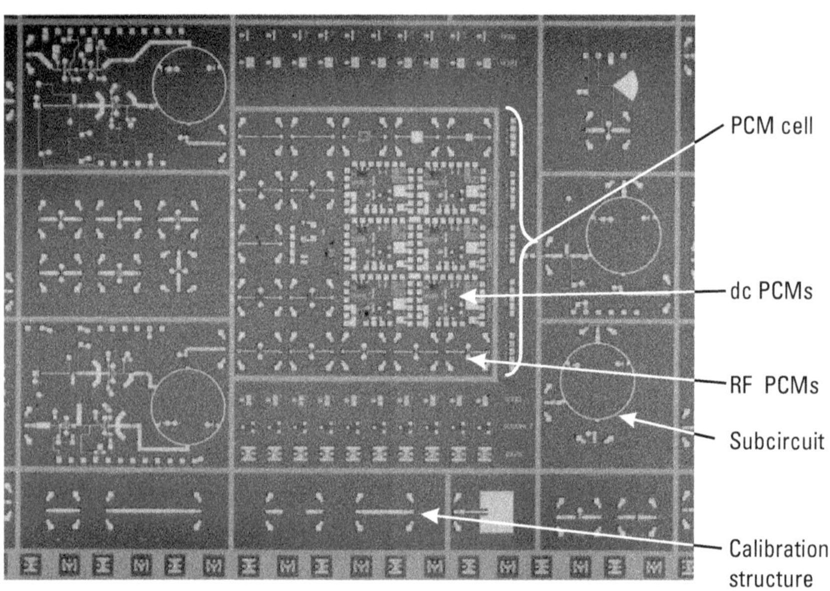

Figure 6.11 Example drop-in array containing the PCM cell, calibration, and other test structures. (*Source:* Bookham Inc., 2006. All Rights Reserved.)

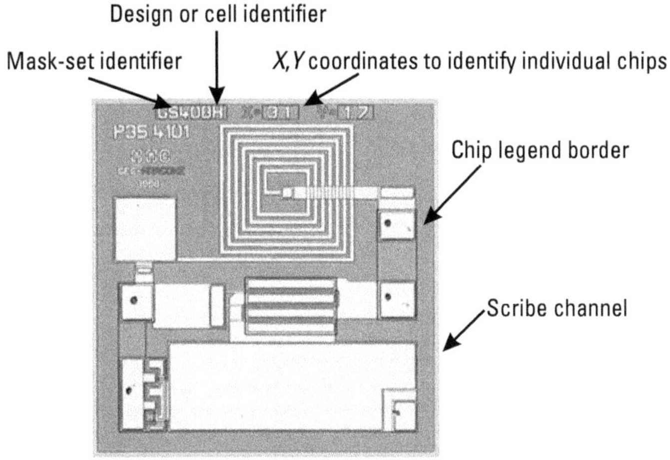

Figure 6.12 Chip identifiers. (*Source:* Bookham Inc., 2006. All Rights Reserved.)

the extremities of the layout features (30 μm to 50 μm larger than the bounding rectangle). The chip identifiers are created as polygonal text within this border, and then scribe/saw channels are placed around the circuit.

6.4.2 Arraying Guidelines

A number of guidelines associated with the arraying process are useful for the MMIC designer to know about, two of which are shown pictorially in Figure 6.13. First, to prevent wasted space, every attempt should be made to use circuit designs of a similar size and aspect ratio to allow the optimum use of the GaAs material. Second, there is likely to be a constraint on the maximum aspect ratio for a chip because long and thin cells are susceptible to breakage while sawing or dicing the wafer. This maximum aspect ratio of cells does depend slightly on the substrate thickness but is typically 3:1. Third, because the circuits may well be tested RFOW, where the probes are stepped from one circuit to the next, the distance between cells may need to be set to a whole number of thousandths of an inch (thou) if the wafer-prober uses imperial units. Fourth, the saw/scribe channels must be continuous across the wafer. Sawing along scribe channels within a subcell is possible but very difficult, so foundries usually will not commit to subscribing themselves.

At the arraying stage it is sometimes tempting to rotate the MMIC layouts to make them fit nicely into the array. If a mesa process is being used,

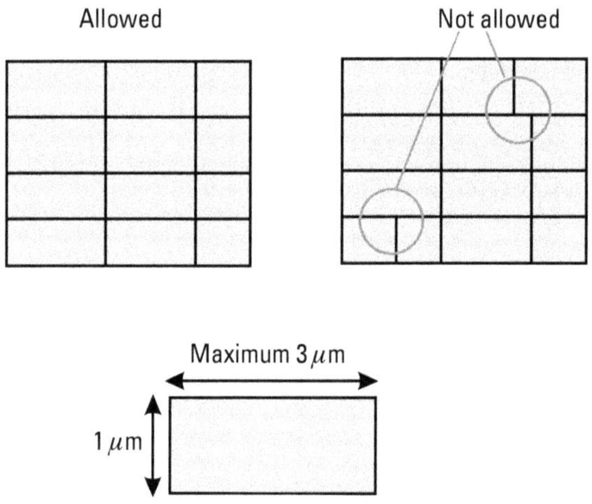

Figure 6.13 Continuous saw-lane and maximum aspect ratio guidelines.

then care must be taken to ensure that the FET/HEMT gate parallel-orientation rule is not forgotten.

6.5 Mask Manufacture

The foundry takes responsibility for the mask manufacture, which may be performed internally or subcontracted out to an external supplier. The photolithography masks are made from quartz, which allows ultraviolet (UV) light to pass through and expose the photoresist on the wafer, and coated with a chrome layer that blocks the UV light where it is not required. The features are patterned on the mask using photoresist and exposed (or written) using an electron beam. Fine features, such as the FET gates, require fine beams to define them, which means that the total writing time is longer, so the gate mask is more expensive. A typical MESFET process requires 11 masks, one for each layout layer.

6.6 Layout Examples

Typical examples of a range of MMIC layouts are shown in Figures 6.14 to 6.17. The simple layout in Figure 6.14 is a layout plot of an X-band VCO, where it can be seen that the different layout layers have different "fill" patterns to distinguish them from each other (e.g., through-substrate via is solid

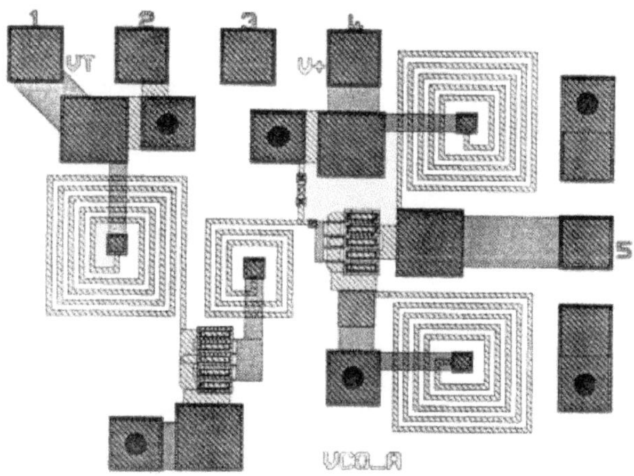

Figure 6.14 Layout of an X-band VCO. (*Source:* Bookham Inc., 2006. All Rights Reserved.)

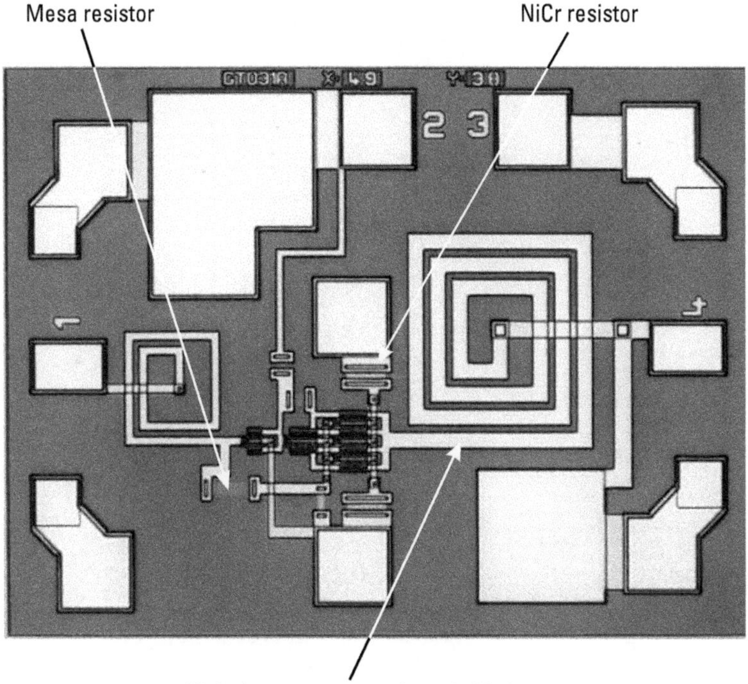

Figure 6.15 Layout of an HBT Darlington amplifier MMIC. (*Source:* Bookham Inc., 2006. All Rights Reserved.)

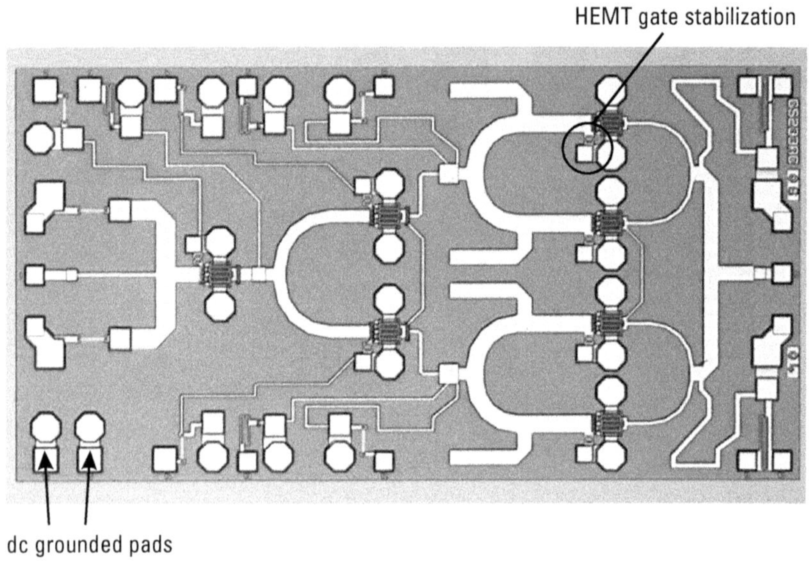

Figure 6.16 Layout of a millimeter-wave HEMT three-stage power amplifier. (*Source:* Bookham Inc., 2006. All Rights Reserved.)

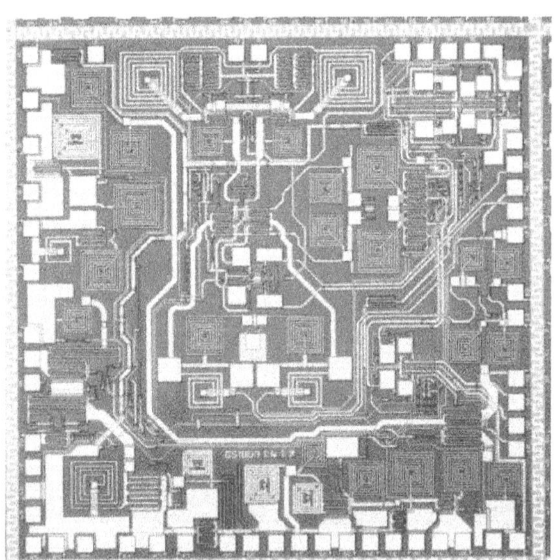

Figure 6.17 Layout of a WLAN transmit/receive MMIC. (*Source:* Bookham Inc., 2006. All Rights Reserved.)

black, metal one is shaded, and metal two is hashed diagonally). Figure 6.15 is a photograph of an actual HBT Darlington amplifier MMIC, where it is hard to distinguish the different metal layers, and the through-substrate vias are not visible. Figure 6.16 is a photograph showing how a more complicated three-stage millimeter-wave HEMT power amplifier MMIC is laid out, with plenty of space between the components to minimize coupling at these high frequencies. In contrast, Figure 6.17 shows how at low frequencies a complex WLAN transmit/receive chip can be laid out with very little space between the individual MMIC components.

Further information on layout can be found in [9, 10].

6.7 Questions

6.1 MMIC layout data files are built up from lines or polygons?

6.2 What does hierarchy mean in the context of MMIC layout?

6.3 Does the MMIC designer construct the layers within the transistors?

6.4 What is the consequence if critical errors are not spotted before the masks are manufactured?

6.5 What determines the layout rules? Give two examples.

What is the general approach for laying out an MMIC?

6.6 What do DRC, LVS, and ERC stand for?

6.7 What is the process of arraying?

6.8 What is the PCM, and where is it placed?

References

[1] Kick, B., et al., "Standard-Cell-Based Design Methodology for High-Performance Support Chips," *IBM J. Research & Development,* Vol. 41, No. 4/5, May 1997.

[2] Calma Corporation, GDS II Stream Format, July 1984.

[3] *GDSII Stream Format Manual,* Release 6.0, G. E. Calma, February 1987.

[4] Cadence Design Systems, 2655 Seely Avenue, San Jose, California, 95134; tel. (408) 943-1234; Web site: htpp://www.cadence.com.

[5] Reich, A. J., et al., "OASIS vs. GDSII Stream Format Efficiency," *23rd Annual BACUS Symposium on Photomask Technology,* ed. Kurt R. Kimmel and Wolfgang Staud, *Proceedings of the SPIE,* Vol. 5256, December 2003, pp. 163–173.

[6] Lavagno, L., G. Martin, and L. Scheffer, *Electronic Design Automation for Integrated Circuits Handbook,* London, U.K.: CRC Taylor & Francis, June 2006.

[7] "New Physical Connectivity Engine for Native High Frequency Interconnect Verification," *Microwave Journal,* August 2004.

[8] Calibre LVS, Mentor Graphics, Wilsonville, Oregon, USA.

[9] Saint, Christopher, *IC Layout Basics: A Practical Guide,* New York: McGraw-Hill, 2001.

[10] Hastings, Alan, *The Art of Analog Layout,* Upper Saddle River, NJ: Prentice Hall, 2000.

7

Processing Technology

Many MMIC design rules and layout constraints are the result of the way the chips are manufactured, in other words, the wafer-processing technology. A basic understanding of the different processes, such as epitaxial growth, photolithography, and metal deposition, enables the MMIC designer to know which rules can be bent during the design phase and which rules cannot be broken under any circumstances. Indeed, it is usually the designers, pushing to improve the performance of their circuit designs, who know which design rules constrain the potential performance of the process and can help direct the process engineers in developing the next generation process. This sort of interaction between the MMIC designers and the foundry process engineers facilitates innovation not just in the development of the next generation process but also in the optimum use of the current process technology.

This chapter gives a brief overview of the typical wafer-processing steps involved in the fabrication of MMICs, from the original semiconductor material all the way through to the final chip. The reader needs to realize that this is just a generic overview, and individual foundries have their own specific process flows and associated design rules, which will be different from those described here. Much of this material is also based on a III–V semiconductor process, which, while generally similar to a silicon fabrication process, will have different associated material systems and processing issues. A summary of these steps is given below:

- *Substrate material growth:* grow material from a single-crystal boule of the semiconductor;
- *Wafer production:* saw the boule into wafers;
- *Surface layers:* produce the active layers;
- *Photolithography:* pattern features on the front of the wafer;
- *Wafer thinning:* lap from the back side;
- *Substrate vias:* etch through grounding vias;
- *Back-face metal:* create the ground plane;
- *Chip separation:* saw or scribe up the wafer into individual chips;
- *Quality assurance:* conduct process control and inspection.

7.1 Substrate Material Growth

The word *monolithic* in the name MMIC means that the circuit is constructed on one solid piece of semiconductor material. The elements making up the semiconductor substrate material determine the potential performance characteristics of the MMIC, primarily due to the insulating properties of the undoped semiconductor and the electron mobility in the doped semiconductor. Table 2.1 compares the properties of the most commonly used semiconductors.

Note that when silicon germanium (SiGe) is used as the semiconductor material for the active device, the SiGe is only an epitaxial layer to provide high electron mobility in the transistors and that the substrate material is silicon (Si); thus, SiGe has much the same benefits and disadvantages as a purely Si process. In a similar way, indium phosphide (InP) is sometimes grown as an epitaxial layer on a gallium arsenide (GaAs) substrate to get the performance of InP transistors on a larger and better-established semiconductor wafer. This type of substrate is known as metamorphic because the lattice constant of the GaAs substrate must be gradually changed through the epitaxial layers to that of InP. InP as a substrate material is more brittle and difficult to handle than GaAs.

The structure of the substrate material is also important because if the transistors fabricated on the substrate surface are going to benefit from the high electron mobility of the material compound, the substrate must be manufactured as a single crystal lattice. If the substrate were amorphous instead of a single crystal lattice, then the material properties would be completely different and unpredictable.

Single-crystal semiconductor substrates are created by taking a seed crystal of the required material and growing it to a much larger size using the same semiconductor material in its molten form. This is generally performed using one of two methods, liquid-encapsulated Czochralski (LEC) [1, 2] or vertical gradient freeze (VGF) [3]. The LEC method, shown in Figure 7.1, uses liquid boron trioxide to cover the molten GaAs and prevent the volatile arsenic from subliming; then, the seed crystal is dipped into the molten GaAs where it grows. The seed crystal is slowly pulled out of the molten GaAs, and the speed of the pulling governs the diameter of the crystal boule.

The VGF method, shown in Figure 7.2, has the molten GaAs in a vertical crucible with the seed crystal at the bottom, and a temperature gradient is moved up the furnace, freeze-forming the single crystal vertically upwards. The diameter of the crystal boule is constrained by the size of the crucible.

The result of both of these techniques is a boule or ingot of the semiconductor substrate material structured as a single crystal with a predetermined diameter. An example of this is the 3-in.-diameter ingot of InP shown in Figure 7.3.

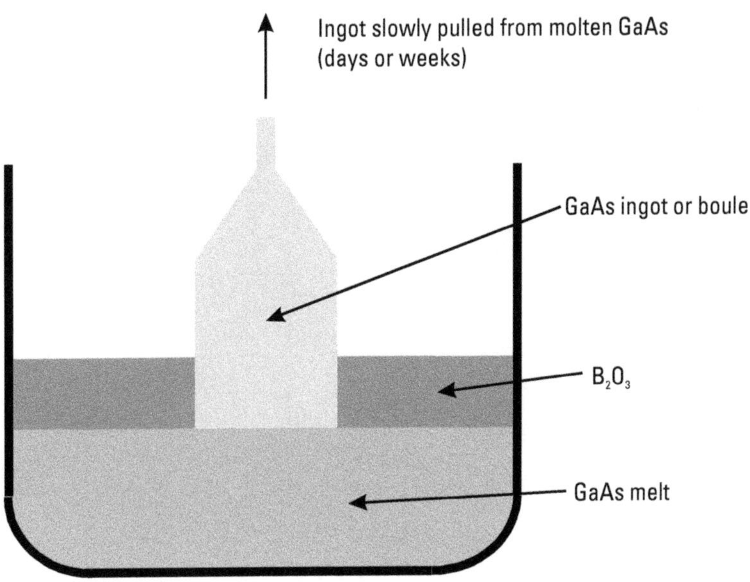

Figure 7.1 LEC single-crystal substrate growth technique.

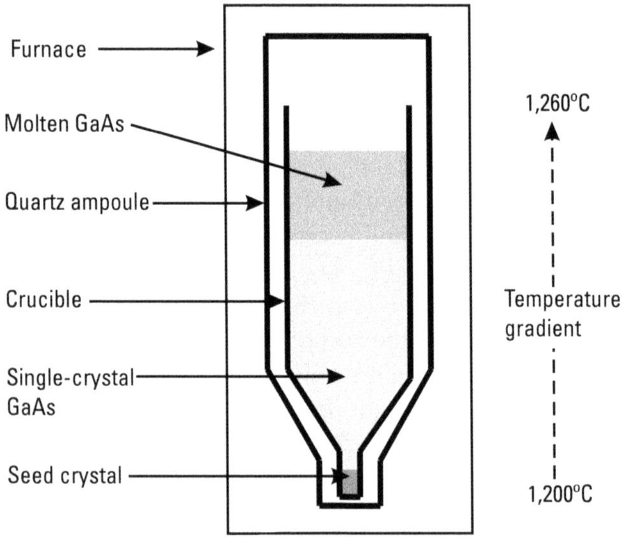

Figure 7.2 VGF single-crystal substrate growth technique.

Figure 7.3 A 3-in. InP ingot grown by LEC. (Courtesy of Wafer Technology Ltd.)

7.2 Wafer Production

The boules of the substrate material are transformed into cylindrical shapes by sawing off the ends, and identifying flats are ground into the sides of the cylinder to identify the crystallographic orientation of the final wafers. The cylindrically shaped substrate material is then sawn into wafers using a saw impregnated with diamond. The sawn wafers are coarsely ground flat, the edges are rounded, and then they are chemically etched by a further 10 μm to remove any defects caused by the sawing and thinning. The wafer surfaces are then polished to a flatness of typically 2 μm. The thickness of a 3-in.-diameter GaAs wafer at this stage is typically 625 μm.

7.3 Surface Layers

The substrate materials discussed are semiconductors, which means their conductivity is greater than that of an insulator but less than that of a good conductor, such as a metal, and can change as a function of the temperature and dopant or impurity levels. At this stage, the wafers are a single crystal purely of the substrate material, and all the electrons are bound up in the crystal lattice bonds, making the substrate close to a good insulator. In this state, the semiconductor is known as semi-insulating, and the higher the resistivity of the semi-insulating substrate, the lower the loss of the circuits designed on it. The resistivities of different semiconductor substrate materials, when in their semi-insulating state, are shown in Table 7.1. This is important for circuit design because lossy substrates reduce the Q-factor of components such as inductors [4, 5] and limit the sharpness of responses achievable, particularly the selectivity of MMIC filters.

Table 7.1
Resistivity of the Different Semiconductor Substrate Materials When in Their Semi-Insulating State

Material	Resistivity (Ω cm)
Silicon	10^3–10^5
Silicon carbide	10^5–10^{12}
Gallium arsenide	10^6–10^9
Gallium nitride	10^8
Indium phosphide	10^6–10^9

The transistors, however, require that the semiconductor conduct current, so dopant elements must be introduced into the crystal lattice that have more or fewer electrons than the element they are replacing. The result of this doping of the semiconductor is that there are now some free electrons or holes to conduct current through the crystal lattice. The semiconductor is usually activated into a conducting state by engineering surface layers, or, in other words, by creating an active layer on the wafer surface. There are three common ways of creating the active layers on the wafer surface: ion implantation, molecular beam epitaxy (MBE), and metal-organic chemical vapor deposition (MOCVD).

Ion implantation works by firing ions of a particular isotope into the wafer surface to a specific dose and with specific energy [6–8]. A typical ion implantation system is shown in Figure 7.4. This positions the donor atoms in the desired position, and then the wafer is heat-treated (annealed) to incorporate the donor atoms into the crystal lattice and make them active. This technique is commonly used in the silicon processing industry and for MESFET processing on GaAs.

MBE and MOCVD create the active layers by growing new layers on the surface of the wafer with precisely controlled compositions. During MBE, molecular beams are targeted onto the surface of the wafer within a very-high-vacuum chamber. An MBE machine for 6-in.-diameter GaAs wafers is shown in Figure 7.5, where the vacuum growth chamber is located on the left side of the image. MBE is particularly suited to producing thin layers with abrupt changes in the material composition as it can be used to grow virtually one atomic layer at a time, and the composition can be

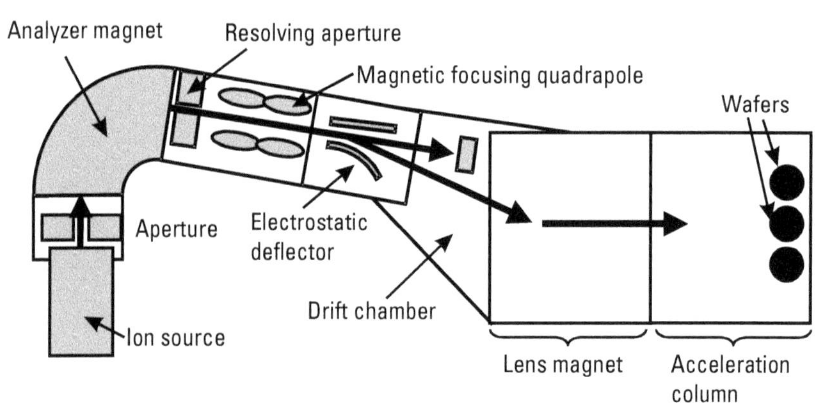

Figure 7.4 Ion implantation system.

Figure 7.5 MBE equipment for 6-in.-diameter GaAs wafers. (*Source:* Bookham Inc., 2006. All Rights Reserved.)

changed very quickly using shutters on the different molecular beams [9, 10]. For this reason, MBE tends to be used for high-electron-mobility transistors (HEMTs) at higher frequencies where all the physical dimensions, including the layer thicknesses, need to be scaled down in size.

MOCVD grows new layers on the surface of the wafer similarly to MBE; however, the material for the new layers is not provided by molecular beams but by precursor organic and hydride molecules, which undergo a controlled gas phase reaction in an open furnace above the wafer. The process performs well for the (Al)GaAs, GaInP, AlInGaP, and InGaN family of compounds and is often used for growing the active layers required for heterojunction bipolar transistors (HBTs) [11].

7.4 Photolithography

The wafers now consist of a semi-insulating substrate with active layers on the surface and are ready to be patterned by photolithography. Photolithography is the process by which the circuit design of transmission lines, metal-insulator-metal (MIM) capacitors, spiral inductors, and transistors are defined on the wafer surface in terms of patterned metal and dielectric layers. As described in Chapter 6, the metal and dielectric features are created as polygons in a CAD system, and this data is used to write a set of mask plates, where the pattern for each layer appears as chrome metal on a quartz mask plate. A liquid photoresist is spun onto the surface of the wafer to a thickness of about 1 μm, baked at 100°C to dry, and exposed through the holes in the chrome mask pattern with ultraviolet (UV) light. The exposure can be via contact lithography, where the mask plate is placed in contact with the wafer during exposure, or by projection lithography, where the mask is held a precise distance away from the wafer during exposure. Contact lithography is the cheaper and simpler of the procedures, but the continual contact means the mask plate will eventually wear out, and the run-out across the width of the mask means it is limited to 3- or 4-in. wafer processing. Projection lithography uses a smaller and less expensive mask, which is stepped across the wafer to expose the whole surface. At each step, the mask is realigned, which eliminates the problem of run-out. The drawback with projection lithography is the capital cost of the stepper equipment (shown in Figure 7.6), which must accurately step and align over a 6-in.-diameter or larger wafer.

Depending on whether the photoresist is positive or negative, the exposed or unexposed resist can be removed with a developing solution, leaving a resist pattern that is a copy of the mask pattern or its inverse. There are two main methods by which the photoresist pattern is used to define the features in the metal and dielectric layers. These are etch and lift-off processes, as shown in Figure 7.7.

An etch process will typically begin with the wafer surface already covered with a layer of the metal to be patterned. The resist pattern is formed to leave developed resist where metal is required, and then the exposed metal is etched, usually using a wet chemical etch. Etching is generally better at handling thick layers, such as plated-up gold, but edge definition can be compromised by etch-undercutting the photoresist pattern.

On the other hand, a lift-off process starts with a bare wafer surface, and the resist pattern is formed directly onto it. The resist pattern is created with an overhang profile, as shown in Figure 7.8, using multiple resist layers or multiple exposures and the metal is deposited anisotropically in a thin layer by techniques such as evaporation. After deposition, the photoresist is dissolved with a solvent, lifting off the metal deposited on top of the resist and leaving behind the metal pattern on the wafer surface. The combination of the overhang resist profile and a thin deposited layer allows the solvents to dissolve the resist under the overhang very effectively and to leave metal patterns on the wafer with very precisely defined edges. Lift-off is not so suitable

Figure 7.6 A 0.5-μm stepper for 6-in.-diameter wafers. (*Source:* Bookham Inc., 2006. All Rights Reserved.)

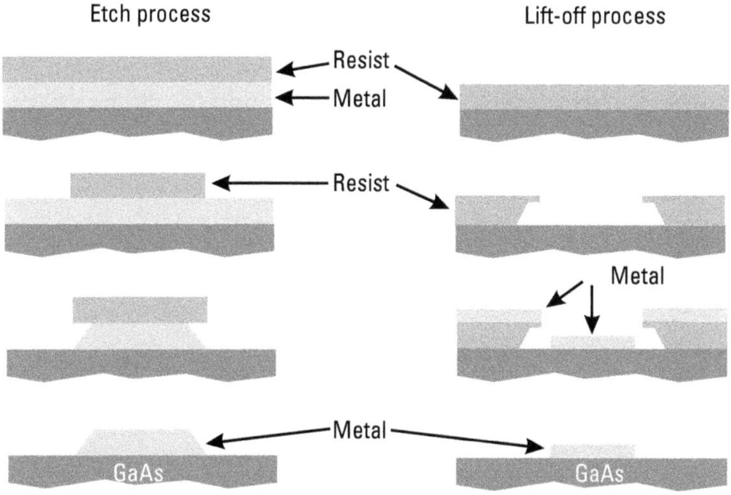

Figure 7.7 Comparison of etch and lift-off photolithography processes.

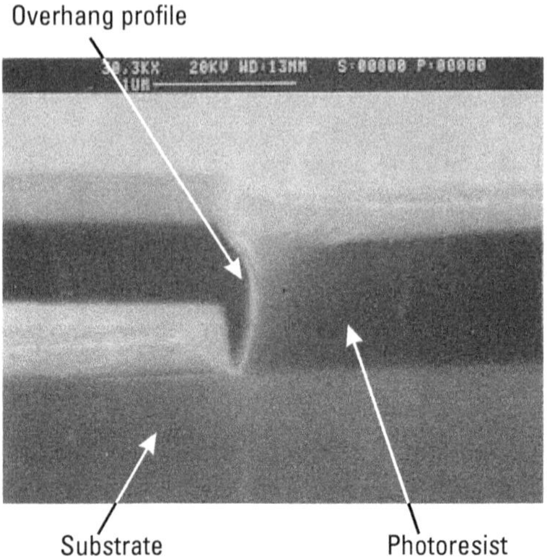

Figure 7.8 Overhang photoresist profile for lift-off photolithography. (*Source:* Bookham Inc., 2006. All Rights Reserved.)

for thick deposited layers because these can obscure the overhung resist edges and prevent the solvent from dissolving the resist.

The minimum size of features that can be defined by photolithography is related to the wavelength of the light used to expose the resist by Rayleigh's equation [12], and the silicon industry is currently able to produce features of the order of 100 nm using special lithography techniques. In practice, the minimum feature size used in optical contact lithography is 500 nm (0.5 μm), and if smaller features are needed, such as the gate of a HEMT, these tend to be directly written into the photoresist using an electron beam.

7.4.1 Typical MMIC Photolithographic Steps

The typical process stages for a MESFET process, shown in Figure 7.9, are as follows:

- Front-face processing;
    - Creation of the active device (MESFET);
      Device isolation (mesa);
      Ohmic contacts;
      Gate contacts;
      First interconnect metal;

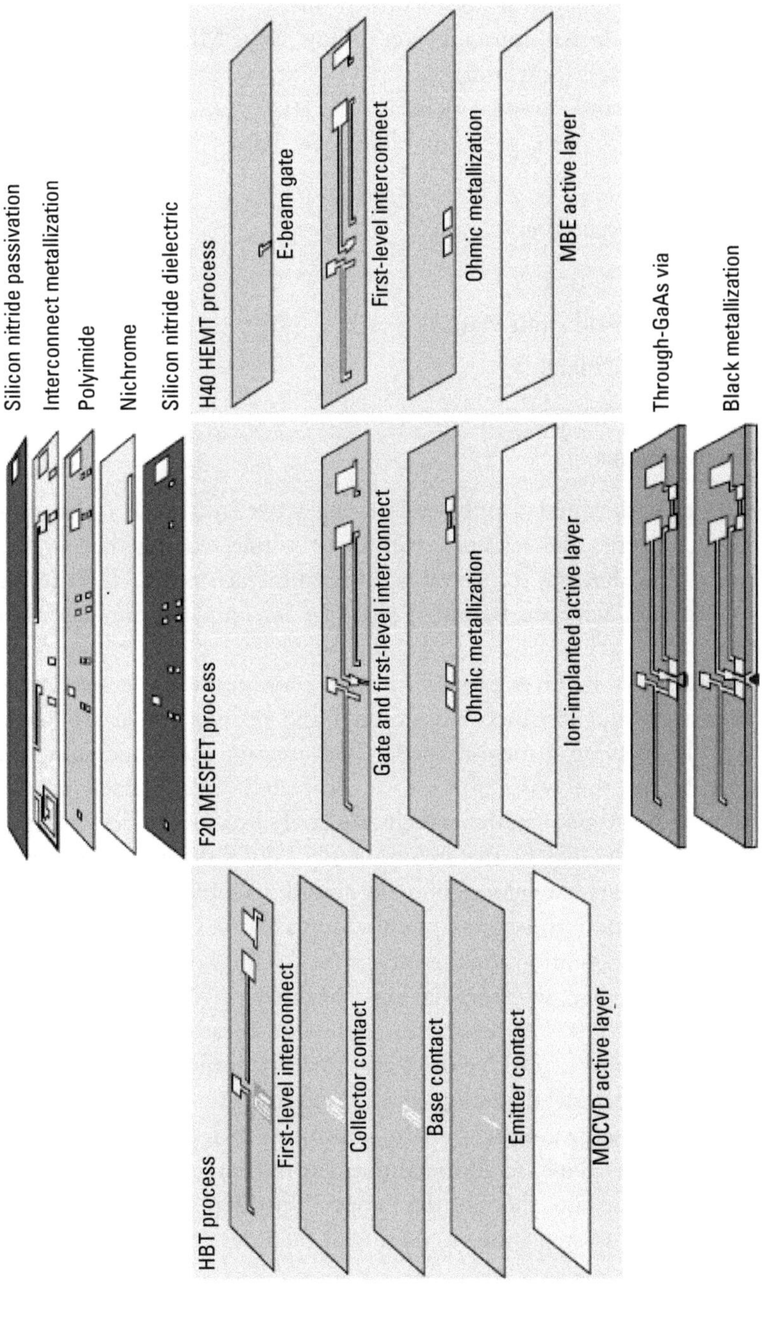

Figure 7.9 Typical MMIC process layers. (*Source:* Bookham Inc., 2006. All Rights Reserved.)

- Creation of passive components;
  - High dielectric-constant layer for high-value MIM capacitors (silicon nitride);
  - Resistive metal for resistors (nichrome);
  - Low dielectric-constant layer for low-value MIM capacitors and cross-overs (polyimide);
  - Second interconnect metal;
  - Dielectric encapsulation and saw lane definition (silicon nitride);
- Back-face processing;
  - Wafer thinning;
  - Through-substrate vias;
  - Back-face metal.

7.4.2 Device Isolation

Devices created with contacts to the conducting active layer, such as the transistors and sometimes the resistors, will all be connected together by the active layer unless they are somehow isolated from each other. This device isolation is usually achieved by either mesa or ion-implantation isolation techniques.

Mesa isolation involves protecting with photoresist the active layers where transistors (and resistors) are required and etching the active layers everywhere else, leaving a mesa-shaped raised area on the semiconductor. The remainder of the wafer surface is completely etched back to the semi-insulating substrate material, which effectively isolates the devices from each other. For GaAs wafers, the etching is typically done with a wet etch that etches at different rates along different crystallographic planes, resulting in sloping sides of the 3-μm-high (typically) mesa. For a GaAs wafer grown with the d crystal orientation, this results in two of the sides' having a positive slope and the other two sides having a negative or re-entrant slope, as indicated by Figure 7.10. The positive slope is useful because metal deposited as a track running up the positive slope will form a continuous path, as, for example, interconnect metal making contact with the gate of a MESFET on the top of the mesa. However, it is almost impossible to deposit a continuous track up the negative slope, so all metal tracks connecting to the gate of the MESFETs must run up the positive slope. Therefore, when designing MMIC on a mesa-isolated process, there will be a design rule that all the

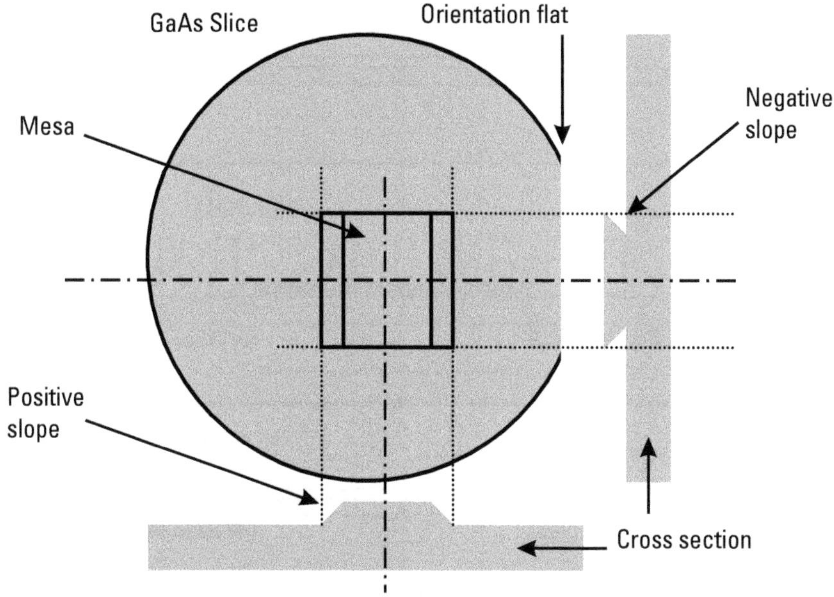

Figure 7.10 Mesa slopes produced when wet etching <100> crystal orientation GaAs wafers.

MESFET gates must be parallel to each other and the mesa etch mask must be aligned correctly to the crystal orientation markers (flats) on the wafer.

Implant isolation uses ions implanted at high energies into the active layers to break up the crystal structure and leave the semiconductor in an amorphous form [13, 14]. Areas that are required to remain active are protected with a mask of gold, which prevents the ions from passing through, as shown in Figure 7.11. The advantage of implant isolation is that the surface is left planar, so there are no mesa edge slopes to constrain the active device orientation. Disadvantages of implant isolation include the fact that the isolation depth is a function of the ion implant energy, and if this is limited, power devices and some HBT devices that use deeper active layers may not be isolated completely.

7.4.3 Ohmic Contacts

Ohmic contacts are used to provide a low-resistance connection between the active semiconductor layers and the interconnecting metal tracks [15]. Ideally, this would be a low-value linear resistance to current flowing in either

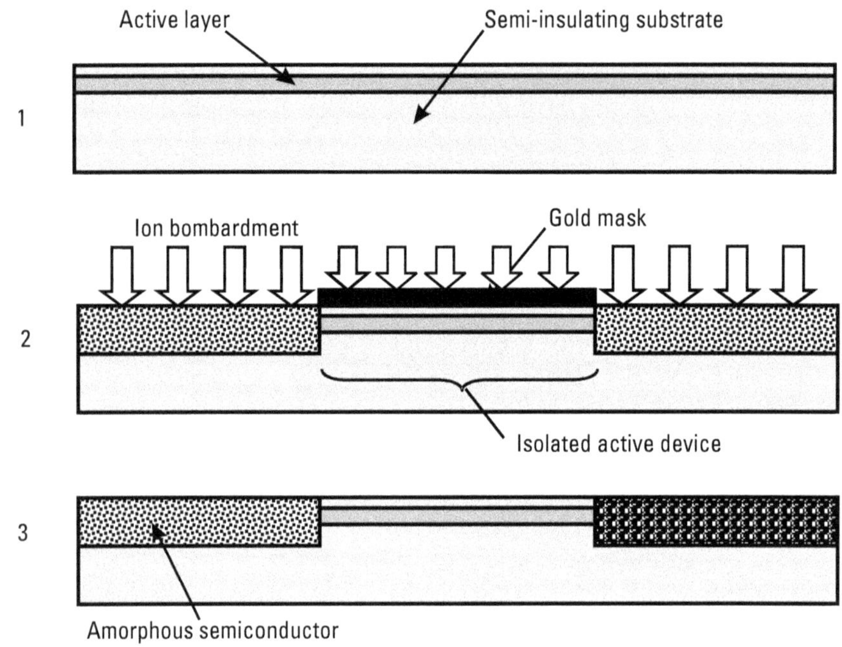

Figure 7.11 Implant isolation process.

direction, but this is not strictly the case. Practically all semiconductor-to-metal junctions form a Schottky barrier with a depletion layer and act as a rectifying junction. However, if the semiconductor layer next to the metal is highly doped (10^{18}–10^{19}/cm^3 for *n*-doped GaAs), conduction across the junction is dominated by quantum tunneling, and a low-resistance contact is formed.

In an MMIC, the ohmic contacts form the source and drain contacts for FETs and the contacts for mesa resistors. Photolithography is used to define an overhang profile resist pattern aligned to the previously created structures (typically the mesas), and metal is deposited in a thin layer by evaporation. Evaporation of metal is where the wafer is placed opposite a crucible or boat containing the metal in a vacuum chamber, and the metal is heated. Metal atoms evaporate from the boat and travel in straight lines across the vacuum chamber to coat the wafer surface. The metal is generally a multilayer deposition of different materials that can help to dope the semiconductor further, act as a metallurgical barrier, and provide a good contact to the interconnecting metal tracks. A gold-based system is typically used for GaAs ohmic contacts. The excess metal is removed using a lift-off process, and the resist is dissolved using solvents. The wafers are then heat-treated in a furnace to alloy the metals and the highly doped semiconductor together. A GaAs FET at this stage is shown in Figure 7.12.

The ohmic metal is a thin layer designed to conduct current vertically into the semiconductor and is not normally used for lateral current conduction, such as interconnecting different components on the wafer surface.

If mesa resistors are required, then they are fabricated at this stage in the process. Their resistance will be a function of their size and the resistivity of the active layer in the semiconductor (typically 300Ω/square for a GaAs wafer). The active layer resistivity may not be grown as accurately as the MMIC requires, so these resistors are often etched slightly until the resistance of a test resistor reaches its correct value, as shown by the example mesa resistors in Figure 7.13. Mesa resistors are ideal for making high-value bias-decoupling resistors, but their resistivity is a strong function of temperature, so should not be used where their value is critical, such as resistors for self-biased transistors.

7.4.4 Gate Contacts

The gate contact for an FET must not be a conducting contact with the semiconductor as any leakage current between the gate and the conducting

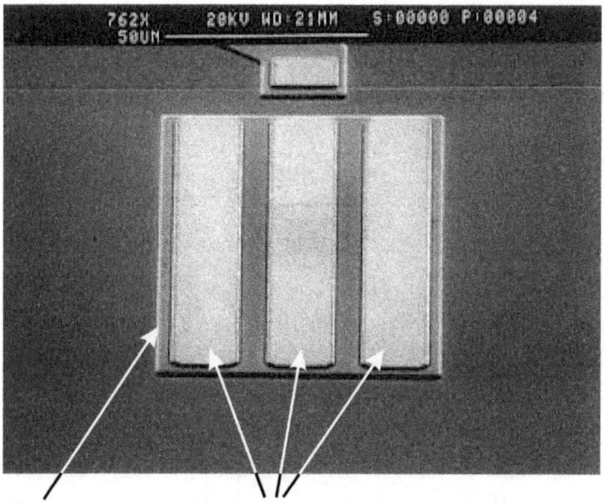

Figure 7.12 Scanning electron microscope (SEM) photo of an FET after formation of the ohmic contacts. (*Source:* Bookham Inc., 2006. All Rights Reserved.)

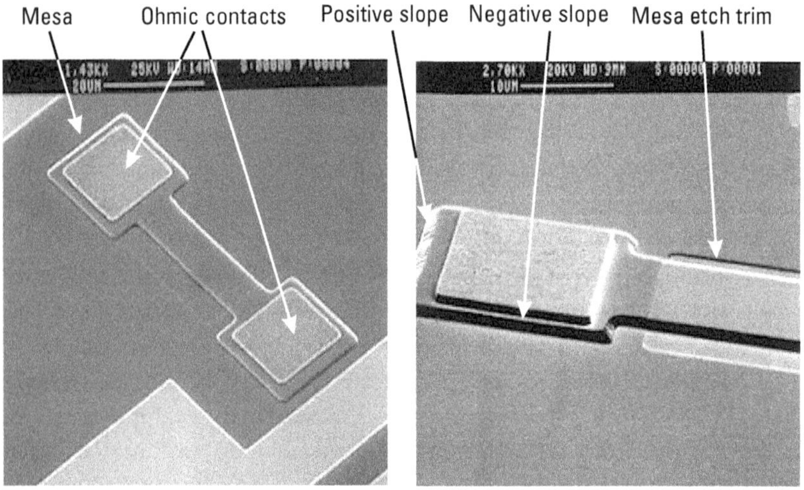

Figure 7.13 SEM photo of mesa resistors after the etch-trim stage. (*Source:* Bookham Inc., 2006. All Rights Reserved.)

semiconductor channel is detrimental to the ideal behavior of the FET, so these contacts are formed to a very-low-doped semiconductor layer to create a good rectifying Schottky contact [16]. The low-doped semiconductor layer is typically the top, or "cap," layer in a GaAs MESFET, and the metal layers of the gate contact are commonly titanium to form the Schottky junction, platinum as a metallurgical barrier to stop the gold from migrating into the GaAs, and finally gold for low resistance.

For a microwave MESFET a single photoresist layer, shown in Figure 7.14(a), and optical lithography are sufficient to define the 0.5-μm-length gate contacts with a lift-off process; however, the length of the FET gate is critical to its frequency response and typically needs to be less than 0.5 μm for good microwave performance and typically 0.25 μm to 0.1 μm for millimeter-wave HEMTs. For this reason, the profile or cross section through the gate contact of a HEMT looks like a "T" or a mushroom, with a short dimension in contact with the semiconductor and a much larger area above for low-loss conduction along the gate finger [Figure 7.15(b)]. This profile may be created with multiple layers of photoresist and exposed using electron beam (e-beam) lithography [Figure 7.15(a)] to achieve accurate lithography at sub-half-micron dimensions [17].

In both cases, after the resist pattern has been developed, a small amount of the GaAs semiconductor is etched through the resist pattern to

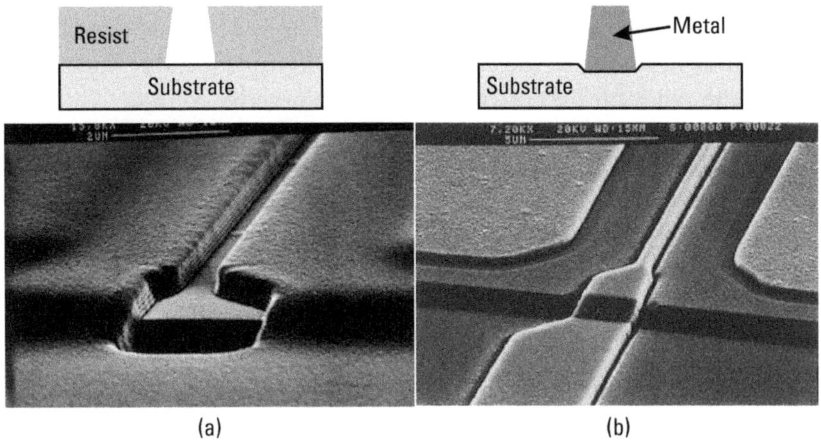

Figure 7.14 (a) MESFET gate photoresist pattern, and (b) gate metal in gate recess. (*Source:* Bookham Inc., 2006. All Rights Reserved.)

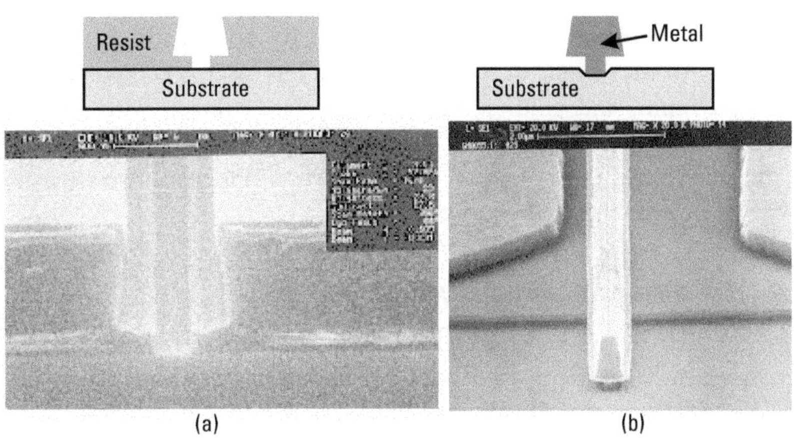

Figure 7.15 (a) HEMT mushroom gate resist profile, and (b) gate metal. (*Source:* Bookham Inc., 2006. All Rights Reserved.)

form a shallow recess in which the gate contact will be deposited. The recess etch provides a clean surface upon which the gate contact is made and is used to refine the eventual saturation current of the transistors. This is achieved by measuring the channel current through a test transistor (simply from the source to the drain contact as there is no gate metal yet) on the wafer during the recess etch and stopping when a predefined current is reached. This ensures a consistent saturation current from wafer to wafer and that the gate metal will be deposited at the correct distance from the conducting channel to be able to pinch off the current at the correct negative voltage.

After the etch, the metals are evaporated onto the wafer, the photoresist is dissolved, and the unwanted metal is lifted off the wafer leaving the gate metal sitting within the gate recess, as shown in Figure 7.14(b). At this stage, it is common to measure the dc response of test transistors on the wafer to check for the expected FET operation (saturation current I_{dss} and pinch-off voltage V_p).

7.4.5 First Interconnect Metal

The first interconnect metal layer is now put onto the wafer (in a MESFET process this is often put down at the same time as the gate contact). This layer is primarily used to join physically separated components electrically and is designed to conduct current laterally across the surface of the chip. It is also usual to use this layer as the lower electrode of metal-insulator-metal (MIM) capacitors and as the bases of dc and RF bond-pads, as well as for connecting to the centers of spiral inductors. The layer is composed of titanium, platinum, and gold, similarly to the gate contact, with an overall thickness of typically 0.5 μm, and it is evaporated onto the wafer and defined by lift-off lithography. Figure 7.16 shows a Pi-style, two-gate-finger MESFET at this stage in the process.

7.4.6 High-Dielectric-Constant Layer

With the first interconnect metal forming the bottom layer of the MIM capacitors, the next step is to deposit a dielectric layer to form the insulator between the two metal capacitor plates. This is normally a high-dielectric-constant material, such as silicon nitride ($\varepsilon_r \sim 7$), which allows the designer to create capacitors with values up to the order of 20 pF.

The dielectric layer covers not only the lower capacitor electrodes but also the other defined features, such as the transistors. The presence of the dielectric on the exposed active layers of the transistor helps to stabilize (passivate) any surface states that may be present and improves the reliability of the device [18]. That is why this dielectric layer is also known as the passivation layer.

The high-dielectric material is usually deposited by plasma electrochemical vapor deposition (PECVD) [19], as shown in Figure 7.17. The wafers are placed in a vacuum chamber, into which reactant gases are fed.

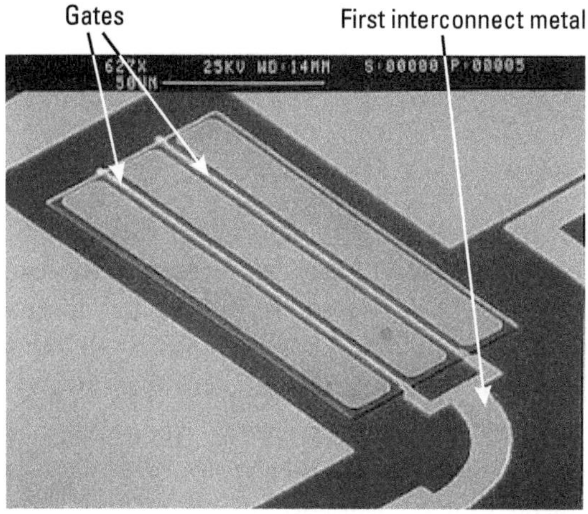

Figure 7.16 A two-gate-finger Pi-style MESFET after gate contact and first interconnect metallization. (*Source:* Bookham Inc., 2006. All Rights Reserved.)

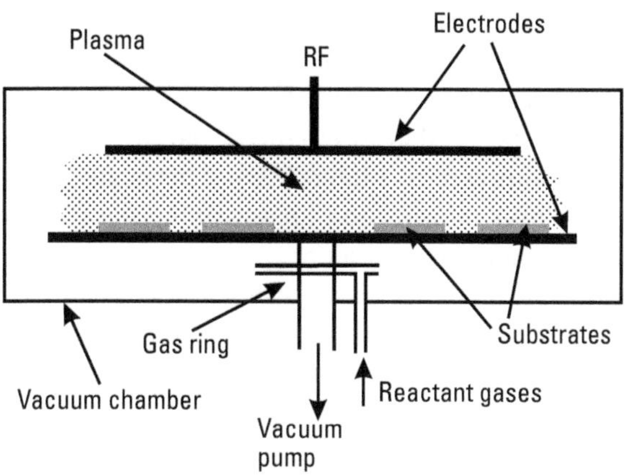

Figure 7.17 Schematic diagram of a PECVD system for coating wafers with silicon nitride.

Radio frequency (RF) signals are applied to the electrodes around where the wafers are sitting to create a plasma of reactive gases, and a solid dielectric layer grows on the surface of the wafers. The layer thickness needs to be high enough to create a pinhole-free layer with a breakdown voltage greater than about 50V and thin enough to produce reasonably valued capacitors. For silicon nitride, this is achieved with layer thickness around 1,000Å. Holes in the dielectric layer for contacting to the lower metal layers are opened using reactive ion etching (RIE) [20].

7.4.7 Resistive Metal

If metal thin-film resistors are required, a thin film of the resistive metal alloy, typically nickel chromium (nichrome) or tantalum nitride [21], can be deposited at this stage. Nichrome resistors have the advantage over mesa resistors because they have virtually zero resistance variation with temperature [22], and the film thickness can be controlled to give a nominal 50 Ω/square resistivity for making accurate low-value resistors [23]. The metal alloy is deposited by sputtering and then annealed to stabilize the film. Patterning of the metal film is invariably by wet etching through developed photoresist.

7.4.8 Low-Dielectric-Constant Layer

A relatively thick low-dielectric-constant layer is the next step in the wafer front-face processing. Its thickness and low dielectric constant mean that further metal interconnect layers will only be loosely coupled electrically to the first interconnect metal (and resistive metal) layers, so different electrical paths can cross without excessive interaction. This is the requirement where dc bias tracks cross over transmission lines carrying the RF signal, as well as for making a connection to the center of spiral inductors.

An example of this low-dielectric layer is a polyimide film.[1] This film is spin-coated onto the wafer and baked to imidise the polymer film. Holes in the polyimide film are patterned by photolithography and etched by RIE. Low-value MIM capacitors can also be formed using the low-dielectric-constant layer as the capacitor dielectric.

An alternative low-dielectric layer is benzocyclobutene (BCB). The advantages of photo-BCB compared to polyimide are its low dielectric constant and dielectric losses [24], the minimal moisture uptake during and after processing, and a very good planarization.

This layer is also usually removed over the transistors to minimize any losses and feedback that may be increased by the presence of a dielectric layer. This is clearly visible as the darker, sunken area of a completed single-gate-finger MESFET shown in Figure 7.18.

1. http://www.hdmicrosystems.com/tech/techinfo.html.

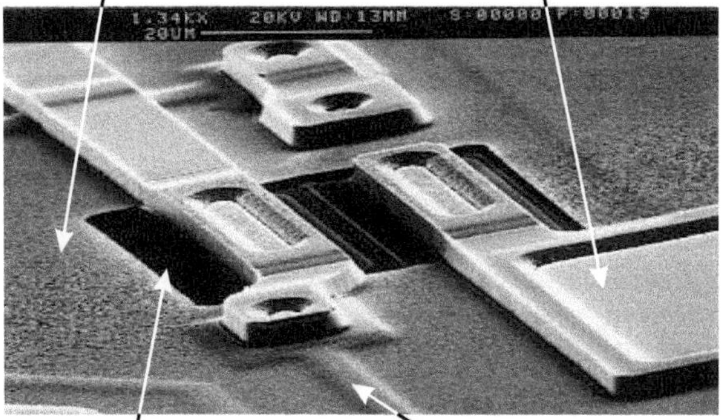

Figure 7.18 MESFET showing the hole in the dielectric layers where the polyimide has been removed. (*Source:* Bookham Inc., 2006. All Rights Reserved.)

7.4.9 Second Interconnect Metal

The second (and usually the top) interconnect metal is normally a thicker metal layer and is used like the first interconnect layer to connect physically separated circuit components laterally across the surface of the chip. It also forms the top plate of the MIM capacitors and the top metal layer of the dc and RF bond-pads. A thick layer of gold is required on the bond-pads to facilitate low-resistance contact with the dc and RF probe needles and to enable wire-bond attachment.

The composition of this metal layer is commonly titanium, platinum, and gold, similarly to the other metal layers, but it has a greater overall thickness (3 μm to 5 μm) and so is deposited by sputtering [25], as shown in Figure 7.19. The wafers are placed in a vacuum chamber opposite a target electrode made of the metal to be deposited, and RF plasma is generated between the electrodes using an inert gas. The RF plasma bombards the pure metal target, dislodging metal atoms, which then coat the wafer surface in a thin layer. The thick metal layer is patterned by photolithography, and the unwanted metal is removed by etching or ion-beam milling [26].

The appearance of the front surface of the wafer at this stage is shown in Figures 7.20 and 7.21. The second interconnect metal forms the most obvious features as it is the top level and a reasonably thick layer. It has depressions in its surface where holes were opened in the two dielectric layers, and it is in

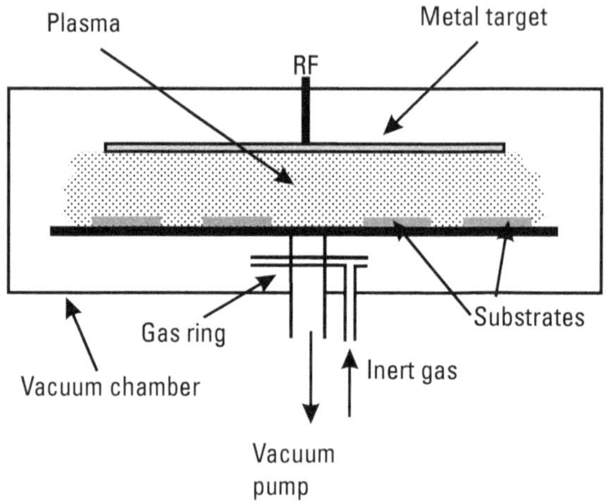

Figure 7.19 Metal sputtering.

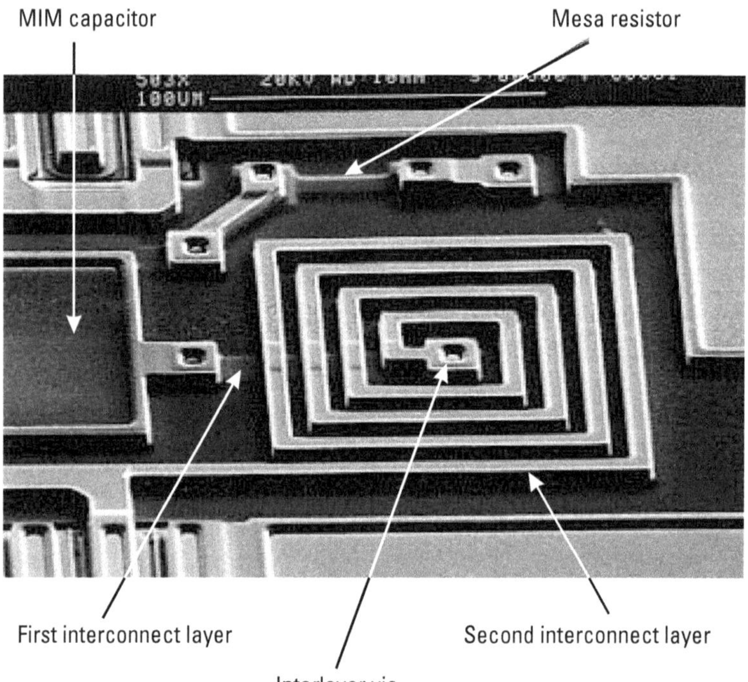

Figure 7.20 Wafer surface after deposition of the second interconnect metal layer. (*Source:* Bookham Inc., 2006. All Rights Reserved.)

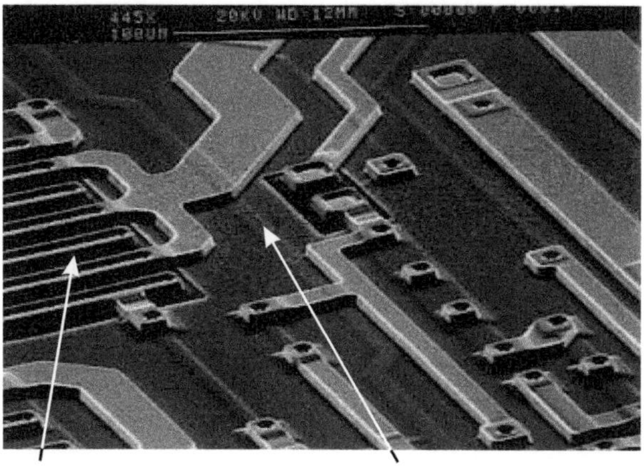

Figure 7.21 MESFET after second interconnect layer showing removal of the low-dielectric layer. (*Source:* Bookham Inc., 2006. All Rights Reserved.)

contact with the first interconnect metal layer, otherwise known as an interlayer via. On the left side of Figure 7.20 is an MIM capacitor where only the low-dielectric layer has been removed to leave the high-dielectric layer between the two metal plates of the capacitor. Small raised areas are also apparent on the second interconnect metal surface at the side of the spiral inductor where a first interconnect metal layer track is passing beneath it.

The top-level interconnect metal is often used as a thermal shunt to help manage the heat generated by the active devices. A thermal shunt is a high-thermal-conductivity material placed on the top surface of the chip to help spread the heat away quickly from the device active area. They are not always constructed from a thick metal interconnect layer but can be formed from diamond films. In HBT amplifiers, they are often used to equalize the temperatures in multifinger devices and help prevent thermal runaway in individual emitter fingers [27, 28].

7.4.10 Dielectric Encapsulation and Saw/Scribe Lane Definition

The last stage of front-face processing is to encapsulate the whole of the chip in another dielectric layer. This improves the reliability of the MMIC by forming a barrier against contamination by water vapor or dust particles and insulates the top interconnect metal layer from accidental contact. In a MESFET process, this encapsulation is typically silicon nitride laid down by PECVD. If dielectric layers are left at the chip edges, they tend to crack when the chips are scribed or sawn apart, and these cracks are seen as a reliability hazard. For this reason, saw/scribe lanes are defined at the chip edges by removing all the dielectric layers down to the semiconductor surface. Holes in the encapsulation layer for the saw lanes and for contacting to the bond-pads are patterned by photolithography, and RIE is used to remove the dielectric layers.

A typical multifinger MESFET at this stage in the process is shown in Figure 7.22.

7.5 Wafer Thinning

When the processing of the front face of the wafer is complete, the wafer is thinned down to a thickness of typically 100 μm or 200 μm [29, 30]. This brings the back face, and ultimately the metal ground plane, to the right distance from the tracks on the front face to create the correct impedances as microstrip transmission lines. This is a primary difference between MMICs and other electronic circuits because the back-face ground plane is not just the current return path, but it is an important part of the transmission lines between individual components. The alternative to microstrip transmission

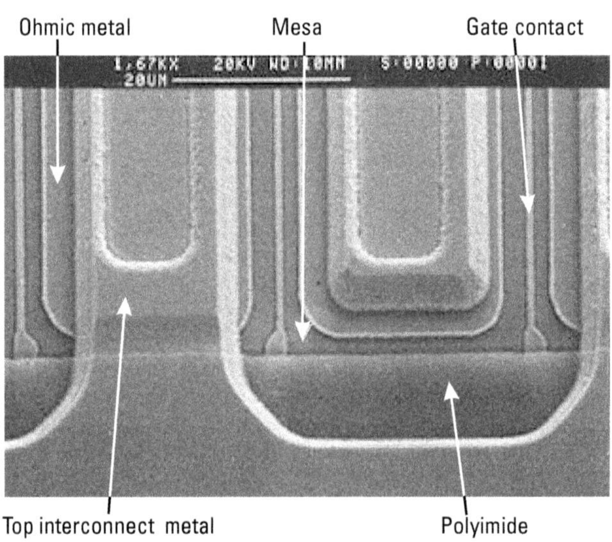

Figure 7.22 Close-up view of a multigate-finger MESFET. (*Source:* Bookham Inc., 2006. All Rights Reserved.)

lines is coplanar waveguide (CPW) transmission lines, and when these are used, the wafer is not necessarily thinned. To thin the wafer, the front face is first protected from mechanical damage with a layer of polyimide or photoresist, and then mounted front-face down onto a glass block using wax to hold it in place. If the finished thickness needs to be 200 μm, the wafer is initially coarse lapped down to about 240 μm, then finely polished to the required 200 μm. A typical wafer lapping and polishing machine capable of handling cassettes of 6-in.-diameter wafers is shown in Figure 7.23 [31]. The back-face surface finish is important because microwave losses in the microstrip transmission lines increase with the surface roughness of the ground plane metal [32].

The factors concerning the final wafer thickness are the frequency of operation and the amount of heat that needs to be sunk out of the backside of the wafer. As the frequency approaches millimeter wavelengths, microstrip propagation can become multimode if the substrate is too thick. Microstrip MMIC processes targeted at RF and microwave frequencies are typically 200-μm thick, processes targeted at millimeter-wave frequencies are typically 100-μm thick, and processes for power devices can be as thin as 50 μm. Power processes require thin substrates to minimize the thermal resistance from the heat sources on the surface, such as the FET channel, down to the system chassis, which forms the heat-sink on the back face of the chip. This is why power devices require very thin substrates. Unfortunately, 50-μm-thick

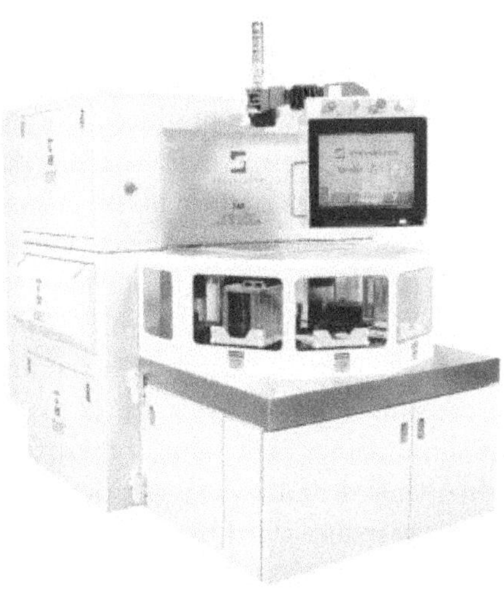

Figure 7.23 Equipment for lapping and polishing 6-in.-diameter GaAs wafers.

wafers are very difficult to handle when they are demounted from the glass block and can lead to a yield hazard particularly during RFOW testing. Some foundries have found a compromise by keeping the wafer reasonably thick and introducing an integral heat-sink into the back of the wafer under every power FET [33]. This bathtub heat-sink (so-called because of its shape) can be produced as low as 50 μm from the FET channel on the surface, giving a high-thermal-conductivity path to the back face and the chassis heat-sink.

7.6 Through-Substrate Vias

Shunt components are useful in MMIC design. As their name implies, they are connected between the microstrip signal path and ground, so they require a low-inductance connection to the back-face ground plane. This is achieved by a metal-plated hole in the substrate, all the way from the front face to the back face. The substrate vias are produced while the wafer is still mounted on the glass block as this gives mechanical stability to the now-very-thin wafer. The back-face of the wafer is coated with photoresist, and the via mask containing the patterns for the vias is aligned to the front-face features through the substrate using an infrared aligner. When the resist is developed, a nickel layer is sputtered onto it, and the excess metal is removed by lift-off. The nickel pattern now on the back face is a protective mask for the majority of the wafer with holes in it where substrate vias are required. RIE is used to etch the GaAs through the holes in the nickel mask until there are holes formed all the way to the front face. The nickel mask is then removed using an acid wet etch. The profile of the hole through the substrate is controlled by the RIE conditions and is commonly produced as a slightly conical shape (as shown in Figure 7.24) to help ensure that the metal deposited will form a continuous layer from the top to the bottom of the hole.

The usual MMIC approach is to have substrate vias as separate components that can be combined with components such as transistors, capacitors, and resistors to form the shunt component to ground. This is usually acceptable at RF and microwave frequencies because the additional inductance to ground generated by the substrate via is typically 15 pH, which does not significantly perturb the shunt component value at these frequencies. However, at millimeter-wave frequencies and for large power transistors, the extra inductance can significantly reduce the device gain and power by introducing negative feedback. Some foundries have found a way around this extra inductance by using individual grounded source vias (IGSVs), which are substrate vias underneath each of the source contacts in the FET, giving them a

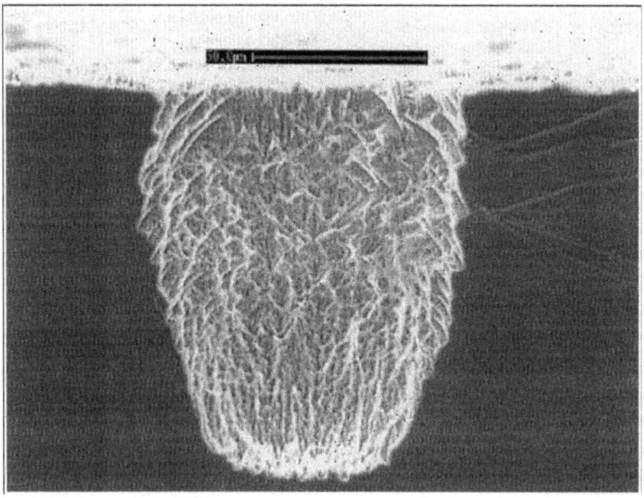

Figure 7.24 Cross section of a GaAs through-substrate via. (*Source:* Bookham Inc., 2006. All Rights Reserved.)

dramatically reduced inductance path to ground. This IGSV process normally requires that the whole wafer be thinned to 50 μm, requiring the thickness accuracy to be ±5 μm, which increases the cost of the processing [34].

7.7 Back-Face Metal

The back-face metal forms the ground plane for the microwave circuit and the material by which the MMIC is normally mounted (except for flip-chip mounting). This needs to be a reasonably thick metal layer with the final layer being gold, particularly if the chip is to be mounted by soldering. The surface of the semiconductor is commonly microroughened using RIE to promote good adhesion of the metal to the back-face of the semiconductor, a process which improves the adhesion but does not make the surface rough enough to increase the microstrip loss. For a GaAs wafer, the next step is to sputter consecutively titanium, platinum, and gold onto back face to form the ground plane and also to make the metal connection to the front face within the substrate via. To finish, the wafer is demounted from the glass block and cleaned with solvents, and the front-face protection layer is removed.

At this stage, the wafer is complete and ready for RFOW testing.

7.8 Chip Separation

The individual MMIC chips are separated from each other by either scribing or sawing. Prior to each of these operations, the wafer is mounted onto a tacky film [35],[2] which is later expanded to space the separated chips physically apart. With scribing, a diamond-tipped scribe is run along the surface of the semiconductor within the saw/scribe lanes, and then the wafer is passed over a rubber former with a small amount of pressure applied to the wafer. Because the wafer is a single crystal, the surface scribe is enough to weaken the crystal, and the pressure causes the crystal to cleave neatly beneath the scribe line [36]. With sawing, a diamond-coated circular saw is passed down all the saw/scribe lanes, and the chips are separated by abrasive sawing [37]. Scribing is normally limited to circuits without a continuous metal ground plane on the back surface; otherwise, as the metal does not cleave like a crystal, it holds the devices together. Circuits that do require a metal ground plane are either sawn, or the back-face ground plane is patterned to produce scribe lanes on the back face as well as the front face. The tacky film is then expanded to allow the separate chips to be picked off and placed in a waffle or gel pack.

7.9 Quality Assurance

The quality of an MMIC process is typically measured in terms of whether the dc and RF electrical characteristics of the common components fall within predetermined limits and based on the visual acceptability of the chips. Some of this is achieved using process control monitors (PCMs), which are sets of the most commonly used components and other test structures, and these are placed on every wafer manufactured by the foundry.

An example PCM is shown in Figure 7.25; it contains six identical dc PCM components within a rectangular set of dc pads and a number of RF-testable PCM components within RFOW input and output pads. Each dc PCM contains at least one of each available foundry component, such as a transistor, capacitor, inductor, resistor, substrate via, interlayer via, interconnect metal layers, and so forth. Typical measured dc parameters are the saturation current of the transistors, the resistance of the interconnects, and the low-frequency (megahertz) capacitance of the MIM capacitors. The RF PCM components are normally several of a standard-size transistor, and one

2. http://www.semicorp.com/Products/tape.htm.

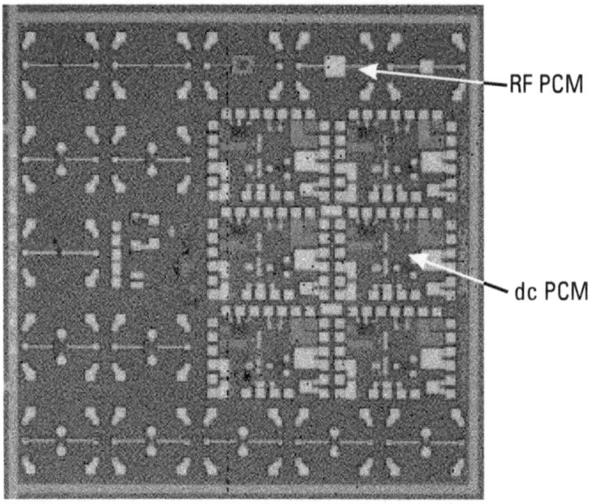

Figure 7.25 PCM. (*Source:* Bookham Inc., 2006. All Rights Reserved.)

of each type of inductor, capacitor, and resistor. Typical measured characteristics of the RF PCM components are capacitance, inductance, and resistance values and the transconductance of the transistors as extracted from an equivalent circuit model fitted to the measured data. The test software analyzes the measurements obtained from each component, calculating the mean and standard deviation for all valid measurements and the percentage yield of measurements within the minimum and maximum specification limits. If the percentages of measurements within the specification limits meet the particular foundry's criteria, then the wafer is classified as an electrically "good" wafer.

Incidentally, the PCM measurements also allow the foundry to gather a large database of the means and standard deviations of the values of all the components. This allows the foundry to use statistical process control (SPC) tools to monitor and correct drifts in the process and keep it centered within the specification limits. The same statistical data is also used to create statistical models for the transistors and other components that are required by the yield-improvement techniques described in Chapter 5.

The final stage of quality assurance is often the visual inspection. This is where the chips are inspected under high-power microscopes for visual defects, such as FETs with missing gates, cracks in the dielectrics or the substrate, debris and extraneous metal, and other blemishes. These defects do not always result in electrical performance deviations but could affect the long-term reliability of the MMIC.

7.10 Questions

7.1 What is meant by the word *monolithic* in the MMIC acronym?

7.2 What is important about the structure of the substrate material?

7.3 How are the wafers produced?

7.4 Why is it useful to use a low-conductivity, semi-insulating substrate material?

7.5 How is the conductivity of a semiconductor increased for the active layers in the transistors?

7.6 What are the different ways to create an active layer?

7.7 What are the two common photoresist processes?

7.8 What are two typical ways to isolate active devices from each other?

7.9 How does mesa isolation impact the orientation of the transistors?

7.10 Can the ohmic contact metallization be used for lateral current conduction?

7.11 Which of the source, gate, and drain contacts of a field effect transistor is a nonconducting Schottky contact?

7.12 What is the high-dielectric layer normally used for?

7.13 What advantages do thin-film metal-alloy resistors (nichrome) have over mesa resistors?

7.14 How is the electromagnetic coupling between crossing tracks on separate interconnecting layers minimized?

7.15 How does the final wafer/chip thickness influence the performance of the MMIC?

7.16 What parasitic impedance do substrate vias add to microstrip shunt components?

7.17 What is the primary function of the back-face metallization?

7.18 How is the quality of a processed wafer assessed?

References

[1] Rudolph, P., et al., "Vapour Pressure Controlled Czochralski (VCz) Growth—A Method to Produce Electronic Materials with Low Dislocation Density," *Crystal Research and Technology,* Vol. 32, 1997, pp. 35–50.

[2] Rudolph, P., and M. Jurich, "Bulk Growth of GaAs—An Overview," *J. Crystal Growth,* Vol. 198, 1999, pp. 325–335.

[3] Pätzold, O., et al., "New Developments in Vertical Gradient Freeze Growth," *Advanced Engineering Materials,* Vol. 6, No. 7, 2004, pp. 554–557.

[4] Yue, C. P., and S. S. Wong, "On-Chip Spiral Inductors with Patterned Ground Shields for Si-Based RF ICs," *IEEE J. Solid-State Circuits,* Vol. 33, No. 5, 1998, pp. 743–752.

[5] Lee, T. H., and S. S. Wong, "CMOS RF Integrated Circuits at 5 GHz and Beyond," *Proc. IEEE,* Vol. 88, No. 10, October 2000.

[6] Shockley, W. B., "Semiconductor Amplifier; Three-Electrode Circuit Element Utilizing Semiconductive Materials," Transistor and Patent Number(s) 2,502,488 and 2,524,035.

[7] Dearnaley, G., et al., *Ion Implantation,* New York: American Elsevier Publishing Co., 1973, pp. 802.

[8] Chason, E., et al., "Ion Beams in Silicon Processing and Characterization," *J. App. Phys.,* Vol. 81, No. 10, 1997, pp. 6513–6561.

[9] Cho, A., "Film Deposition by Molecular Beam Techniques," *J. Vac. Sci. Tech.,* Vol. 8, 1971, pp. S31–S38.

[10] Cho, A., and J. Arthur, "Molecular Beam Epitaxy," *Proc. Solid-State Chem.,* Vol. 10, 1975, pp. 157–192.

[11] Christiansen, K., et al., "Advances in MOCVD Technology for Research, Development and Mass Production of Compound Semiconductor Devices," *Opto-Electron. Rev.,* Vol. 10, No. 4, 2002, pp. 237–242.

[12] Liebmann, L. W., et al., "TCAD Development for Lithography Resolution Enhancement," *IBM Journal of Research and Development,* Vol. 45, No. 5, 2001.

[13] Sealy, B. J., "Implantation into III–V Compound Semiconductors," *Proc. Institute of Physics Conference on Ion Implantation,* Aston, UK, December 1996.

[14] Sealy, B. J., "Ion Implantation in GaAs: An Overview," *Properties of Gallium Arsenide,* ed. M. R. Brozel and G. E. Stillman, 3rd ed., EMIS Datareviews Series 16, London, United Kingdom: IEEE, 1996, pp. 765–769.

[15] Rideout, V. L., "A Review of the Theory and Technology for Ohmic Contacts to Group III–V Compound Semiconductors," *Solid State Electronics,* Vol. 18, 1975, pp. 541–550.

[16] Kahen, K. B., "Theory of Schottky-Contact Formation on GaAs(110)," *Phys Rev B Condens Matter,* May 1991, Vol. 43, No. 14, pp. 11,745–11,753.

[17] Via, D., et al., "'Safe' Solvent Resist Process for Sub-Quarter Micron T-Gates," *J. Vacuum Science & Technology B: Microelectronics and Nanometer Structures,* Vol. 15, No. 6, November 1997, pp. 2916–2920.

[18] Dalton, J. V., and J. Drobek, "Structure and Sodium Migration in Silicon Nitride Films," *J. Electrochemical Society,* Vol. 115, August 1968, pp. 865–868.

[19] Cote, D. R., et al., "Plasma-Assisted Chemical Vapor Deposition of Dielectric Thin Films for ULSI Semiconductor Circuits," *IBM J. Research and Development,* Vol. 43, 1999.

[20] Law, V. J., and G. A. C. Jones, "Chloromethane-Based Reactive Ion Etching of GaAs and InP," *Semicond. Sci. Tech.,* 1992, Vol. 7, pp. 281–283.

[21] Mellberg, A., et al., "Fabrication and Characterization of Reactively Sputtered TaN Thin-Film Resistors for Millimeter Wave Applications," *Electrochemical and Solid-State Letters,* Vol. 7, No. 11, 2004, pp. G261–G263.

[22] Zelenka, J., V. Chudoba, and J. Rehak, "Thin Resistive Film with Temperature Coefficient of Resistance Close to Zero," *Thin Solid Films,* Vol. 200, 1991, pp. 239–246.

[23] Maissel, L. I., and R. Glang, *Handbook of Thin Film Technology,* New York: McGraw-Hill, 1970.

[24] Janezic, M. D., et al., "Permittivity Characterisation of Low-*k* Thin-Films from Transmission Line Measurements," *IEEE Trans. Microwave Theory Tech.,* Vol. 51, No. 1, January 2003, pp. 132–136.

[25] Wilson, B., "Applying Metal/Sputtering," Number m1039, Document type -//CNX// DTD CNXML 0.5 plus MathML//EN, Creative Commons Attribution License, Updates July 3, 2003.

[26] Giannuzzi, L. A., and F. A. Stevie, *Introduction to Focused Ion Beam: Instrumentation, Theory, Applications and Practice,* New York: Springer Verlag, 2005.

[27] Bozada, C. A., et al, "Microwave Power Heterojunction Bipolar Transistors Fabricated with Thermal Shunt and Bathtub," *Proc. GaAs IC Symposium,* San Diego, CA, October 29–November 1, 1995, pp. 155–158.

[28] Dhondt, F., J. Barrette, and P. A. Rolland, "3D Finite-Difference Electrothermal Model for Multifinger HBTs with Thermal Shunt and Emitter Ballast Resistance," *Solid-State Electronics,* Vol. 42, No. 9, September 1999, pp. 1723–1729.

[29] Hazeldine, T., and J. Rubin, "Rapid Routes to Planar Polishing," *Materials World,* Vol. 6, No. 2, Bristol, U.K.: Institute of Materials, February 1998.

[30] Fynn, G. W., and I. A. Powell, *Cutting and Polishing Optical and Electronic Materials,* 2nd ed., Bristol, U.K.: Adam Hilger, 1988.

[31] Strasbaugh, San Luis Obispo, CA, USA, at http://www.strasbaugh.com.

[32] Hammerstad, E., and Ø. Jensen, "Accurate Models for Microstrip Computer-Aided Design," *Proc. IEEE MTT Symposium,* June 1980, pp. 407–409.

[33] Davies, I., et al., "The Development of a Foundry Compatible Power MMIC Production Technology," *Proc. 12th GaAs IC Symposium,* New Orleans, LA, October 7–10, 1990, pp. 71–72.

[34] Alavi, K., et al., "A Very High Performance, High Yield, and High Throughput Millimeter Wave Power PHEMT Process Technology," *Proc. GaAs MANTECH,* Las Vegas, NV, May 2001.

[35] Moore, D., "Sticky Issues with Semiconductor Processing Tape," *Semiconductor International Magazine,* May 1996.

[36] Gogotsi, Y., and V. Domnich, "High-Pressure Surface Science and Engineering," Bristol, U.K.: Institue of Physics Publishing, August 2004.

[37] Hazeldine, T., *Annular Sawing,* Angel Publishing, October 1997.

8

Test

This chapter is concerned only with the tests and measurements that are performed while the wafer is still whole. After these tests, the MMICs are diced or sawn into individual chips or packaged, then jig-tested, a process that is outside the scope of this design book.

The whole wafers are tested at dc because these tests are used for the process control monitoring (PCM) function to ensure that the process was performed correctly and that the components are within their allowed specification ranges. The RF-on-wafer (RFOW) tests also confirm that the RF performance of the components is within the allowed ranges and can be used to check the performance of the MMIC designs. RFOW measurement of the MMICs, before dicing and assembly, allows quicker testing of prototype circuits and enables out-of-spec wafers to be rejected before costly dicing operations are performed. It also allows only known-good dies to be selected for production assembly, which reduces costly rework at later stages.

8.1 Process Control and Monitoring (PCM)

As mentioned in the layout chapter, process control monitors [1] are placed either in each array or in the drop-in arrays, and they contain at least one of each type of component from resistors to transistors, including chains of metal-layer interconnects and through-substrate vias. The PCMs have a standard pad layout and are generally autoprobed using a probe card with numerous probe needles and a wafer probe station (Figure 8.1). The probe card is positioned over the first PCM and is lowered to contact the pads so that the measurements can be performed. The probe card is then stepped to the next PCM, and the measurements are repeated. As the probe needles are stepped across the wafer, there is usually some run-out due to imperfect alignment of the wafer to the probe station axes, but the size of the MMIC pads is designed so that the needles should still contact the pads.

8.2 The dc Test and Stability Problems

The dc test performed on MMICs usually just applies the dc bias voltages to the chip and verifies that the correct currents are drawn and that the chip is not oscillating. The voltages are applied using probe cards (described in Chapter 5), the current is measured, and an oscilloscope is used to check for ac signals on the bias lines, which are a sign that the chip is oscillating. Oscillation due to inadequate RF decoupling is the most common cause of problems during RFOW testing. The designer should ensure either that there is high series impedance, such as a spiral inductor or a large resistor between the

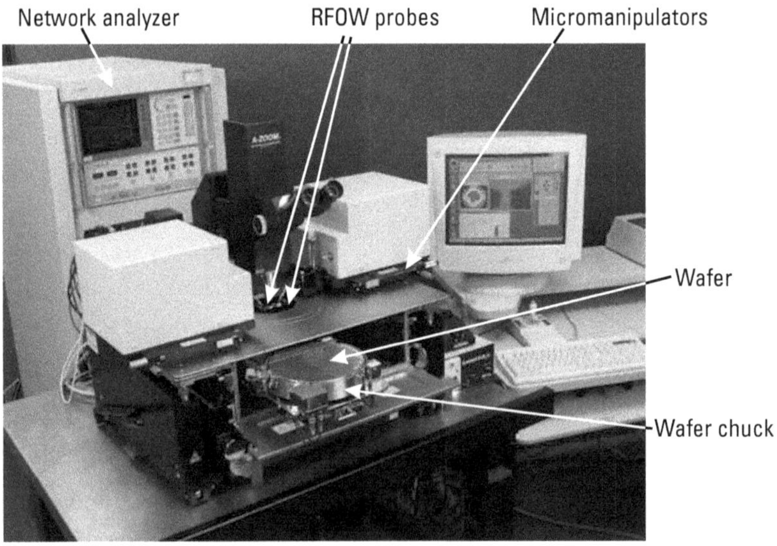

Figure 8.1 RFOW probe station. (Courtesy of Cascade Microtech.)

RF signals and the dc bias pads, or that there is low impedance to ground near the pads, such as a large capacitor that is well grounded, or preferably both. Careful on-chip decoupling is not always adequate to prevent oscillation via the bias lines because of the limited impedance values of the MMIC component elements. Where the transistors have high and broadband gain (such as HEMTs), additional decoupling is sometimes required off chip in the form of MIM capacitors mounted next to the chip. This is not possible during RFOW because the MMIC is surrounded by many other chips, so the additional decoupling needs to be added to the probe needles. Note that the additional capacitive decoupling needs to be well grounded to the MMIC ground plane, which is the back of the wafer, and the lowest impedance path to that ground is through a needle contacting a through-substrate via and not by an inductive-looking wire connected to the wafer chuck of the probe station (which could be ~100 mm away). This is why the designer should provide dc pads connected to through-substrate vias next to each bias supply dc pad. Extra series impedance decoupling can be achieved by putting ferrous beads onto the probe needles after the capacitors.

Take care that electrostatic discharge (ESD) precautions are used and that the maximum current ratings of the dc pins (typically 0.5A per pin) are not exceeded.

8.3 RF Testing and Calibration

During RFOW testing, the RF signal is applied to the chip using a transmission line of known impedance (usually 50Ω) so that the *s*-parameters of the design can be recorded. This means that both the signal and ground return connections must be carefully controlled to minimize the transmission-line discontinuities and allow accurate calibration. The convention is to use three connections, one signal connection between two ground connections, known as ground-signal-ground (G-S-G) RFOW pads, as in Figure 8.2. Two connections, known as ground-signal (G-S), can be used, but the fields around the signal track are symmetric, and doing away with one of the ground connections adds around 100-pH inductance in the ground return path [2], so G-S connections are not used much above 5 GHz.

These G-S-G RFOW pads are contacted using RFOW probes like those in Figure 8.3, which are then connected by coax cables to the measurement equipment, such as a network analyzer.

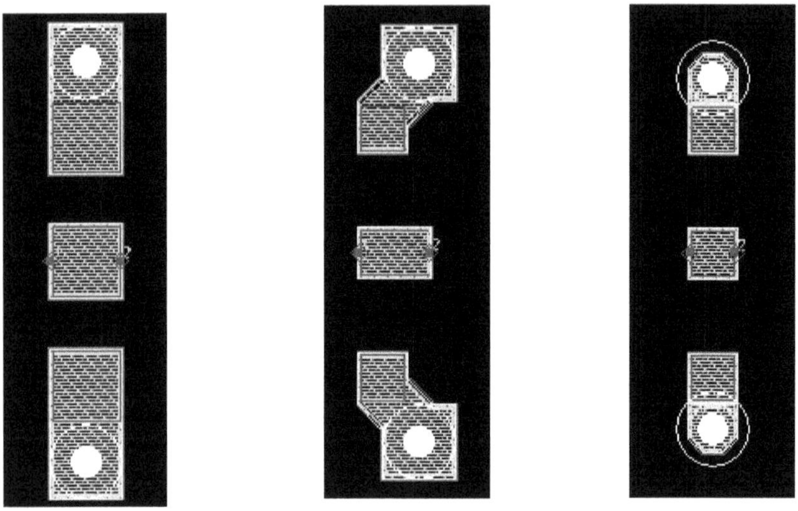

Figure 8.2 Three examples of G-S-G RFOW pads. (*Source:* Bookham Inc., 2006. All Rights Reserved.)

Ideally, the G-S-G contact pads need to be made from the same metal layers so that they have the same height and physical resilience to the probe contacts. This is because the probes are brought down into contact with the pads at an acute angle, and they skate or slide over the surface of the pad by a small distance according to the contact pressure the probe has with the wafer. Lack of planarity of the three probe contacts with the pad surface can lead to poor connections, bad calibration, and erroneous measurement data [3]. Through-substrate via pads should not be used as the ground-connection pads because the metal layers will be pushed into the via hole by the probe ground contacts. This has the effect of making poor ground connections and disturbs the planarity of the probe connections because the signal contacts are pushed up by the signal pad resting on the substrate, while the ground contact is pushed into the via holes.

Calibration enables the network analyzer to place the reference plane of the *s*-parameter measurements at any position along the RF signal path. The position of this reference plane is typically at the RFOW probe tips when measuring circuits and just after the RFOW pads when characterizing individual components. The calibration method requires specific measurements of either off-wafer standard calibration structures [e.g., impedance standard substrate (ISS) [4] structures for calibrating to the probe tips] or standard structures placed in the drop-in array (when calibrating after the RFOW

Figure 8.3 RFOW probes. (Courtesy of Cascade Microtech.)

pads). These standard calibration structures consist of transmission lines and open and short circuits.

There are three popular calibration methods known as short-open-load-thru (SOLT), line-reflect-match (LRM), and thru-reflect-line (TRL). SOLT is commonly used for coaxial calibration of network analyzers using manufacturer-supplied coaxial standards [5, 6]. SOLT calibration tends to be sensitive to the probe placement and is very dependant on the standard definitions for an accurate calibration [7]. This type of calibration is available on nearly all network analyzers but is used less for MMIC calibration because the standards are difficult to achieve accurately on wafer. LRM is useful for automated calibration because the probe separation distance can be kept constant during the calibration process. This technique is broadband but requires that the line delay and match resistance be known. The match resistance value is especially important as the measurements are referenced to this value of resistance. A variation on this calibration is the line-reflect-reflect-match (LRRM) method, which includes an open circuit reflect measurement, as well as a short circuit reflect measurement [8]. TRL uses multiple transmission lines as measurement standards so the measurements are referenced to the line impedance, but it has a relatively limited frequency range [9]. In general, TRL can achieve the best absolute accuracy, but LRRM is also nearly as accurate for MMIC measurements. Further details on RFOW calibration and verification methods can be found in [10–15].

Once an accurate reference plane has been achieved through the calibration method, the data can be de-embedded to a different position along the reference transmission line either using the built-in de-embedding feature within the network analyzer or afterwards by inserting negative lengths of the reference transmission line in series with the measured data within an RF simulator [16].

When the dc probe cards are supplying the dc bias to the MMIC, and the RFOW probes enable coax connection to the RF ports of the MMIC, any measurements performed in a typical microwave laboratory on a coax-connected component can be carried out on the chip while it is part of a whole wafer. Mixed signal (dc and RF) measurements can also be undertaken using multiple dc pins and microcoax probes (Figure 8.4) for multifunction MMICs, such as mixers and transmit/receive chips.

Figure 8.4 Multiple RF microcoax and dc on-wafer probing of a complex multifunction MMIC. (*Source:* Bookham Inc., 2006. All Rights Reserved.)

When analyzing the RFOW measurements, bear in mind that the chip-substrate or chip-package interface parasitics, such as bond-wire inductances, were probably incorporated into the design so that the specification would be met when fully assembled, but as they are not present during RFOW measurement, the RF performance will be different. This should be expected, and the assembled performance can be predicted by taking the RFOW measurement data back into an RF simulator and adding on the expected parasitics. Obviously, the final check would be to remeasure the RF performance of the assembled/packaged MMIC. Figure 8.5 shows me in my younger days loading a 3-in. GaAs wafer onto an RFOW probe station.

8.4 Questions

8.1 What else should be provided at the side of the chip, as well as dc bias pads, to aid decoupling?

8.2 How much current can a typical probe needle feed onto the chip?

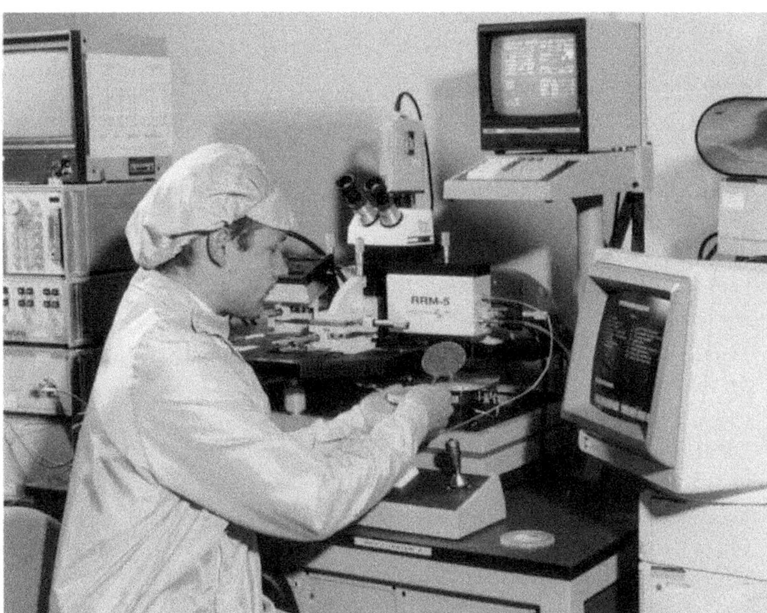

Figure 8.5 The author (in earlier days) loading a 3-in. GaAs wafer onto a RFOW probe station. (*Source:* Bookham Inc., 2006. All Rights Reserved.)

8.3 What does RFOW stand for, and what are its advantages?

8.4 How many contacts does an RFOW probe have, and what are they for?

8.5 What does the calibration do?

References

[1] Miller, D., "Deterministic Process Control Using a Multivariate Model," *Proc. GaAsMANTECH,* Scottsdale, AZ, May 22, 2003.

[2] Godshalk, E., and G. Sundberg, "Modeling Ground Signal Probe Invasiveness and Its Effect on Extracted Inductor Q," *Proc. 59th ARFTG Conference,* Seattle, Washington, June 2002.

[3] Wartenberg, S., "RF Coplanar Probe Basics," *Microwave Journal,* March 2003.

[4] ISS Series of Impedance Standard Substrates, Cascade Microtech, Document ISS-DS-0205, at http://www.cmicro.com/pubs/issfam-ds.pdf.

[5] Rytting, D., "An Analysis of Vector Measurement Accuracy Enhancement Techniques," *Proc. RF & Microwave Measurement Symposium,* Santa Rosa, CA, March 1982.

[6] "Applying Error Correction to Network Analyzer Measurements," Agilent Application Note 1287–3.

[7] Walker, D. K., and D. F. Williams, "Comparison of SOLR and TRL Calibrations," *Proc. 51st ARFTG Conference,* Baltimore, MD, June 1998, pp. 83–87.

[8] Davidson, A., K. Jones, and E. Strid, "LRM and LRRM Calibrations with Automatic Determination of Load Inductance," *Proc. 36th ARFTG Conference,* Monterey, CA, November 1990, pp. 57–63.

[9] Engen, G. F., and C. A. Hoer, "Thru-Reflect-Line: An Improved Technique for Calibrating the Dual Six-Port Automatic Network Analyser," *IEEE Trans. Microwave Theory Tech.,* Vol. 27, December 1979, pp. 987–993.

[10] Marks, R., "A Multiline Method of Network Analyser Calibration," *IEEE Trans. Microwave Theory Tech.,* Vol. 39, 1991, pp. 1205–121.

[11] Cho, H., and D. Burk, "A Three Step Method for De-embedding of High Frequency *S*-Parameter Measurements," *IEEE Trans. Electron Devices,* Vol. 38, No. 6, June 1991, pp. 1371–1375.

[12] Koolen, M., "An Improved De-Embedding Technique for On-Wafer High Frequency Characterization," *IEEE Bipolar Circuits and Tech. Meeting Digest,* Minneapolis, MN, September 1991, pp. 188–191.

[13] "On Wafer Vector Network Analyzer Calibration and Measurements," Application Note, Cascade Microtech Inc., Beaverton, OR, at http://www.cascademicrotech.com.

[14] Williams, D. F., C. M. Wang, and U. Arz, "An Optimal Vector-Network-Analyzer Calibration Algorithm," *IEEE Trans. Microwave Theory Tech.,* Vol. 51, No. 12, December 2003, pp. 2391–2401.

[15] Marks, R. B., and D. F. Williams, "Verification of Commercial Probe-Tip Calibrations," *Proc. 42nd ARFTG Conference,* San Jose, CA, December 1993, pp. 37–44.

[16] "Embedding/De-embedding," Application Note 11410-00278, Anritsu Microwave Measurements Division, Morgan Hill, California, at http://www.eu.anritsu.com/files/11410–00278B.pdf.

Appendix: Answers to Questions

Chapter 2

2.1 The electron peak velocity and the electron mobility.

2.2 Silicon carbide and gallium nitride.

2.3 Gate voltage of −0.5V and drain current of half the drain saturation current ($I_{DSS}/2$).

2.4 V_{BE} forward biased at typically +0.7V.

2.5 The distance between components on an MMIC can approach significant fractions of the signal wavelength, so the phase and amplitude vary along the connections as functions of time and distance.

2.6 The characteristic impedance becomes lower as the track width is increased because the track looks more capacitive.

2.7 125Ω.

2.8 The 30-μm silicon nitride capacitor is 0.5 pF, which is 25 times larger than the same size polyimide capacitor.

2.9 The width of the metal track used for constructing the spiral inductor.

2.10 At RF and low microwave frequencies.

Chapter 4

4.1 It ignores close components that are not connected to its ports, and extrapolation of the data beyond the frequencies measured is unpredictable.

4.2 The components are produced with multiple geometries between RFOW pads and lengths of feed transmission line using a characterization mask-set, and their *s*-parameters are measured RFOW.

4.3 The dimensions of the component are less than one-tenth of the wavelength of the highest signal frequency in the circuit.

4.4 The prime element is the dominant characteristic of the component, and the parasitic elements represent second-order effects, such as loss and fringing capacitance.

4.5 The models must be valid at dc and at the higher harmonic frequencies at which the nonlinear simulation operates.

4.6 To predict coupling effects between unconnected components, wavelength-related effects, and other modes of propagation.

4.7 It will change the effective electrical length of the transmission line.

Chapter 5

5.1 HBT and bipolar transistors.

5.2 HEMT.

5.3 InP HEMT and InP HBT.

5.4 10 dB.

5.5 Between −10 and −15 dB.

5.6 At the side of the chip, and their size and pitch.

5.7 The first and its gain.

5.8 Series inductive feedback.

5.9 Stability.

5.10 Placing stabilizing resistors at the ground end of shunt inductive matching elements.

5.11 From the input to the input-artificial-transmission-line terminating resistor, and from the output-artificial-transmission-line terminating resistor to the output.

5.12 An inductor of value 250 pH.

5.13 Higher gain, wider bandwidth, and higher output impedance.

5.14 Stack bias.

5.15 Self-bias.

5.16 6.25Ω.

5.17 The performance will vary with the resistor process variations.

5.18 A *linear device* is defined as a device with properties (e.g., resistance, transconductance) that are independent of the voltage or current applied to the device.

5.19 Large-signal operation assumes that the voltage and current signals applied to the circuit are large enough that the devices may become nonlinear.

5.20 Weak nonlinearities.

5.21 Less power in the fundamental as the input power increases corresponds to compression of the device gain. The dc component produced can change the dc bias point of the device. Increasing power at the harmonic frequencies distorts the output waveform. The generation of power at other frequencies can have serious consequences on the system performance.

5.22 Architecture design, small-signal design, and large-signal optimization.

5.23 GaAs HBT, and power processes are rated in terms of watts per millimeter of gate width or emitter-finger length.

5.24 As FET device size increases, output power increases, and the gain decreases. The optimum device is one that gives the maximum amount of power and still has useable signal gain over the specified frequency range. The rule of thumb is to select the largest device that still exhibits a G_{max} of 10 dB at the high end of the specified frequency range.

5.25 The output (last) stage.

5.26 All the stages before the last stage should be backed off by 3 dB below their P_{1dB} output power level.

5.27 Large power devices will have too little gain or get too hot if just one large device is used.

5.28 Wilkinson splitters and bus-bar and parallel-matching networks.

5.29 The completed power budget is shown in Figure A.1. The input power needs to be +16.5 dBm. The output power from FET *A* in the first stage is +27 dBm, but the P_{1dB} of FET *A* is +29 dBm; therefore, linear behavior (3-dB back-off from P_{1dB}) is limited up to +26 dBm, so this does not quite meet the required specification.

5.30 Small-signal simulation.

5.31 dc bias paths.

5.32 Gain.

5.33 The Cripps load line technique.

5.34 The load-pull technique.

5.35 Large-signal simulation using nonlinear models.

5.36 Hard.

5.37 HBT.

5.38 MESFET.

5.39 Resonator, negative-resistance circuit, and load.

5.40 The resonator.

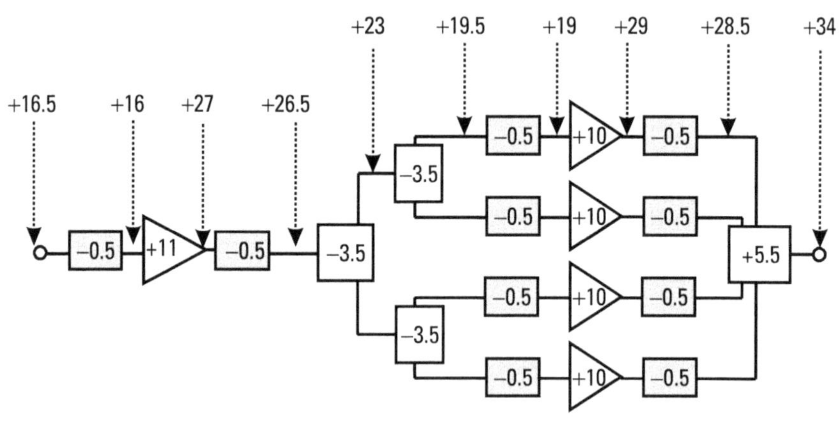

Figure A.1 Power amplifier power budget for answer 5.29.

5.41 The active devices (transistors).

5.42 Frequency conversion (to higher or lower frequencies) while retaining any modulation data on the signal.

5.43 Local oscillator, radio frequency, and intermediate frequency.

5.44 All the sum and difference products ($n \times \omega_{LO} \pm m \times \omega_{RF}$ where n and m are integers).

5.45 Biased transistors in an active mixer, and unbiased transistors and diodes in a passive mixer.

5.46 Passive mixers are reciprocal, active mixers can have conversion gain, and diode mixers are easiest to bias.

5.47 Additional suppression of the even harmonics of the LO frequency.

5.48 Four.

5.49 In a subharmonic mixer.

5.50 The image frequency is the unwanted sideband in a single-sideband mixer, which, if the mixer is designed to reject it, reduces the mixer noise figure by 3 dB.

5.51 Active and passive modes.

5.52 Passive FET mixers give the best linearity, and active FET mixers can give gain.

5.53 A Gilbert-cell mixer is a balanced mixer.

5.54 A balun is used to convert a balanced to an unbalanced signal.

5.55 0V and –5V.

5.56 Near-zero power consumption and easy control bias via the gate contact.

5.57 Higher frequency, shorter wavelength, parasitics have a larger effect, and 3D simulation may be required.

5.58 The self-resonant frequency.

5.59 8.4 to 8.5.

5.60 0.25 mm.

5.61 10Ω and a radial stub.

5.62 Microstrip and CPW.

5.63 CPW.

5.64 Microstrip because the thermal dissipation is better through a thinner substrate and the high dc bias currents are easy to return to ground using vias.

5.65 CPW has fewer shunt parasitics, so the active devices have more gain.

5.66 Microstrip.

5.67 Microstrip.

5.68 One-quarter of the wavelength.

5.69 The short wavelength means that radiation is more likely, and coupling between close components is also possible.

Chapter 6

6.1 Polygons.

6.2 Hierarchy means that a group of individually defined polygons (on any number of layers) can be grouped into one object known as a cell. The cell can then be copied and reproduced at many points in the layout, and the link to the original cell is retained so that when it is changed, all copies are changed as well.

6.3 No, the foundry supplies the complete transistor layouts, and these are just connected together with transmission lines by the MMIC designer.

6.4 Large amounts of time and money will be wasted before the error is brought to light by nonfunctioning designs found during RFOW testing.

6.5 The processing methods and technology determine the layout rules. For example, a mesa isolation process requires that all FET gates be parallel to each other so that the gate feed lines are fabricated up a positive slope of the mesa, and closed structures cannot be made with metal layers because they do not fabricate cleanly when using a lift-off process.

6.6 DRC stands for design rule checking, LVS stands for layout versus schematic checking, and ERC stands for electrical rule checking.

6.7 Arraying is the process of taking multiple MMIC design layouts and constructing an orthogonal array or reticule of the designs.

6.8 PCMs are the process control monitors, and they are placed within the drop-in arrays.

Chapter 7

7.1 The word *monolithic* in the name MMIC means that the circuit is constructed on one solid piece of semiconductor material.

7.2 The structure of the substrate material must be manufactured as a single crystal lattice if the transistors fabricated on the substrate surface are going to benefit from the high electron mobility of the substrate semiconductor material.

7.3 Wafers are produced by sawing the single-crystal boule into slices and then polishing the surfaces.

7.4 Higher-conductivity (lossy) substrates reduce the Q-factor of components, such as inductors, and limit the sharpness of filter responses.

7.5 Dopant atoms are introduced into the semiconductor lattice that have more or fewer electrons than the atoms they are replacing.

7.6 Implantation and epitaxy.

7.7 Etch and lift-off lithography.

7.8 Mesa and implant isolation.

7.9 The metal tracks to the transistor gate contacts must run off the positive slope of the mesa, so all transistors must be aligned with their gate contacts parallel to each other.

7.10 No, the ohmic contact metallization is normally only used to conduct current vertically into the semiconductor.

7.11 The gate contact.

7.12 The high dielectric layer is normally used for the dielectric insulator for MIM capacitors and for passivation of the semiconductor surface.

7.13 Nichrome thin-film resistors have virtually zero variation of resistance with temperature, while mesa resistors vary considerably.

7.14 A relatively thick layer of the low-dielectric material helps to minimize the coupling.

7.15 The chip thickness affects the number of modes that can propagate along a microstrip transmission line and the amount of heat that can be conducted out from the back face of the device.

7.16 Primarily inductance.

7.17 The back-face metallization forms the ground plane for microstrip transmission lines.

7.18 Test devices on the wafer are measured, and their parameters are compared to predetermined limits set by the foundry.

Chapter 8

8.1 A dc pad connected to a through-substrate via, preferably one for each bias supply.

8.2 0.5A.

8.3 Radio frequency on Wafer, which allows the rapid testing of all the circuits before scribing so that only good dies are assembled into systems.

8.4 Three: one central signal contact and two ground contacts on either side.

8.5 Calibration brings the measurement reference plane to a known point within the measurement system, typically to the RFOW probe tips.

Glossary

$\sqrt{\ }$	Square root
$\|\Gamma\|$	Magnitude of the reflection coefficient
$1/f$ noise	Noise that varies inversely to frequency
3D	Three dimensional
A	Loss per unit length of a transmission line
ac	Alternating current
admittance (Y)	Reciprocal of impedance
array	Multiple MMIC design layouts assembled orthogonally
balun	Converts a balanced signal into an unbalanced signal or visa versa
BCB	Benzocyclobutene
boule	A large, cylindrical, single crystal of semiconductor material pulled from the melt, which is cut into semiconductor wafers

bus-bar	Output combiner that uses a wide track to feed bias current directly to the outputs of a power amplifier stage
C	Shunt capacitance per unit length of a transmission line
CAD	Computer-aided design
cascode topology	Common-emitter FET followed by a common-base FET
characteristic admittance (Y_0)	The ratio of the current to the voltage at every point on a transmission line
characteristic impedance (Z_0)	Ratio of the voltage to the current at every point on a transmission line
CMOS	Complementary metal-oxide semiconductor
conductance (G)	Real part of admittance
conjugate (*)	Equal real part, opposite imaginary part
CPW	Coplanar waveguide transmission line
CW	Continuous wave
dc	Direct current
depletion region	Region of a doped semiconductor where there are no free carriers
dispersion	Performance variation with frequency
DOE	Design of experiments
dopant	Element or impurity added to a semiconductor to change its conductivity
drain efficiency	Output power divided by the drain dc bias power
DRC	Design-rule check
DRO	Dielectric resonator oscillator

early voltage	Bipolar devices have a linear increase in current while in saturation. A line extension with the slope of this increase intercepts the voltage axis at a negative voltage called the Early Voltage
e-beam	Electron beam
EM	Electromagnetic
ERC	Electric-rule check
ESD	Electrostatic discharge
FBAR	Film bulk acoustic resonator
FET	Field effect transistor
F_{min}	Minimum noise figure
G	Shunt conductance per unit length of a transmission line
GaAs	Gallium arsenide semiconductor
gamma opt. (Γ_{opt})	Optimum reflection coefficient for minimum noise figure
GaN	Gallium nitride semiconductor
G_{ass}	Gain associated with the minimum noise figure
GCPW	Grounded coplanar waveguide transmission line
Ge	Germanium
GHz	Gigahertz (10^9 hertz)
G_m	Transconductance of an FET
G_{max}	Maximum stable gain
GPS	Global positioning system
G-S	Ground-signal contact pad
G-S-G	Ground-signal-ground contact pad
GSM	Global System for Mobile Communications

HBT	Heterojunction bipolar transistor
HEMT	High-electron-mobility transistor
hertz	Cycles per second
HFET	Heterojunction FET
hybrid coupler	Transmitted and coupled signals are 90° out of phase
I/V	Current-voltage characteristic
IC	Integrated circuit
I_{DSS}	Drain-to-source saturation current
IF	Intermediate frequency
IGSV	Individual-grounded source vias
III–V semiconductor	Semiconductor using elements from columns III and V of the periodic table
I_{MAX}	Drain-to-source current with positive voltage on the gate (V_{gs})
impedance (*Z*)	Complex ratio of voltage to current
InP	Indium phosphide semiconductor
ISS	Impedance standard substrate
K	Stability factor
k	Boltzmann's constant
kHz	Kilohertz (10^3 hertz)
L	Inductance per unit length of a transmission line
LAN	Local-area network
lapping	Reduction of a wafer thickness by mechanical abrasion
LDMOS	Lateral double diffuse metal-oxide semiconductor

LEC	Liquid-encapsulated Czochralski
LNA	Low-noise amplifier
LO	Local oscillator
load line	Line followed by the drain voltage and current on an *I*/*V* plot as determined by the load resistance on the drain terminal of an FET
lossy	Attribute of a component, such as an inductor or transmission line, which has series resistance or shunt conductance that introduces significant loss to signals passing though it
LRM	Line-reflect-match calibration technique
LRRM	Line-reflect-reflect-match calibration technique
LVS	Layout versus schematic
$\angle\Gamma$	Angle of the reflection coefficient
MAG	Maximum available gain
MBE	Molecular beam epitaxy
MEMS	Microelectromechanical systems
mesa	Raised area of semiconductor with clifflike edges
MESFET	Metal semiconductor field effect transistor
mHEMT	Metamorphic HEMT
MHz	Megahertz (10^6 hertz)
MIC	Microwave integrated circuit typically made from packaged transistors mounted on an alumina passive circuit
MIM	Metal-insulator-metal
MMIC	Monolithic microwave integrated circuit
MOCVD	Metal organic chemical vapor deposition

MODFET	Modulation-doped FET
MSG	Maximum stable gain
NDA	Nondisclosure agreement
NEPER	Natural log of a ratio of two amplitudes (= 8.686 dB)
NiCr	Nickel chromium alloy resistive metal layer
noise factor	Ratio of the output signal-to-noise ratio to the input signal-to-noise ratio
noise figure	Noise factor expressed in decibels [$10\log_{10}$(*noise factor*)]
normalized	Relative to the characteristic impedance of the system
NPN	Negative-positive-negative doping in a bipolar transistor
ohmic contact	Low-resistance contact between metal and semiconductor
P_{1dB}	Output power level when the gain is 1 dB less than the small-signal gain
PAE	Power-added efficiency
PbSe	Lead selenium
PbTe	Lead tellurium
PCB	Printed circuit board
PCM	Process control monitor
PECVD	Plasma electrochemical vapor deposition
pHEMT	Pseudomorphic high-electron-mobility transistor
PIN diode	Positive-intrinsic-negative doped diode
PNP	Positive-negative-positive doping in a bipolar transistor

P_{SAT}	Saturated output power
P_{TOI}	Third-order intercept point
puck	Cylindrically shaped dielectric resonator
quadrature	Through and coupled signals are 90° out of phase
quality factor (Q)	Ratio of the imaginary part over the real part of the impedance of a component
R	Resistance per unit length of a transmission line
reactance (X)	Imaginary part of impedance
reflection coefficient (Γ)	Complex ratio of reflected voltage to incident voltage
resistance (R)	Real part of impedance
reticule	Multiple MMIC design layouts assembled orthogonally
RF	Radio frequency
RFIC	Radio frequency integrated circuit
RFOW	RF-on-wafer
RIE	Reactive ion etching
rms	Root mean square
run-out	Deviation from true alignment across a large diameter wafer
Rx	Receive
S_{11}	Input reflection scattering parameter
S_{12}	Reverse transmission scattering parameter
S_{21}	Forward transmission scattering parameter
S_{22}	Output reflection scattering parameter
Schottky barrier	Potential barrier at a metal-semiconductor junction

SDH	Synchronous digital hierarchy
SDL	Schematic-driven layout
SEM	Scanning electron microscope
semiconductor	Any of various solid crystalline substances having greater electrical conductivity than insulators but less than good conductors
Si	Silicon
SiC	Silicon carbide semiconductor
SiGe	Silicon germanium semiconductor
SOLT	Short-open-load-thru calibration technique
SPC	Statistical process control
SPDT	Single-pole double-throw
SPST	Single-pole single-throw
susceptance (B)	Imaginary part of admittance
TE	Transverse-electric
TEM	Transverse-electromagnetic
thou	Thousandths of an inch
TM	Transverse-magnetic
TRL	Thru-reflect-line calibration technique
Tx	Transmit
Tx/Rx	Transmit and receive
UV	Ultraviolet
varactor	Voltage-controllable capacitor
VCO	Voltage-controlled oscillator
VGF	Vertical gradient freeze
VHF	Very high frequency

V_{JBE}	Base-emitter junction potential
V_p	Pinch-off voltage
VSWR	Voltage standing wave ratio
WLAN	Wireless local area network
YIG	Yttrium iron garnet
Z_0	Characteristic impedance
Z_N	Normalized impedance
β	Beta; ratio of the collector current to the base current in a bipolar transistor, also known as the current gain
Γ	Reflection coefficient
η	Efficiency
λ	Wavelength
ν	Phase velocity
ξ	Dielectric constant
ξ_0	Dielectric constant of free space
ξ_{eff}	Effective relative dielectric constant of a transmission line
ξ_R	Relative dielectric constant

About the Author

Steve Marsh graduated from the University of Bath in 1985 with a first-class honors degree in physics with physical electronics. In 1989, he received his Ph.D. from the same establishment on the subject of the design and optimization of a planar Schottky diode 183-GHz subharmonic mixer. After this, he spent some time at Plessey 3–5 working on hybrid amplifier design and microstrip alumina circuit component characterization, and then at British Aerospace Dynamics developing millimeter-wave components and systems for radar and satellite applications.

Dr. Marsh worked for 13 years with Bookham Technology, Caswell, Northamptonshire, in the GaAs MMIC division. He has worked on the development of HEMT processes for 40-GHz and 60-GHz operation, designing distributed amplifier, LNA, and multiplier MMICs. His experience also includes the design and development of ultrabroadband-tunable active band-stop filter MMICs. He has also designed high-power MESFET MMICs at 14 GHz for a European power amplifier project, and while assisting in the development of an HBT process, has designed demonstrator HBT power amplifiers to produce 10W at X-band frequencies.

Dr. Marsh has published over 40 papers in the field of microwave and MMIC design, is honorary secretary of the Administrative Committee of the

Institute of Electrical and Electronics Engineers (IEEE), United Kingdom, and Republic of Ireland microwave theory and techniques/electron/devices/ antennas and propagation/laser and electro optics (MTT/ED/AP/LEO) joint chapter, sits on the IEEE Micorwave Theory and Techniques (MTT-6) Technical Committee, and is associate editor of *IEEE Transactions on Microwave Theory and Techniques.* He is currently working as a consultant in his own company, MIDAS Consulting, http://www.midasconsulting.co.uk.

Index

Recent Titles in the Artech House Microwave Library

Active Filters for Integrated-Circuit Applications, Fred H. Irons

Advanced Techniques in RF Power Amplifier Design, Steve C. Cripps

Automated Smith Chart, Version 4.0: Software and User's Manual, Leonard M. Schwab

Behavioral Modeling of Nonlinear RF and Microwave Devices, Thomas R. Turlington

Broadband Microwave Amplifiers, Bal S. Virdee, Avtar S. Virdee, and Ben Y. Banyamin

Classic Works in RF Engineering: Combiners, Couplers, Transformers, and Magnetic Materials, John L. B. Walker, Daniel P. Myer, Frederick H. Raab, and Chris Trask, editors

Computer-Aided Analysis of Nonlinear Microwave Circuits, Paulo J. C. Rodrigues

Design of FET Frequency Multipliers and Harmonic Oscillators, Edmar Camargo

Design of Linear RF Outphasing Power Amplifiers, Xuejun Zhang, Lawrence E. Larson, and Peter M. Asbeck

Design of RF and Microwave Amplifiers and Oscillators, Pieter L. D. Abrie

Distortion in RF Power Amplifiers, Joel Vuolevi and Timo Rahkonen

EMPLAN: Electromagnetic Analysis of Printed Structures in Planarly Layered Media, Software and User's Manual, Noyan Kinayman and M. I. Aksun

FAST: Fast Amplifier Synthesis Tool—Software and User's Guide, Dale D. Henkes

Feedforward Linear Power Amplifiers, Nick Pothecary

Generalized Filter Design by Computer Optimization, Djuradj Budimir

High-Linearity RF Amplifier Design, Peter B. Kenington

High-Speed Circuit Board Signal Integrity, Stephen C. Thierauf

Integrated Circuit Design for High-Speed Frequency Synthesis, John Rogers, Calvin Plett, and Foster Dai

Intermodulation Distortion in Microwave and Wireless Circuits, José Carlos Pedro and Nuno Borges Carvalho

Lumped Elements for RF and Microwave Circuits, Inder Bahl

Microwave Circuit Modeling Using Electromagnetic Field Simulation, Daniel G. Swanson, Jr. and Wolfgang J. R. Hoefer

Microwave Component Mechanics, Harri Eskelinen and Pekka Eskelinen

Microwave Engineers' Handbook, Two Volumes, Theodore Saad, editor

Microwave Filters, Impedance-Matching Networks, and Coupling Structures, George L. Matthaei, Leo Young, and E.M.T. Jones

Microwave Materials and Fabrication Techniques, Second Edition, Thomas S. Laverghetta

Microwave Mixers, Second Edition, Stephen A. Maas

Microwave Radio Transmission Design Guide, Trevor Manning

Microwaves and Wireless Simplified, Thomas S. Laverghetta

Modern Microwave Circuits, Noyan Kinayman and M. I. Aksun

Neural Networks for RF and Microwave Design, Q. J. Zhang and K. C. Gupta

Nonlinear Microwave and RF Circuits, Second Edition, Stephen A. Maas

QMATCH: Lumped-Element Impedance Matching, Software and User's Guide, Pieter L. D. Abrie

Practical Analog and Digital Filter Design, Les Thede

Practical MMIC Design, Steve Marsh

Practical RF Circuit Design for Modern Wireless Systems, Volume I: Passive Circuits and Systems, Les Besser and Rowan Gilmore

Practical RF Circuit Design for Modern Wireless Systems, Volume II: Active Circuits and Systems, Rowan Gilmore and Les Besser

Production Testing of RF and System-on-a-Chip Devices for Wireless Communications, Keith B. Schaub and Joe Kelly

Radio Frequency Integrated Circuit Design, John Rogers and Calvin Plett

RF Design Guide: Systems, Circuits, and Equations, Peter Vizmuller

RF Measurements of Die and Packages, Scott A. Wartenberg

The RF and Microwave Circuit Design Handbook, Stephen A. Maas

RF and Microwave Coupled-Line Circuits, Rajesh Mongia, Inder Bahl, and Prakash Bhartia

RF and Microwave Oscillator Design, Michal Odyniec, editor

RF Power Amplifiers for Wireless Communications, Steve C. Cripps

RF Systems, Components, and Circuits Handbook, Second Edition, Ferril A. Losee

Stability Analysis of Nonlinear Microwave Circuits, Almudena Suárez and Raymond Quéré

TRAVIS 2.0: Transmission Line Visualization Software and User's Guide, Version 2.0, Robert G. Kaires and Barton T. Hickman

Understanding Microwave Heating Cavities, Tse V. Chow Ting Chan and Howard C. Reader

For further information on these and other Artech House titles, including previously considered out-of-print books now available through our In-Print-Forever® (IPF®) program, contact:

Artech House
685 Canton Street
Norwood, MA 02062
Phone: 781-769-9750
Fax: 781-769-6334
e-mail: artech@artechhouse.com

Artech House
46 Gillingham Street
London SW1V 1AH UK
Phone: +44 (0)20 7596-8750
Fax: +44 (0)20 7630 0166
e-mail: artech-uk@artechhouse.com

Find us on the World Wide Web at: www.artechhouse.com

www.ingramcontent.com/pod-product-compliance
Ingram Content Group UK Ltd.
Pitfield, Milton Keynes, MK11 3LW, UK
UKHW020222290926
15470UKWH00001B/320

9 781596 930360